Introduction to Optics

Introduction to Optics

Edited by
Albie Reid

Larsen & Keller
www.larsen-keller.com

Introduction to Optics
Edited by Albie Reid
ISBN: 978-1-63549-208-8 (Hardback)

© 2017 Larsen & Keller

Larsen & Keller

Published by Larsen and Keller Education,
5 Penn Plaza,
19th Floor,
New York, NY 10001, USA

Cataloging-in-Publication Data

Introduction to optics / edited by Albie Reid.
    p. cm.
Includes bibliographical references and index.
ISBN 978-1-63549-208-8
1. Optics. 2. Optical instruments. 3. Optics--History.
4. Optical engineering. I. Reid, Albie.
QC355.3 .I58 2017
535--dc23

The publisher's policy is to use permanent paper from mills that operate a sustainable forestry policy. Furthermore, the publisher ensures that the text paper and cover boards used have met acceptable environmental accreditation standards.

Printed and bound in the United States of America.

For more information regarding Larsen and Keller Education and its products, please visit the publisher's website www.larsen-keller.com

Table of Contents

Permissions

Index

Preface

This book provides comprehensive insights into the field of optics. It provides detailed information about the various advancements within this field. Optics is the study of light. It also discusses the design and manufacture of technology related to the manipulation and control of sources of light. It is used in many fields like photography, astronomy, optometry, ophthalmology and many other engineering fields. The aim of this textbook is to provide thorough insights to readers about this subject. Most of the topics introduced in it cover fundamentals techniques and the applications of optics. This book, with its detailed analyses and data, will prove immensely beneficial students involved in this area at various levels.

A detailed account of the significant topics covered in this book is provided below:

Chapter 1- The branch of physics that involves light, its properties and its interaction with matter is known as optics. Optics as a subject concentrates on the behavior of visible, ultraviolet and infrared light. The chapter on optics offers an insightful focus, keeping in mind the subject matter.

Chapter 2- Optics evolved from the development of lenses. The history of optics is divided into two halves, classical optics and modern optics. Modern optics has developed in the 20th century and covers topics such as wave optics and quantum optics. The following section will not only provide an overview, it will delve deep into the topics related to it.

Chapter 3- The branches of optics discussed in this section are geometrical optics, physical optics, quantum optics, nonimaging optics and nonlinear optics. Geometrical optics defines light propagation in terms of rays whereas physics optics is a branch of optics that concerns itself with diffraction and polarization. This chapter is a compilation of the various branches of optics that form an integral part of the broader subject matter.

Chapter 4- The significant topics of optics explained in this section are optical instruments and optical engineering. An optical instrument processes light waves to improve an image for viewing and optical engineering is the study of optics. The section strategically encompasses and incorporates the major components and significant topics of optics, providing a complete understanding.

Chapter 5- The concepts of optics explained in the following text are photometry, exposure, atmospheric optics, optical vortex, focus and dispersion. Photometry is the study of measuring light; there is a difference between photometry and radiometry. Photometry mainly studies the sensitivity of visible light on the human eye; it measures this by calculating the power of each wavelength by representing the sensitivity of eye at that particular wavelength. This section elucidates the crucial theories of optics.

Chapter 6- Optics is used for scientific and commercial use. Fiber- optic communication is the technique of conveying information from one place to another. This is done by sending pulses of light through an optical fiber. Alternatively the other uses of optics are electromagnetically induced transparency, superlens, optical fiber, spectrometer and optical computing. This section has been carefully written to provide an easy understanding of the theory of optics.

It gives me an immense pleasure to thank our entire team for their efforts. Finally in the end, I would like to thank my family and colleagues who have been a great source of inspiration and support.

Editor

Introduction to Optics in Engineering

The branch of physics that involves light, its properties and its interaction with matter is known as optics. Optics as a subject concentrates on the behavior of visible, ultraviolet and infrared light. The chapter on optics offers an insightful focus, keeping in mind the subject matter.

Optics is the branch of physics which involves the behaviour and properties of light, including its interactions with matter and the construction of instruments that use or detect it. Optics usually describes the behaviour of visible, ultraviolet, and infrared light. Because light is an electromagnetic wave, other forms of electromagnetic radiation such as X-rays, microwaves, and radio waves exhibit similar properties.

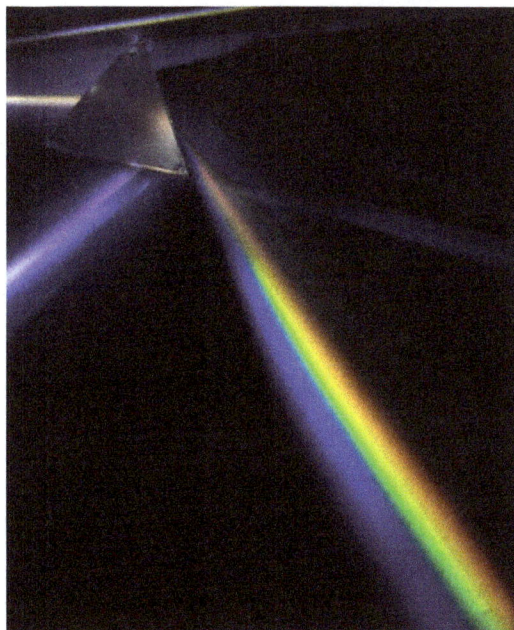

Optics includes study of dispersion of light.

Most optical phenomena can be accounted for using the classical electromagnetic description of light. Complete electromagnetic descriptions of light are, however, often difficult to apply in practice. Practical optics is usually done using simplified models. The most common of these, geometric optics, treats light as a collection of rays that travel in straight lines and bend when they pass through or reflect from surfaces. Physical optics is a more comprehensive model of light, which includes wave effects such as diffraction and interference that cannot be accounted for in geometric optics. Historically, the ray-based model of light was developed first, followed by the wave model of light. Progress in electromagnetic theory in the 19th century led to the discovery that light waves were in fact electromagnetic radiation.

Some phenomena depend on the fact that light has both wave-like and particle-like properties.

Explanation of these effects requires quantum mechanics. When considering light's particle-like properties, the light is modelled as a collection of particles called "photons". Quantum optics deals with the application of quantum mechanics to optical systems.

Optical science is relevant to and studied in many related disciplines including astronomy, various engineering fields, photography, and medicine (particularly ophthalmology and optometry). Practical applications of optics are found in a variety of technologies and everyday objects, including mirrors, lenses, telescopes, microscopes, lasers, and fibre optics.

History

The Nimrud lens

Optics began with the development of lenses by the ancient Egyptians and Mesopotamians. The earliest known lenses, made from polished crystal, often quartz, date from as early as 700 BC for Assyrian lenses such as the Layard/Nimrud lens. The ancient Romans and Greeks filled glass spheres with water to make lenses. These practical developments were followed by the development of theories of light and vision by ancient Greek and Indian philosophers, and the development of geometrical optics in the Greco-Roman world. The word *optics* comes from the ancient Greek word πτικὴ (*optikē*), meaning "appearance, look".

Greek philosophy on optics broke down into two opposing theories on how vision worked, the "intromission theory" and the "emission theory". The intro-mission approach saw vision as coming from objects casting off copies of themselves (called eidola) that were captured by the eye. With many propagators including Democritus, Epicurus, Aristotle and their followers, this theory seems to have some contact with modern theories of what vision really is, but it remained only speculation lacking any experimental foundation.

Plato first articulated the emission theory, the idea that visual perception is accomplished by rays emitted by the eyes. He also commented on the parity reversal of mirrors in *Timaeus*. Some hundred years later, Euclid wrote a treatise entitled *Optics* where he linked vision to geometry, creating *geometrical optics*. He based his work on Plato's emission theory wherein he described the mathematical rules of perspective and described the effects of refraction qualitatively, although he questioned that a beam of light from the eye could instantaneously light up the stars every time someone blinked. Ptolemy, in his treatise *Optics*, held an extramission-intromission theory of vision: the rays (or flux) from the eye formed a cone, the vertex being within the eye, and the base defining the visual field. The rays were sensitive, and conveyed information back to the observer's intellect about the distance and orientation of surfaces. He summarised much of Euclid and went on to describe a way to measure the angle of refraction, though he failed to notice the empirical relationship between it and the angle of incidence.

Alhazen (Ibn al-Haytham), "the father of Optics"

During the Middle Ages, Greek ideas about optics were resurrected and extended by writers in the Muslim world. One of the earliest of these was Al-Kindi (c. 801–73) who wrote on the merits of Aristotelian and Euclidean ideas of optics, favouring the emission theory since it could better quantify optical phenomenon. In 984, the Persian mathematician Ibn Sahl wrote the treatise "On burning mirrors and lenses", correctly describing a law of refraction equivalent to Snell's law. He used this law to compute optimum shapes for lenses and curved mirrors. In the early 11th century, Alhazen (Ibn al-Haytham) wrote the *Book of Optics (Kitab al-manazir)* in which he explored reflection and refraction and proposed a new system for explaining vision and light based on observation and experiment. He rejected the "emission theory" of Ptolemaic optics with its rays being emitted by the eye, and instead put forward the idea that light reflected in all directions in straight lines from all points of the objects being viewed and then entered the eye, although he was unable to correctly explain how the eye captured the rays. Alhazen's work was largely ignored in the Arabic world but it was anonymously translated into Latin around 1200 A.D. and further summarised and expanded on by the Polish monk Witelo making it a standard text on optics in Europe for the next 400 years.

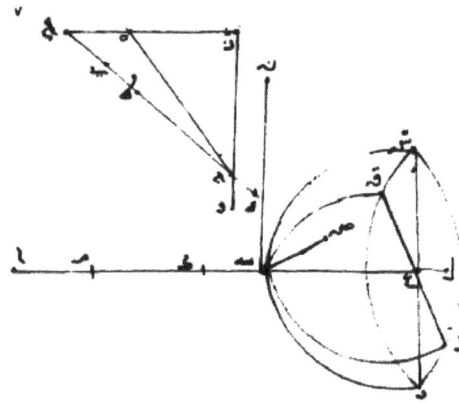

كان ان ما ته عليها سطح مستوى غيره فلان هذا السطح يقطع سطح بن ج
على نقطة ب فلابد من ان يقطع احد خطى ب ن بعد فلكن ذلك
الخط بتر والفصل المشترك بين هذا السطح وبين سطح قطع فرد
خط بتر فلان هذا السطح ثابت من سبط ب على نقطة ب لخط
ب خط ان قطع ف ب د على نقطة ت وكل خط بتر وهذا بالحال
فلابا ان سبط ت على نقطة ت سطح مستوى غير سطح ب ن تص ٥

Reproduction of a page of Ibn Sahl's manuscript showing his knowledge of the law of refraction.

In the 13th century in medieval Europe, English bishop Robert Grosseteste wrote on a wide range of scientific topics, and discussed light from four different perspectives: an epistemology of light, a metaphysics or cosmogony of light, an etiology or physics of light, and a theology of light, basing it on the works Aristotle and Platonism. Grosseteste's most famous disciple, Roger Bacon, wrote works citing a wide range of recently translated optical and philosophical works, including those of Alhazen, Aristotle, Avicenna, Averroes, Euclid, al-Kindi, Ptolemy, Tideus, and Constantine the African. Bacon was able to use parts of glass spheres as magnifying glasses to demonstrate that light reflects from objects rather than being released from them.

The first wearable eyeglasses were invented in Italy around 1286. This was the start of the optical industry of grinding and polishing lenses for these "spectacles", first in Venice and Florence in the thirteenth century, and later in the spectacle making centres in both the Netherlands and Germany. Spectacle makers created improved types of lenses for the correction of vision based more on empirical knowledge gained from observing the effects of the lenses rather than using the rudimentary optical theory of the day (theory which for the most part could not even adequately explain how spectacles worked). This practical development, mastery, and experimentation with lenses led directly to the invention of the compound optical microscope around 1595, and the refracting telescope in 1608, both of which appeared in the spectacle making centres in the Netherlands.

In the early 17th century Johannes Kepler expanded on geometric optics in his writings, covering lenses, reflection by flat and curved mirrors, the principles of pinhole cameras, inverse-square law governing the intensity of light, and the optical explanations of astronomical phenomena such as lunar and solar eclipses and astronomical parallax. He was also able to correctly deduce the role of the retina as the actual organ that recorded images, finally being able to scientifically quantify the effects of different types of lenses that spectacle makers had been observing over the previous 300 years. After the invention of the telescope Kepler set out the theoretical basis on how they worked and described an improved version, known as the *Keplerian telescope*, using two convex lenses to produce higher magnification.

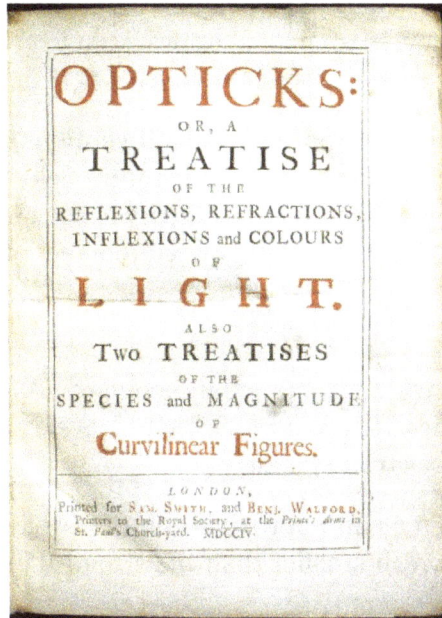

Cover of the first edition of Newton's *Opticks*

Optical theory progressed in the mid-17th century with treatises written by philosopher René Descartes, which explained a variety of optical phenomena including reflection and refraction by assuming that light was emitted by objects which produced it. This differed substantively from the ancient Greek emission theory. In the late 1660s and early 1670s, Isaac Newton expanded Descartes' ideas into a corpuscle theory of light, famously determining that white light was a mix of colours which can be separated into its component parts with a prism. In 1690, Christiaan Huygens proposed a wave theory for light based on suggestions that had been made by Robert Hooke in 1664. Hooke himself publicly criticised Newton's theories of light and the feud between the two lasted until Hooke's death. In 1704, Newton published *Opticks* and, at the time, partly because of his success in other areas of physics, he was generally considered to be the victor in the debate over the nature of light.

Newtonian optics was generally accepted until the early 19th century when Thomas Young and Augustin-Jean Fresnel conducted experiments on the interference of light that firmly established light's wave nature. Young's famous double slit experiment showed that light followed the law of superposition, which is a wave-like property not predicted by Newton's corpuscle theory. This work led to a theory of diffraction for light and opened an entire area of study in physical optics. Wave optics was successfully unified with electromagnetic theory by James Clerk Maxwell in the 1860s.

The next development in optical theory came in 1899 when Max Planck correctly modelled black-body radiation by assuming that the exchange of energy between light and matter only occurred in discrete amounts he called *quanta*. In 1905 Albert Einstein published the theory of the photo-electric effect that firmly established the quantization of light itself. In 1913 Niels Bohr showed that atoms could only emit discrete amounts of energy, thus explaining the discrete lines seen in emission and absorption spectra. The understanding of the interaction between light and matter which followed from these developments not only formed the basis of quantum optics but also was crucial for the development of quantum mechanics as a whole. The ultimate culmination, the theory of quantum electrodynamics, explains all optics and electromagnetic processes in general as the result of the exchange of real and virtual photons.

Quantum optics gained practical importance with the inventions of the maser in 1953 and of the laser in 1960. Following the work of Paul Dirac in quantum field theory, George Sudarshan, Roy J. Glauber, and Leonard Mandel applied quantum theory to the electromagnetic field in the 1950s and 1960s to gain a more detailed understanding of photodetection and the statistics of light.

Classical Optics

Classical optics is divided into two main branches: geometrical (or ray) optics and physical (or wave) optics. In geometrical optics, light is considered to travel in straight lines, while in physical optics, light is considered as an electromagnetic wave.

Geometrical optics can be viewed as an approximation of physical optics that applies when the wavelength of the light used is much smaller than the size of the optical elements in the system being modelled.

Geometrical Optics

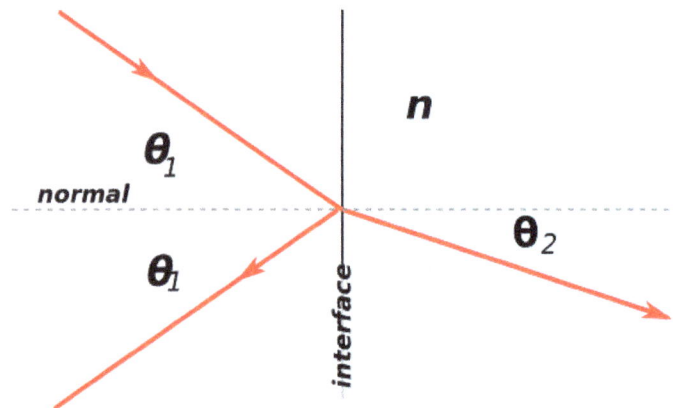

Geometry of reflection and refraction of light rays

Geometrical optics, or *ray optics*, describes the propagation of light in terms of "rays" which travel in straight lines, and whose paths are governed by the laws of reflection and refraction at interfaces between different media. These laws were discovered empirically as far back as 984 AD and have been used in the design of optical components and instruments from then until the present day. They can be summarised as follows:

When a ray of light hits the boundary between two transparent materials, it is divided into a reflected and a refracted ray.

> The law of reflection says that the reflected ray lies in the plane of incidence, and the angle of reflection equals the angle of incidence.

The law of refraction says that the refracted ray lies in the plane of incidence, and the sine of the angle of refraction divided by the sine of the angle of incidence is a constant:

$$\frac{\sin \theta_1}{\sin \theta_2} = n,$$

where n is a constant for any two materials and a given colour of light. If the first material is air or vacuum, n is the refractive index of the second material.

The laws of reflection and refraction can be derived from Fermat's principle which states that *the path taken between two points by a ray of light is the path that can be traversed in the least time.*

Approximations

Geometric optics is often simplified by making the paraxial approximation, or "small angle approximation". The mathematical behaviour then becomes linear, allowing optical components and systems to be described by simple matrices. This leads to the techniques of Gaussian optics and *paraxial ray tracing*, which are used to find basic properties of optical systems, such as approximate image and object positions and magnifications.

Reflections

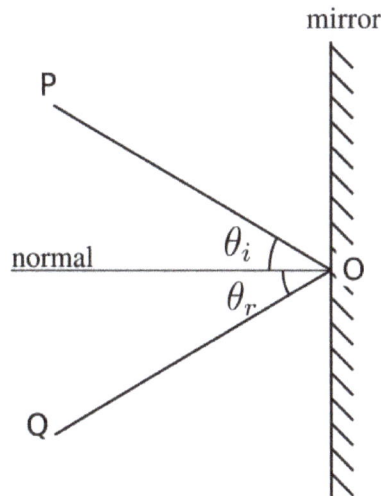

Diagram of specular reflection

Reflections can be divided into two types: specular reflection and diffuse reflection. Specular reflection describes the gloss of surfaces such as mirrors, which reflect light in a simple, predictable way. This allows for production of reflected images that can be associated with an actual (real) or extrapolated (virtual) location in space. Diffuse reflection describes non-glossy materials, such as paper or rock. The reflections from these surfaces can only be described statistically, with the exact distribution of the reflected light depending on the microscopic structure of the material. Many diffuse reflectors are described or can be approximated by Lambert's cosine law, which describes surfaces that have equal luminance when viewed from any angle. Glossy surfaces can give both specular and diffuse reflection.

In specular reflection, the direction of the reflected ray is determined by the angle the incident ray makes with the surface normal, a line perpendicular to the surface at the point where the ray hits. The incident and reflected rays and the normal lie in a single plane, and the angle between the reflected ray and the surface normal is the same as that between the incident ray and the normal. This is known as the Law of Reflection.

For flat mirrors, the law of reflection implies that images of objects are upright and the same distance behind the mirror as the objects are in front of the mirror. The image size is the same as the object size. The law also implies that mirror images are parity inverted, which we perceive as a left-right inversion. Images formed from reflection in two (or any even number of) mirrors are not parity inverted. Corner reflectors retroreflect light, producing reflected rays that travel back in the direction from which the incident rays came.

Mirrors with curved surfaces can be modelled by ray-tracing and using the law of reflection at each point on the surface. For mirrors with parabolic surfaces, parallel rays incident on the mirror produce reflected rays that converge at a common focus. Other curved surfaces may also focus light, but with aberrations due to the diverging shape causing the focus to be smeared out in space. In particular, spherical mirrors exhibit spherical aberration. Curved mirrors can form images with magnification greater than or less than one, and the magnification can be negative, indicating that the image is inverted. An upright image formed by reflection in a mirror is always virtual, while an inverted image is real and can be projected onto a screen.

Refraction occurs when light travels through an area of space that has a changing index of refraction; this principle allows for lenses and the focusing of light. The simplest case of refraction occurs when there is an interface between a uniform medium with index of refraction n_1 and another medium with index of refraction n_2. In such situations, Snell's Law describes the resulting deflection of the light ray:

$$n_1 \sin \theta_1 = n_2 \sin \theta_2$$

where θ_1 and θ_2 are the angles between the normal (to the interface) and the incident and refracted waves, respectively.

The index of refraction of a medium is related to the speed, v, of light in that medium by

$$n = c / v ,$$

where c is the speed of light in vacuum.

Snell's Law can be used to predict the deflection of light rays as they pass through linear media as long as the indexes of refraction and the geometry of the media are known. For example, the propagation of light through a prism results in the light ray being deflected depending on the shape and orientation of the prism. In most materials, the index of refraction varies with the frequency of the light. Taking this into account, Snell's Law can be used to predict how a prism will disperse light into a spectrum. The discovery of this phenomenon when passing light through a prism is famously attributed to Isaac Newton.

Some media have an index of refraction which varies gradually with position and, thus, light rays in the medium are curved. This effect is responsible for mirages seen on hot days: a change in index of refraction air with height causes light rays to bend, creating the appearance of specular reflections in the distance (as if on the surface of a pool of water). Optical materials with varying index of refraction are called gradient-index (GRIN) materials. Such materials are used to make gradient-index optics.

For light rays travelling from a material with a high index of refraction to a material with a low index of refraction, Snell's law predicts that there is no θ_2 when θ_1 is large. In this case, no transmission occurs; all the light is reflected. This phenomenon is called total internal reflection and allows for fibre optics technology. As light travels down an optical fibre, it undergoes total internal reflection allowing for essentially no light to be lost over the length of the cable.

Lenses

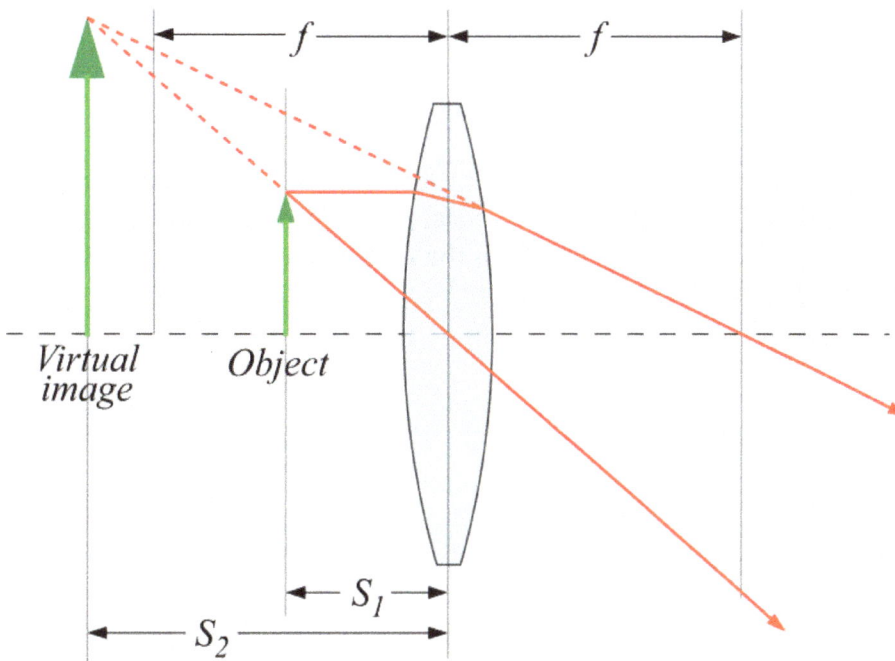

A ray tracing diagram for a converging lens.

A device which produces converging or diverging light rays due to refraction is known as a *lens*. Lenses are characterized by their focal length: a converging lens has positive focal length, while a diverging lens has negative focal length. Smaller focal length indicates that the lens has a stronger converging or diverging effect. The focal length of a simple lens in air is given by the lensmaker's equation.

Ray tracing can be used to show how images are formed by a lens. For a thin lens in air, the location of the image is given by the simple equation

$$\frac{1}{S_1} + \frac{1}{S_2} = \frac{1}{f},$$

where S_1 is the distance from the object to the lens, S_2 is the distance from the lens to the image, and f is the focal length of the lens. In the sign convention used here, the object and image distances are positive if the object and image are on opposite sides of the lens.

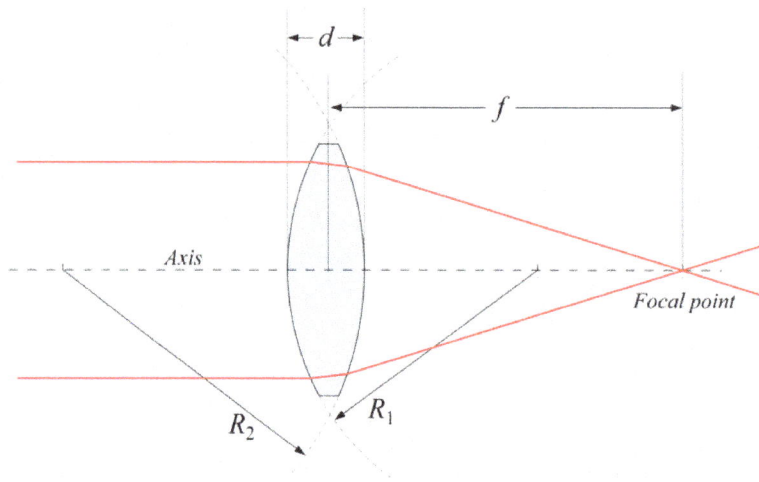

Positive (converging) lens

Incoming parallel rays are focused by a converging lens onto a spot one focal length from the lens, on the far side of the lens. This is called the rear focal point of the lens. Rays from an object at finite distance are focused further from the lens than the focal distance; the closer the object is to the lens, the further the image is from the lens.

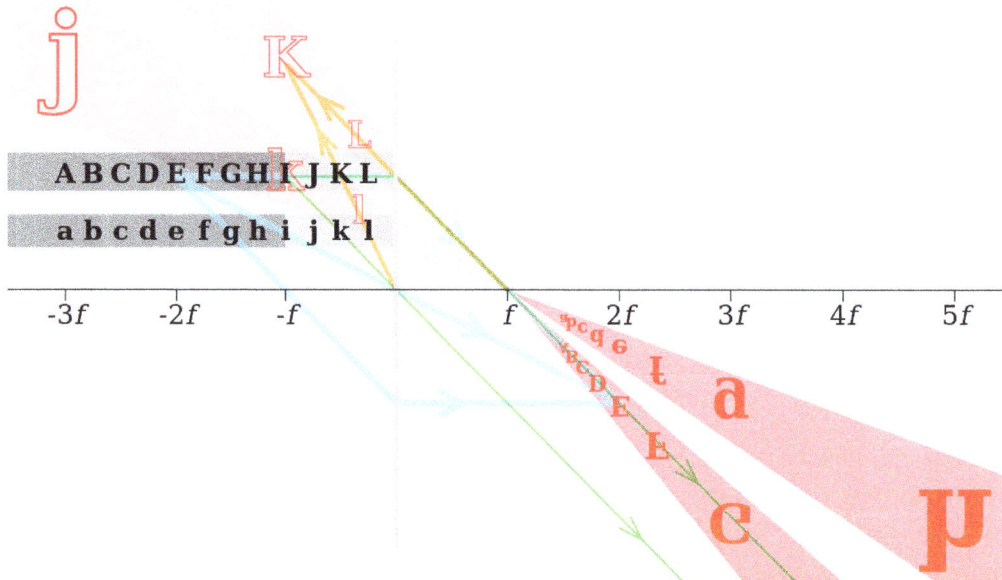

Images of black letters in a thin convex lens of focal length f are shown in red. Selected rays are shown for letters E, I and K in blue, green and orange, respectively. Note that E (at $2f$) has an equal-size, real and inverted image; I (at f) has its image at infinity; and K (at $f/2$) has a double-size, virtual and upright image.

With diverging lenses, incoming parallel rays diverge after going through the lens, in such a way that they seem to have originated at a spot one focal length in front of the lens. This is the lens's front focal point. Rays from an object at finite distance are associated with a virtual image that is closer to the lens than the focal point, and on the same side of the lens as the object. The closer the object is to the lens, the closer the virtual image is to the lens. As with mirrors, upright images produced by a single lens are virtual, while inverted images are real.

Lenses suffer from aberrations that distort images. *Monochromatic aberrations* occur because the geometry of the lens does not perfectly direct rays from each object point to a single point on the image, while chromatic aberration occurs because the index of refraction of the lens varies with the wavelength of the light.

Physical Optics

In physical optics, light is considered to propagate as a wave. This model predicts phenomena such as interference and diffraction, which are not explained by geometric optics. The speed of light waves in air is approximately 3.0×10^8 m/s (exactly 299,792,458 m/s in vacuum). The wavelength of visible light waves varies between 400 and 700 nm, but the term "light" is also often applied to infrared (0.7–300 µm) and ultraviolet radiation (10–400 nm).

The wave model can be used to make predictions about how an optical system will behave without requiring an explanation of what is "waving" in what medium. Until the middle of the 19th century, most physicists believed in an "ethereal" medium in which the light disturbance propagated. The existence of electromagnetic waves was predicted in 1865 by Maxwell's equations. These waves propagate at the speed of light and have varying electric and magnetic fields which are orthogonal to one another, and also to the direction of propagation of the waves. Light waves are now generally treated as electromagnetic waves except when quantum mechanical effects have to be considered.

Modelling and Design of Optical Systems Using Physical Optics

Many simplified approximations are available for analysing and designing optical systems. Most of these use a single scalar quantity to represent the electric field of the light wave, rather than using a vector model with orthogonal electric and magnetic vectors. The Huygens–Fresnel equation is one such model. This was derived empirically by Fresnel in 1815, based on Huygens' hypothesis that each point on a wavefront generates a secondary spherical wavefront, which Fresnel combined with the principle of superposition of waves. The Kirchhoff diffraction equation, which is derived using Maxwell's equations, puts the Huygens-Fresnel equation on a firmer physical foundation. Examples of the application of Huygens–Fresnel principle can be found in the sections on diffraction and Fraunhofer diffraction.

More rigorous models, involving the modelling of both electric and magnetic fields of the light wave, are required when dealing with the detailed interaction of light with materials where the interaction depends on their electric and magnetic properties. For instance, the behaviour of a light wave interacting with a metal surface is quite different from what happens when it interacts with a dielectric material. A vector model must also be used to model polarised light.

Numerical modeling techniques such as the finite element method, the boundary element method and the transmission-line matrix method can be used to model the propagation of light in systems which cannot be solved analytically. Such models are computationally demanding and are normally only used to solve small-scale problems that require accuracy beyond that which can be achieved with analytical solutions.

All of the results from geometrical optics can be recovered using the techniques of Fourier optics which apply many of the same mathematical and analytical techniques used in acoustic engineering and signal processing.

Gaussian beam propagation is a simple paraxial physical optics model for the propagation of coherent radiation such as laser beams. This technique partially accounts for diffraction, allowing accurate calculations of the rate at which a laser beam expands with distance, and the minimum size to which the beam can be focused. Gaussian beam propagation thus bridges the gap between geometric and physical optics.

Superposition and Interference

In the absence of nonlinear effects, the superposition principle can be used to predict the shape of interacting waveforms through the simple addition of the disturbances. This interaction of waves to produce a resulting pattern is generally termed "interference" and can result in a variety of outcomes. If two waves of the same wavelength and frequency are *in phase*, both the wave crests and wave troughs align. This results in constructive interference and an increase in the amplitude of the wave, which for light is associated with a brightening of the waveform in that location. Alternatively, if the two waves of the same wavelength and frequency are out of phase, then the wave crests will align with wave troughs and vice versa. This results in destructive interference and a decrease in the amplitude of the wave, which for light is associated with a dimming of the waveform at that location. See below for an illustration of this effect.

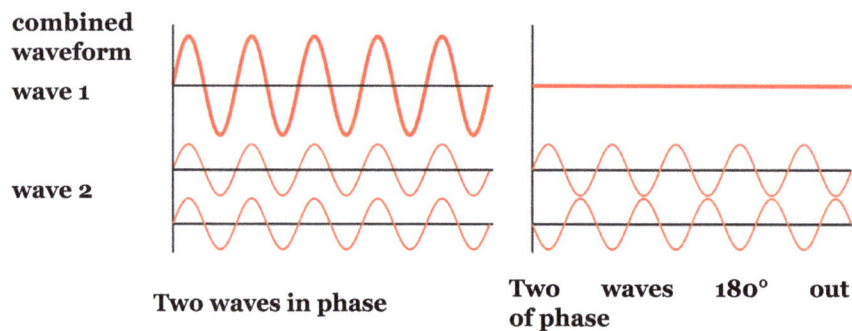

combined waveform

wave 1

wave 2

Two waves in phase   **Two waves 180° out of phase**

Since the Huygens–Fresnel principle states that every point of a wavefront is associated with the production of a new disturbance, it is possible for a wavefront to interfere with itself constructively or destructively at different locations producing bright and dark fringes in regular and predictable patterns. Interferometry is the science of measuring these patterns, usually as a means of making precise determinations of distances or angular resolutions. The Michelson interferometer was a famous instrument which used interference effects to accurately measure the speed of light.

The appearance of thin films and coatings is directly affected by interference effects. Antireflective coatings use destructive interference to reduce the reflectivity of the surfaces they coat, and can be used to minimise glare and unwanted reflections. The simplest case is a single layer with thickness one-fourth the wavelength of incident light. The reflected wave from the top of the film and the reflected wave from the film/material interface are then exactly 180° out of phase, causing destructive interference. The waves are only exactly out of phase for one wavelength, which would typically be chosen to be near the centre of the visible spectrum, around 550 nm. More complex designs using multiple layers can achieve low reflectivity over a broad band, or extremely low reflectivity at a single wavelength.

When oil or fuel is spilled, colourful patterns are formed by thin-film interference.

Constructive interference in thin films can create strong reflection of light in a range of wavelengths, which can be narrow or broad depending on the design of the coating. These films are used to make dielectric mirrors, interference filters, heat reflectors, and filters for colour separation in colour television cameras. This interference effect is also what causes the colourful rainbow patterns seen in oil slicks.

Diffraction and Optical Resolution

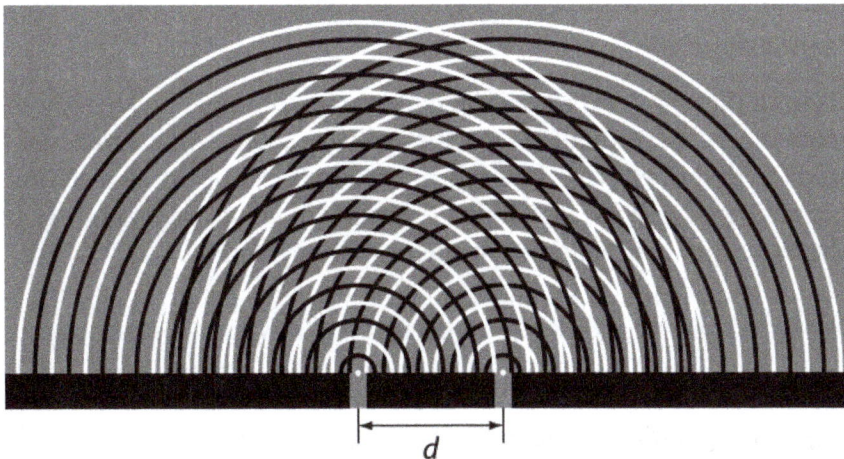

Diffraction on two slits separated by distance d. The bright fringes occur along lines where black lines intersect with black lines and white lines intersect with white lines. These fringes are separated by angle θ and are numbered as order n.

Diffraction is the process by which light interference is most commonly observed. The effect was first described in 1665 by Francesco Maria Grimaldi, who also coined the term from the Latin *diffringere*, 'to break into pieces'. Later that century, Robert Hooke and Isaac Newton also described phenomena now known to be diffraction in Newton's rings while James Gregory recorded his observations of diffraction patterns from bird feathers.

The first physical optics model of diffraction that relied on the Huygens–Fresnel principle was developed in 1803 by Thomas Young in his interference experiments with the interference patterns of two closely spaced slits. Young showed that his results could only be explained if the two slits acted as two unique sources of waves rather than corpuscles. In 1815 and 1818, Augustin-Jean Fresnel firmly established the mathematics of how wave interference can account for diffraction.

The simplest physical models of diffraction use equations that describe the angular separation of light and dark fringes due to light of a particular wavelength (λ). In general, the equation takes the form

$$m\lambda = d \sin\theta$$

where d is the separation between two wavefront sources (in the case of Young's experiments, it was two slits), θ is the angular separation between the central fringe and the m th order fringe, where the central maximum is $m = 0$.

This equation is modified slightly to take into account a variety of situations such as diffraction through a single gap, diffraction through multiple slits, or diffraction through a diffraction grating that contains a large number of slits at equal spacing. More complicated models of diffraction require working with the mathematics of Fresnel or Fraunhofer diffraction.

X-ray diffraction makes use of the fact that atoms in a crystal have regular spacing at distances that are on the order of one angstrom. To see diffraction patterns, x-rays with similar wavelengths to that spacing are passed through the crystal. Since crystals are three-dimensional objects rather than two-dimensional gratings, the associated diffraction pattern varies in two directions according to Bragg reflection, with the associated bright spots occurring in unique patterns and d being twice the spacing between atoms.

Diffraction effects limit the ability for an optical detector to optically resolve separate light sources. In general, light that is passing through an aperture will experience diffraction and the best images that can be created (as described in diffraction-limited optics) appear as a central spot with surrounding bright rings, separated by dark nulls; this pattern is known as an Airy pattern, and the central bright lobe as an Airy disk. The size of such a disk is given by

$$\sin\theta = 1.22\frac{\lambda}{D}$$

where θ is the angular resolution, λ is the wavelength of the light, and D is the diameter of the lens aperture. If the angular separation of the two points is significantly less than the Airy disk angular radius, then the two points cannot be resolved in the image, but if their angular separation is much greater than this, distinct images of the two points are formed and they can therefore be resolved. Rayleigh defined the somewhat arbitrary "Rayleigh criterion" that two points whose angular separation is equal to the Airy disk radius (measured to first null, that is, to the first place where no light is seen) can be considered to be resolved. It can be seen that the greater the diameter of the lens or its aperture, the finer the resolution. Interferometry, with its ability to mimic extremely large baseline apertures, allows for the greatest angular resolution possible.

For astronomical imaging, the atmosphere prevents optimal resolution from being achieved in the visible spectrum due to the atmospheric scattering and dispersion which cause stars to twinkle. Astronomers refer to this effect as the quality of astronomical seeing. Techniques known as adaptive optics have been used to eliminate the atmospheric disruption of images and achieve results that approach the diffraction limit.

Dispersion and Scattering

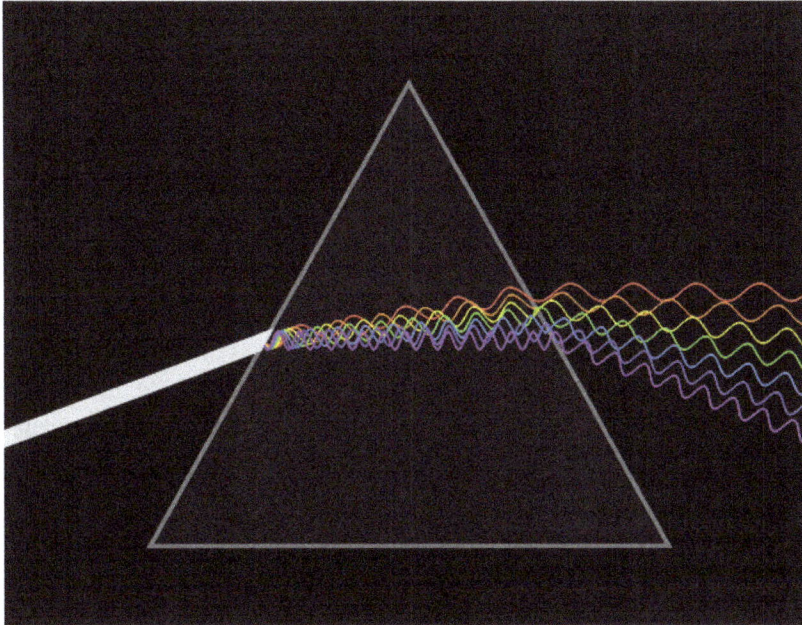

Conceptual animation of light dispersion through a prism. High frequency (blue) light is deflected the most, and low frequency (red) the least.

Refractive processes take place in the physical optics limit, where the wavelength of light is similar to other distances, as a kind of scattering. The simplest type of scattering is Thomson scattering which occurs when electromagnetic waves are deflected by single particles. In the limit of Thomson scattering, in which the wavelike nature of light is evident, light is dispersed independent of the frequency, in contrast to Compton scattering which is frequency-dependent and strictly a quantum mechanical process, involving the nature of light as particles. In a statistical sense, elastic scattering of light by numerous particles much smaller than the wavelength of the light is a process known as Rayleigh scattering while the similar process for scattering by particles that are similar or larger in wavelength is known as Mie scattering with the Tyndall effect being a commonly observed result. A small proportion of light scattering from atoms or molecules may undergo Raman scattering, wherein the frequency changes due to excitation of the atoms and molecules. Brillouin scattering occurs when the frequency of light changes due to local changes with time and movements of a dense material.

Dispersion occurs when different frequencies of light have different phase velocities, due either to material properties (*material dispersion*) or to the geometry of an optical waveguide (*waveguide dispersion*). The most familiar form of dispersion is a decrease in index of refraction with increasing wavelength, which is seen in most transparent materials. This is called "normal dispersion". It occurs in all dielectric materials, in wavelength ranges where the material does not absorb light. In wavelength ranges where a medium has significant absorption, the index of refraction can increase with wavelength. This is called "anomalous dispersion".

The separation of colours by a prism is an example of normal dispersion. At the surfaces of the prism, Snell's law predicts that light incident at an angle θ to the normal will be refracted at an angle arcsin(sin (θ) / n). Thus, blue light, with its higher refractive index, is bent more strongly than red light, resulting in the well-known rainbow pattern.

Dispersion: two sinusoids propagating at different speeds make a moving interference pattern. The red dot moves with the phase velocity, and the green dots propagate with the group velocity. In this case, the phase velocity is twice the group velocity. The red dot overtakes two green dots, when moving from the left to the right of the figure. In effect, the individual waves (which travel with the phase velocity) escape from the wave packet (which travels with the group velocity).

Material dispersion is often characterised by the Abbe number, which gives a simple measure of dispersion based on the index of refraction at three specific wavelengths. Waveguide dispersion is dependent on the propagation constant. Both kinds of dispersion cause changes in the group characteristics of the wave, the features of the wave packet that change with the same frequency as the amplitude of the electromagnetic wave. "Group velocity dispersion" manifests as a spreading-out of the signal "envelope" of the radiation and can be quantified with a group dispersion delay parameter:

$$\sin \theta = 1.22 \frac{\lambda}{D}$$

where v_g is the group velocity. For a uniform medium, the group velocity is

$$v_g = c \left(n - \lambda \frac{dn}{d\lambda} \right)^{-1}$$

where n is the index of refraction and c is the speed of light in a vacuum. This gives a simpler form for the dispersion delay parameter:

$$D = -\frac{\lambda}{c} \frac{d^2 n}{d\lambda^2}.$$

If D is less than zero, the medium is said to have *positive dispersion* or normal dispersion. If D is greater than zero, the medium has *negative dispersion*. If a light pulse is propagated through a normally dispersive medium, the result is the higher frequency components slow down more than the lower frequency components. The pulse therefore becomes *positively chirped*, or *up-chirped*, increasing in frequency with time. This causes the spectrum coming out of a prism to appear with red light the least refracted and blue/violet light the most refracted. Conversely, if a pulse travels through an anomalously (negatively) dispersive medium, high frequency components travel faster than the lower ones, and the pulse becomes *negatively chirped*, or *down-chirped*, decreasing in frequency with time.

The result of group velocity dispersion, whether negative or positive, is ultimately temporal spreading of the pulse. This makes dispersion management extremely important in optical communications systems based on optical fibres, since if dispersion is too high, a group of pulses representing information will each spread in time and merge, making it impossible to extract the signal.

Polarization

Polarization is a general property of waves that describes the orientation of their oscillations. For transverse waves such as many electromagnetic waves, it describes the orientation of the oscillations in the plane perpendicular to the wave's direction of travel. The oscillations may be oriented in a single direction (linear polarization), or the oscillation direction may rotate as the wave travels (circular or elliptical polarization). Circularly polarised waves can rotate rightward or leftward in the direction of travel, and which of those two rotations is present in a wave is called the wave's chirality.

The typical way to consider polarization is to keep track of the orientation of the electric field vector as the electromagnetic wave propagates. The electric field vector of a plane wave may be arbitrarily divided into two perpendicular components labeled x and y (with z indicating the direction of travel). The shape traced out in the x-y plane by the electric field vector is a Lissajous figure that describes the *polarization state*. The following figures show some examples of the evolution of the electric field vector (blue), with time (the vertical axes), at a particular point in space, along with its x and y components (red/left and green/right), and the path traced by the vector in the plane (purple): The same evolution would occur when looking at the electric field at a particular time while evolving the point in space, along the direction opposite to propagation.

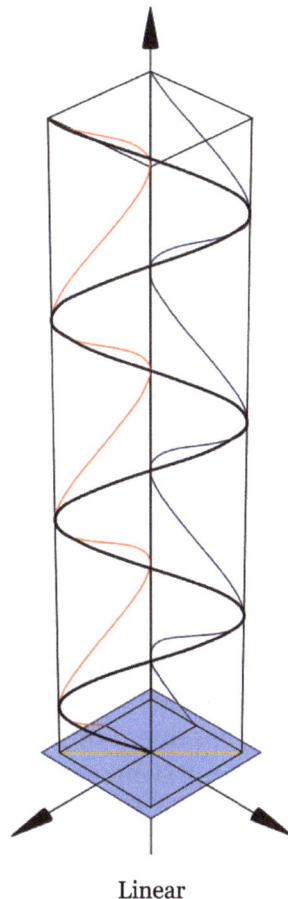

Linear

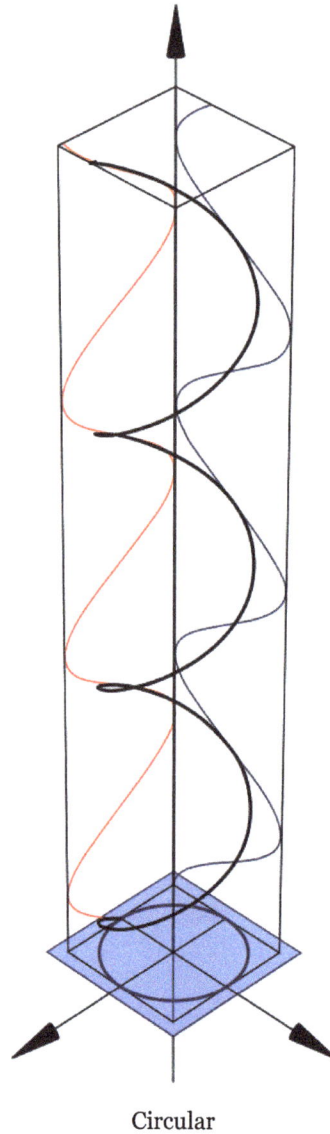

Circular

In the leftmost figure above, the x and y components of the light wave are in phase. In this case, the ratio of their strengths is constant, so the direction of the electric vector (the vector sum of these two components) is constant. Since the tip of the vector traces out a single line in the plane, this special case is called linear polarization. The direction of this line depends on the relative amplitudes of the two components.

In the middle figure, the two orthogonal components have the same amplitudes and are 90° out of phase. In this case, one component is zero when the other component is at maximum or minimum amplitude. There are two possible phase relationships that satisfy this requirement: the x component can be 90° ahead of the y component or it can be 90° behind the y component. In this special case, the electric vector traces out a circle in the plane, so this polarization is called circular polarization. The rotation direction in the circle depends on which of the two phase relationships exists and corresponds to *right-hand circular polarization* and *left-hand circular polarization*.

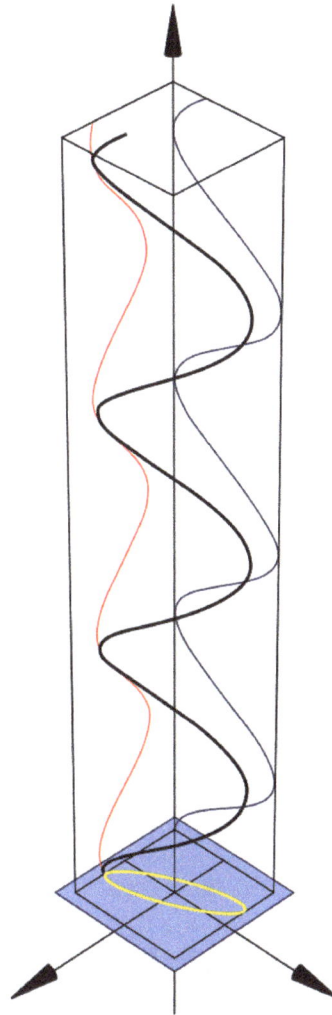

Elliptical polarization

In all other cases, where the two components either do not have the same amplitudes and/or their phase difference is neither zero nor a multiple of 90°, the polarization is called elliptical polarization because the electric vector traces out an ellipse in the plane (the *polarization ellipse*). This is shown in the above figure on the right. Detailed mathematics of polarization is done using Jones calculus and is characterised by the Stokes parameters.

Changing Polarization

Media that have different indexes of refraction for different polarization modes are called *birefringent*. Well known manifestations of this effect appear in optical wave plates/retarders (linear modes) and in Faraday rotation/optical rotation (circular modes). If the path length in the birefringent medium is sufficient, plane waves will exit the material with a significantly different propagation direction, due to refraction. For example, this is the case with macroscopic crystals of calcite, which present the viewer with two offset, orthogonally polarised images of whatever is viewed through them. It was this effect that provided the first discovery of polarization, by Erasmus Bartholinus in 1669. In addition, the phase shift, and thus the change in polarization state, is usually frequency dependent, which, in combination with dichroism, often gives rise to bright colours and rainbow-like effects. In mineralogy, such properties, known as pleochroism, are frequently exploited for the purpose of identifying minerals using polarization microscopes. Additionally, many plastics that are not normally birefringent will become so when subject to mechanical stress, a phenomenon which is the basis of photoelasticity. Non-birefringent methods, to rotate the linear polarization of light beams, include the use of prismatic polarization rotators which use total internal reflection in a prism set designed for efficient collinear transmission.

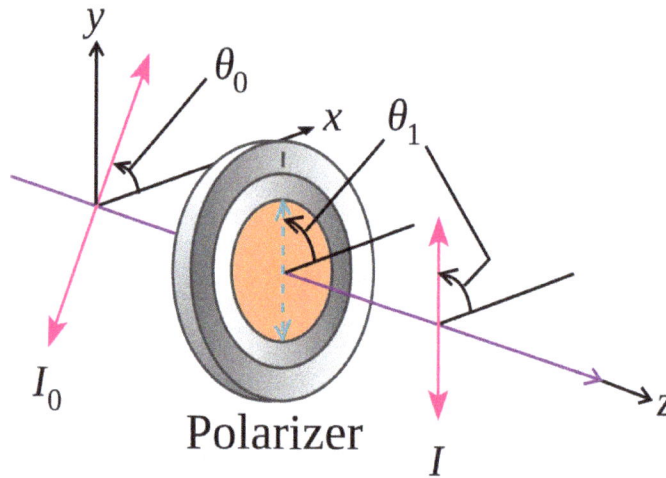

A polariser changing the orientation of linearly polarised light.
In this picture, $\theta_1 - \theta_0 = \theta_i$.

Media that reduce the amplitude of certain polarization modes are called *dichroic*, with devices that block nearly all of the radiation in one mode known as *polarizing filters* or simply "polarisers". Malus' law, which is named after Étienne-Louis Malus, says that when a perfect polariser is placed in a linear polarised beam of light, the intensity, I, of the light that passes through is given by

$$I = I_0 \cos^2 \theta_i ,$$

where

I_0 is the initial intensity,

and θ_i is the angle between the light's initial polarization direction and the axis of the polariser.

A beam of unpolarised light can be thought of as containing a uniform mixture of linear polarizations at all possible angles. Since the average value of $\cos^2 \theta$ is 1/2, the transmission coefficient becomes

$$\frac{I}{I_0} = \frac{1}{2}$$

In practice, some light is lost in the polariser and the actual transmission of unpolarised light will be somewhat lower than this, around 38% for Polaroid-type polarisers but considerably higher (>49.9%) for some birefringent prism types.

In addition to birefringence and dichroism in extended media, polarization effects can also occur at the (reflective) interface between two materials of different refractive index. These effects are treated by the Fresnel equations. Part of the wave is transmitted and part is reflected, with the ratio depending on angle of incidence and the angle of refraction. In this way, physical optics recovers Brewster's angle. When light reflects from a thin film on a surface, interference between the reflections from the film's surfaces can produce polarization in the reflected and transmitted light.

Natural Light

The effects of a polarising filter on the sky in a photograph. Left picture is taken without polariser. For the right picture, filter was adjusted to eliminate certain polarizations of the scattered blue light from the sky.

Most sources of electromagnetic radiation contain a large number of atoms or molecules that emit light. The orientation of the electric fields produced by these emitters may not be correlated, in which case the light is said to be *unpolarised*. If there is partial correlation between the emitters, the light is *partially polarised*. If the polarization is consistent across the spectrum of the source, partially polarised light can be described as a superposition of a completely unpolarised component, and a completely polarised one. One may then describe the light in terms of the degree of polarization, and the parameters of the polarization ellipse.

Light reflected by shiny transparent materials is partly or fully polarised, except when the light is normal (perpendicular) to the surface. It was this effect that allowed the mathematician Étienne-Louis Malus to make the measurements that allowed for his development of the first mathematical models for polarised light. Polarization occurs when light is scattered in the atmosphere. The scattered light produces the brightness and colour in clear skies. This partial polarization of scattered light can be taken advantage of using polarizing filters to darken the sky in photographs. Optical polarization is principally of importance in chemistry due to circular dichroism and optical rotation ("*circular birefringence*") exhibited by optically active (chiral) molecules.

Modern Optics

Modern optics encompasses the areas of optical science and engineering that became popular in the 20th century. These areas of optical science typically relate to the electromagnetic or quantum properties of light but do include other topics. A major subfield of modern optics, quantum optics, deals with specifically quantum mechanical properties of light. Quantum optics is not just theoretical; some modern devices, such as lasers, have principles of operation that depend on quantum mechanics. Light detectors, such as photomultipliers and channeltrons, respond to individual photons. Electronic image sensors, such as CCDs, exhibit shot noise corresponding to the statistics of individual photon events. Light-emitting diodes and photovoltaic cells, too, cannot be understood without quantum mechanics. In the study of these devices, quantum optics often overlaps with quantum electronics.

Specialty areas of optics research include the study of how light interacts with specific materials as in crystal optics and metamaterials. Other research focuses on the phenomenology of electromagnetic waves as in singular optics, non-imaging optics, non-linear optics, statistical optics, and radiometry. Additionally, computer engineers have taken an interest in integrated optics, machine vision, and photonic computing as possible components of the "next generation" of computers.

Today, the pure science of optics is called optical science or optical physics to distinguish it from applied optical sciences, which are referred to as optical engineering. Prominent subfields of optical engineering include illumination engineering, photonics, and optoelectronics with practical applications like lens design, fabrication and testing of optical components, and image processing. Some of these fields overlap, with nebulous boundaries between the subjects terms that mean slightly different things in different parts of the world and in different areas of industry. A professional community of researchers in nonlinear optics has developed in the last several decades due to advances in laser technology.

Lasers

Experiments such as this one with high-power lasers are part of the modern optics research.

A laser is a device that emits light (electromagnetic radiation) through a process called *stimulated emission*. The term *laser* is an acronym for *Light Amplification by Stimulated Emission of Radiation*. Laser light is usually spatially coherent, which means that the light either is emitted in a narrow, low-divergence beam, or can be converted into one with the help of optical components such as lenses. Because the microwave equivalent of the laser, the *maser*, was developed first, devices that emit microwave and radio frequencies are usually called *masers*.

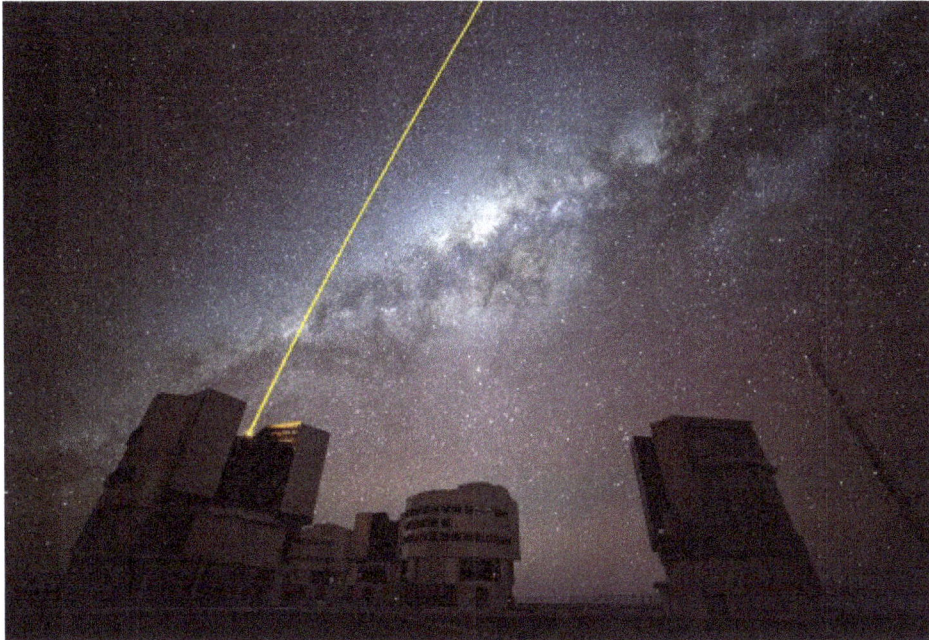

VLT's laser guided star.

The first working laser was demonstrated on 16 May 1960 by Theodore Maiman at Hughes Research Laboratories. When first invented, they were called "a solution looking for a problem". Since then, lasers have become a multibillion-dollar industry, finding utility in thousands of highly varied applications. The first application of lasers visible in the daily lives of the general population was the supermarket barcode scanner, introduced in 1974. The laserdisc player, introduced in 1978, was the first successful consumer product to include a laser, but the compact disc player was the first laser-equipped device to become truly common in consumers' homes, beginning in 1982. These optical storage devices use a semiconductor laser less than a millimetre wide to scan the surface of the disc for data retrieval. Fibre-optic communication relies on lasers to transmit large amounts of information at the speed of light. Other common applications of lasers include laser printers and laser pointers. Lasers are used in medicine in areas such as bloodless surgery, laser eye surgery, and laser capture microdissection and in military applications such as missile defence systems, electro-optical countermeasures (EOCM), and lidar. Lasers are also used in holograms, bubblegrams, laser light shows, and laser hair removal.

Kapitsa–dirac Effect

The Kapitsa–Dirac effect causes beams of particles to diffract as the result of meeting a standing wave of light. Light can be used to position matter using various phenomena.

Applications

Optics is part of everyday life. The ubiquity of visual systems in biology indicates the central role optics plays as the science of one of the five senses. Many people benefit from eyeglasses or contact lenses, and optics are integral to the functioning of many consumer goods including cameras. Rainbows and mirages are examples of optical phenomena. Optical communication provides the backbone for both the Internet and modern telephony.

Human Eye

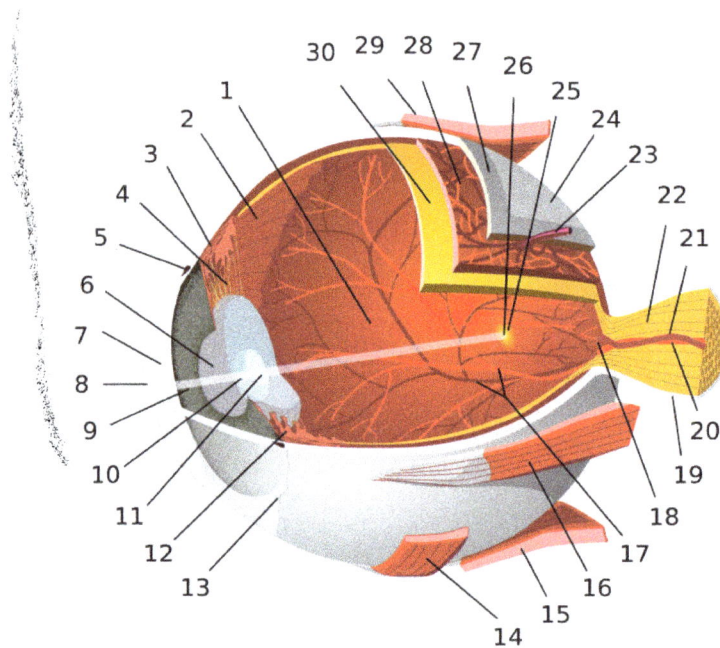

Model of a human eye. Features mentioned in this article are 3. ciliary muscle, 6. pupil, 8. cornea, 10. lens cortex, 22. optic nerve, 26. fovea, 30. retina

The human eye functions by focusing light onto a layer of photoreceptor cells called the retina, which forms the inner lining of the back of the eye. The focusing is accomplished by a series of transparent media. Light entering the eye passes first through the cornea, which provides much of the eye's optical power. The light then continues through the fluid just behind the cornea—the anterior chamber, then passes through the pupil. The light then passes through the lens, which focuses the light further and allows adjustment of focus. The light then passes through the main body of fluid in the eye—the vitreous humour, and reaches the retina. The cells in the retina line the back of the eye, except for where the optic nerve exits; this results in a blind spot.

There are two types of photoreceptor cells, rods and cones, which are sensitive to different aspects of light. Rod cells are sensitive to the intensity of light over a wide frequency range, thus are responsible for black-and-white vision. Rod cells are not present on the fovea, the area of the retina responsible for central vision, and are not as responsive as cone cells to spatial and temporal changes in light. There are, however, twenty times more rod cells than cone cells in the retina because the rod cells are present across a wider area. Because of their wider distribution, rods are responsible for peripheral vision.

In contrast, cone cells are less sensitive to the overall intensity of light, but come in three varieties that are sensitive to different frequency-ranges and thus are used in the perception of colour and photopic vision. Cone cells are highly concentrated in the fovea and have a high visual acuity meaning that they are better at spatial resolution than rod cells. Since cone cells are not as sensitive to dim light as rod cells, most night vision is limited to rod cells. Likewise, since cone cells are in the fovea, central vision (including the vision needed to do most reading, fine detail work such as sewing, or careful examination of objects) is done by cone cells.

Ciliary muscles around the lens allow the eye's focus to be adjusted. This process is known as accommodation. The near point and far point define the nearest and farthest distances from the eye at which an object can be brought into sharp focus. For a person with normal vision, the far point is located at infinity. The near point's location depends on how much the muscles can increase the curvature of the lens, and how inflexible the lens has become with age. Optometrists, ophthalmologists, and opticians usually consider an appropriate near point to be closer than normal reading distance—approximately 25 cm.

Defects in vision can be explained using optical principles. As people age, the lens becomes less flexible and the near point recedes from the eye, a condition known as presbyopia. Similarly, people suffering from hyperopia cannot decrease the focal length of their lens enough to allow for nearby objects to be imaged on their retina. Conversely, people who cannot increase the focal length of their lens enough to allow for distant objects to be imaged on the retina suffer from myopia and have a far point that is considerably closer than infinity. A condition known as astigmatism results when the cornea is not spherical but instead is more curved in one direction. This causes horizontally extended objects to be focused on different parts of the retina than vertically extended objects, and results in distorted images.

All of these conditions can be corrected using corrective lenses. For presbyopia and hyperopia, a converging lens provides the extra curvature necessary to bring the near point closer to the eye while for myopia a diverging lens provides the curvature necessary to send the far point to infinity. Astigmatism is corrected with a cylindrical surface lens that curves more strongly in one direction than in another, compensating for the non-uniformity of the cornea.

The optical power of corrective lenses is measured in diopters, a value equal to the reciprocal of the focal length measured in metres; with a positive focal length corresponding to a converging lens and a negative focal length corresponding to a diverging lens. For lenses that correct for astigmatism as well, three numbers are given: one for the spherical power, one for the cylindrical power, and one for the angle of orientation of the astigmatism.

Visual Effects

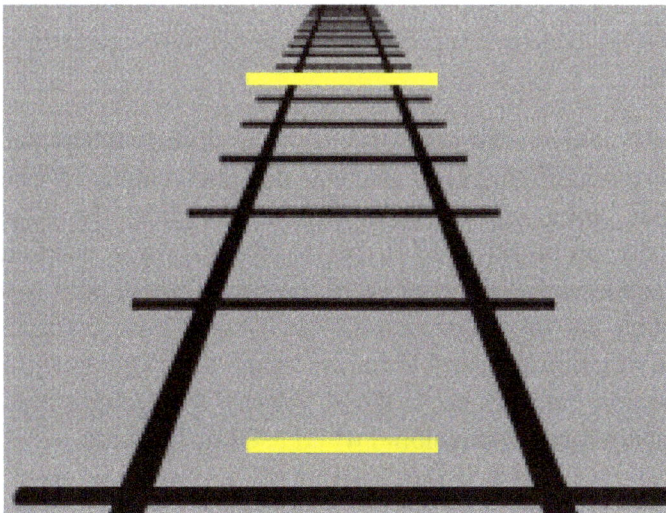

The Ponzo Illusion relies on the fact that parallel lines appear to converge as they approach infinity.

Optical illusions (also called visual illusions) are characterized by visually perceived images that differ from objective reality. The information gathered by the eye is processed in the brain to give a percept that differs from the object being imaged. Optical illusions can be the result of a variety of phenomena including physical effects that create images that are different from the objects that make them, the physiological effects on the eyes and brain of excessive stimulation (e.g. brightness, tilt, colour, movement), and cognitive illusions where the eye and brain make unconscious inferences.

Cognitive illusions include some which result from the unconscious misapplication of certain optical principles. For example, the Ames room, Hering, Müller-Lyer, Orbison, Ponzo, Sander, and Wundt illusions all rely on the suggestion of the appearance of distance by using converging and diverging lines, in the same way that parallel light rays (or indeed any set of parallel lines) appear to converge at a vanishing point at infinity in two-dimensionally rendered images with artistic perspective. This suggestion is also responsible for the famous moon illusion where the moon, despite having essentially the same angular size, appears much larger near the horizon than it does at zenith. This illusion so confounded Ptolemy that he incorrectly attributed it to atmospheric refraction when he described it in his treatise, *Optics*.

Another type of optical illusion exploits broken patterns to trick the mind into perceiving symmetries or asymmetries that are not present. Examples include the café wall, Ehrenstein, Fraser spiral, Poggendorff, and Zöllner illusions. Related, but not strictly illusions, are patterns that occur due to the superimposition of periodic structures. For example, transparent tissues with a grid structure produce shapes known as moiré patterns, while the superimposition of periodic transparent patterns comprising parallel opaque lines or curves produces line moiré patterns.

Optical Instruments

Single lenses have a variety of applications including photographic lenses, corrective lenses, and magnifying glasses while single mirrors are used in parabolic reflectors and rear-view mirrors. Combining a number of mirrors, prisms, and lenses produces compound optical instruments which have practical uses. For example, a periscope is simply two plane mirrors aligned to allow for viewing around obstructions. The most famous compound optical instruments in science are the microscope and the telescope which were both invented by the Dutch in the late 16th century.

Microscopes were first developed with just two lenses: an objective lens and an eyepiece. The objective lens is essentially a magnifying glass and was designed with a very small focal length while the eyepiece generally has a longer focal length. This has the effect of producing magnified images of close objects. Generally, an additional source of illumination is used since magnified images are dimmer due to the conservation of energy and the spreading of light rays over a larger surface area. Modern microscopes, known as *compound microscopes* have many lenses in them (typically four) to optimize the functionality and enhance image stability. A slightly different variety of microscope, the comparison microscope, looks at side-by-side images to produce a stereoscopic binocular view that appears three dimensional when used by humans.

The first telescopes, called *refracting telescopes* were also developed with a single objective and eyepiece lens. In contrast to the microscope, the objective lens of the telescope was designed with a large focal length to avoid optical aberrations. The objective focuses an image of a distant object at its focal point which is adjusted to be at the focal point of an eyepiece of a much smaller focal length. The main goal of a telescope is not necessarily magnification, but rather collection of light which is determined by the physical size of the objective lens. Thus, telescopes are normally indicated by the diameters of their objectives rather than by the magnification which can be changed by switching eyepieces. Because the magnification of a telescope is equal to the focal length of the objective divided by the focal length of the eyepiece, smaller focal-length eyepieces cause greater magnification.

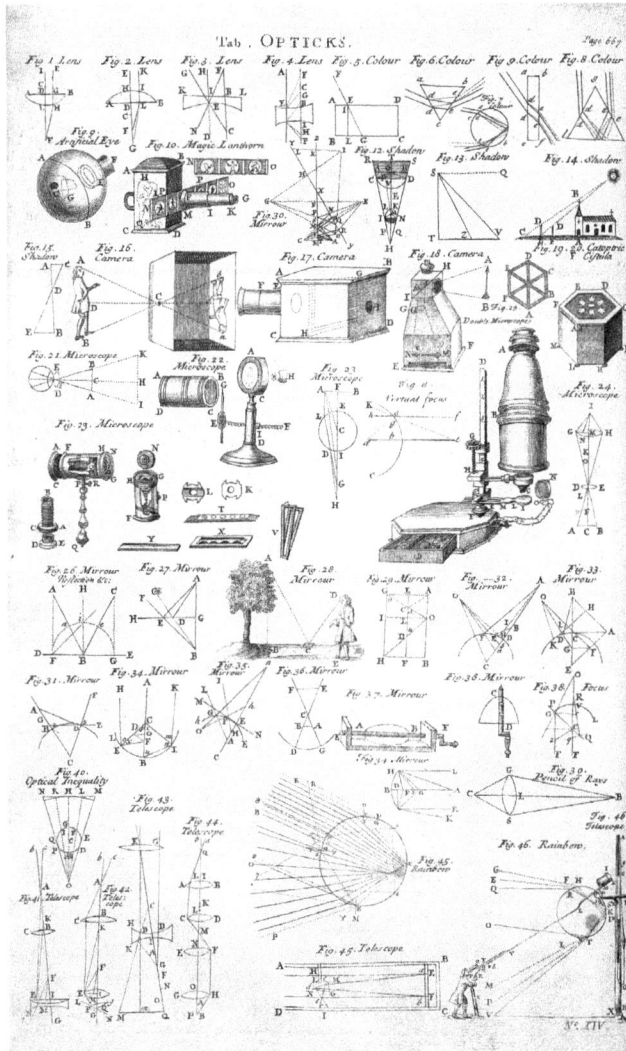

Illustrations of various optical instruments from the 1728 *Cyclopaedia*

Since crafting large lenses is much more difficult than crafting large mirrors, most modern telescopes are *reflecting telescopes*, that is, telescopes that use a primary mirror rather than an objective lens. The same general optical considerations apply to reflecting telescopes that applied to refracting telescopes, namely, the larger the primary mirror, the more light collected, and the magnification is still equal to the focal length of the primary mirror divided by the focal length of the eyepiece. Professional telescopes generally do not have eyepieces and instead place an instrument (often a charge-coupled device) at the focal point instead.

Photography

Photograph taken with aperture $f/32$

The optics of photography involves both lenses and the medium in which the electromagnetic radiation is recorded, whether it be a plate, film, or charge-coupled device. Photographers must consider the reciprocity of the camera and the shot which is summarized by the relation

$$\text{Exposure} \propto \text{ApertureArea} \times \text{ExposureTime} \times \text{SceneLuminance}$$

Photograph taken with aperture $f/5$

In other words, the smaller the aperture (giving greater depth of focus), the less light coming in, so the length of time has to be increased (leading to possible blurriness if motion occurs). An example of the use of the law of reciprocity is the Sunny 16 rule which gives a rough estimate for the settings needed to estimate the proper exposure in daylight.

A camera's aperture is measured by a unitless number called the f-number or f-stop, $f/\#$, often notated as N, and given by

$$f\,/\,\# = N = \frac{f}{D}$$

where f is the focal length, and D is the diameter of the entrance pupil. By convention, "$f/\#$" is treated as a single symbol, and specific values of $f/\#$ are written by replacing the number sign with the value. The two ways to increase the f-stop are to either decrease the diameter of the entrance pupil or change to a longer focal length (in the case of a zoom lens, this can be done by simply adjusting the lens). Higher f-numbers also have a larger depth of field due to the lens approaching the limit of a pinhole camera which is able to focus all images perfectly, regardless of distance, but requires very long exposure times.

The field of view that the lens will provide changes with the focal length of the lens. There are three basic classifications based on the relationship to the diagonal size of the film or sensor size of the camera to the focal length of the lens:

- Normal lens: angle of view of about 50° (called *normal* because this angle considered roughly equivalent to human vision) and a focal length approximately equal to the diagonal of the film or sensor.

- Wide-angle lens: angle of view wider than 60° and focal length shorter than a normal lens.

- Long focus lens: angle of view narrower than a normal lens. This is any lens with a focal length longer than the diagonal measure of the film or sensor. The most common type of long focus lens is the telephoto lens, a design that uses a special *telephoto group* to be physically shorter than its focal length.

Modern zoom lenses may have some or all of these attributes.

The absolute value for the exposure time required depends on how sensitive to light the medium being used is (measured by the film speed, or, for digital media, by the quantum efficiency). Early photography used media that had very low light sensitivity, and so exposure times had to be long even for very bright shots. As technology has improved, so has the sensitivity through film cameras and digital cameras.

Other results from physical and geometrical optics apply to camera optics. For example, the maximum resolution capability of a particular camera set-up is determined by the diffraction limit associated with the pupil size and given, roughly, by the Rayleigh criterion.

Atmospheric Optics

The unique optical properties of the atmosphere cause a wide range of spectacular optical phenomena. The blue colour of the sky is a direct result of Rayleigh scattering which redirects higher frequency (blue) sunlight back into the field of view of the observer. Because blue light is scattered more easily than red light, the sun takes on a reddish hue when it is observed through a thick atmosphere, as during a sunrise or sunset. Additional particulate matter in the sky can scatter different colours at different angles creating colourful glowing skies at dusk and dawn. Scattering off of ice crystals and other particles in the atmosphere are responsible for halos, afterglows, coronas, rays of sunlight, and sun dogs. The variation in these kinds of phenomena is due to different particle sizes and geometries.

Mirages are optical phenomena in which light rays are bent due to thermal variations in the refraction index of air, producing displaced or heavily distorted images of distant objects. Other dramatic optical phenomena associated with this include the Novaya Zemlya effect where the sun appears to rise earlier than predicted with a distorted shape. A spectacular form of refraction occurs with a temperature inversion called the Fata Morgana where objects on the horizon or even beyond the horizon, such as islands, cliffs, ships or icebergs, appear elongated and elevated, like "fairy tale castles".

Rainbows are the result of a combination of internal reflection and dispersive refraction of light in raindrops. A single reflection off the backs of an array of raindrops produces a rainbow with an angular size on the sky that ranges from 40° to 42° with red on the outside. Double rainbows are produced by two internal reflections with angular size of 50.5° to 54° with violet on the outside. Because rainbows are seen with the sun 180° away from the centre of the rainbow, rainbows are more prominent the closer the sun is to the horizon.

References

- Euclid (1999). Elaheh Kheirandish, ed. The Arabic version of Euclid's optics = Kitāb Uqlīdis fi ikhtilāf al-manāir. New York: Springer. ISBN 0-387-98523-9.

- Ptolemy (1996). A. Mark Smith, ed. Ptolemy's theory of visual perception: an English translation of the Optics with introduction and commentary. DIANE Publishing. ISBN 0-87169-862-5.

- Adamson, Peter (2006). "Al-Kindi¯ and the reception of Greek philosophy». In Adamson, Peter; Taylor, R.. The Cambridge companion to Arabic philosophy. Cambridge University Press. p. 45. ISBN 978-0-521-52069-0.

- Elena Agazzi; Enrico Giannetto; Franco Giudice (2010). Representing Light Across Arts and Sciences: Theories and Practices. V&R unipress GmbH. p. 42. ISBN 978-3-89971-735-8.

- Paul S. Agutter; Denys N. Wheatley (2008). Thinking about Life: The History and Philosophy of Biology and Other Sciences. Springer. p. 17. ISBN 978-1-4020-8865-0.

- Vincent, Ilardi (2007). Renaissance Vision from Spectacles to Telescopes. Philadelphia, PA: American Philosophical Society. pp. 4–5. ISBN 978-0-87169-259-7.

- N. Taylor (2000). LASER: The inventor, the Nobel laureate, and the thirty-year patent war. New York: Simon & Schuster. ISBN 0-684-83515-0.

- Ariel Lipson; Stephen G. Lipson; Henry Lipson (28 October 2010). Optical Physics. Cambridge University Press. p. 48. ISBN 978-0-521-49345-1. Retrieved 12 July 2012.

- Alastair D. McAulay (16 January 1991). Optical computer architectures: the application of optical concepts to next generation computers. Wiley. ISBN 978-0-471-63242-9. Retrieved 12 July 2012.

- C. H. Townes (2003). "The first laser". In Laura Garwin; Tim Lincoln. A Century of Nature: Twenty-One Discoveries that Changed Science and the World. University of Chicago Press. pp. 107–12. ISBN 0-226-28413-1.

- J. Wilson & J.F.B. Hawkes (1987). Lasers: Principles and Applications, Prentice Hall International Series in Optoelectronics. Prentice Hall. ISBN 0-13-523697-5.

- E. R. Kandel; J. H. Schwartz; T. M. Jessell (2000). Principles of Neural Science (4th ed.). New York: McGraw-Hill. pp. 507–513. ISBN 0-8385-7701-6.

- A. K. Jain; M. Figueiredo; J. Zerubia (2001). Energy Minimization Methods in Computer Vision and Pattern Recognition. Springer. ISBN 978-3-540-42523-6.

Evolution of Optics

Optics evolved from the development of lenses. The history of optics is divided into two halves, classical optics and modern optics. Modern optics has developed in the 20th century and covers topics such as wave optics and quantum optics. The following section will not only provide an overview, it will delve deep into the topics related to it.

Optics began with the development of lenses by the ancient Egyptians and Mesopotamians, followed by theories on light and vision developed by ancient Greek philosophers, and the development of geometrical optics in the Greco-Roman world. Optics was significantly reformed by the developments in the medieval Islamic world, such as the beginnings of physical and physiological optics, and then significantly advanced in early modern Europe, where diffractive optics began. These earlier studies on optics are now known as "classical optics". The term "modern optics" refers to areas of optical research that largely developed in the 20th century, such as wave optics and quantum optics.

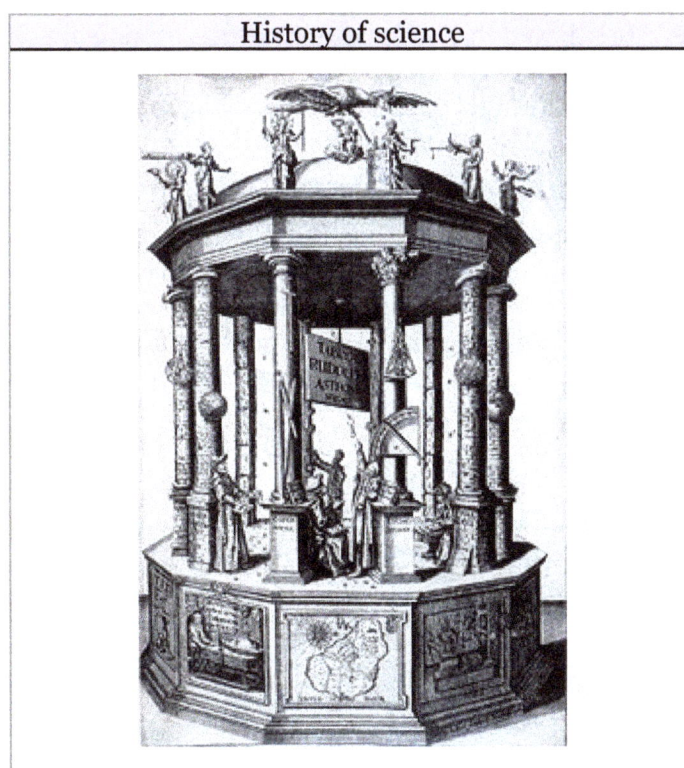

History of science

Early History of Optics

The earliest known lenses were made from polished crystal, often quartz, and have been dated as early as 750 BC for Assyrian lenses such as the Nimrud / Layard lens. There are many similar lenses from ancient Egypt, Greece and Babylon. The ancient Romans and Greeks filled glass spheres with water to make lenses. However, glass lenses were not thought of until the Middle Ages.

Some lenses fixed in ancient Egyptian statues are much older than those mentioned above. There is some doubt as to whether or not they qualify as lenses, but they are undoubtedly glass and served at least ornamental purposes. The statues appear to be anatomically correct schematic eyes

In ancient India, the philosophical schools of Samkhya and Vaisheshika, from around the 6th–5th century BC, developed theories on light. According to the Samkhya school, light is one of the five fundamental "subtle" elements (*tanmatra*) out of which emerge the gross elements.

In contrast, the Vaisheshika school gives an atomic theory of the physical world on the non-atomic ground of ether, space and time. The basic atoms are those of earth (*prthivı*), water (*apas*), fire (*tejas*), and air (*vayu*), that should not be confused with the ordinary meaning of these terms. These atoms are taken to form binary molecules that combine further to form larger molecules. Motion is defined in terms of the movement of the physical atoms. Light rays are taken to be a stream of high velocity of *tejas* (fire) atoms. The particles of light can exhibit different characteristics depending on the speed and the arrange-ments of the *tejas* atoms. Around the first century BC, the *Vishnu Purana* refers to sunlight as "the seven rays of the sun".

In the fifth century BC, Empedocles postulated that everything was composed of four elements; fire, air, earth and water. He believed that Aphrodite made the human eye out of the four elements and that she lit the fire in the eye which shone out from the eye making sight possible. If this were true, then one could see during the night just as well as during the day, so Empedocles postulated an interaction between rays from the eyes and rays from a source such as the sun.

In his *Optics* Greek mathematician Euclid observed that "things seen under a greater angle appear greater, and those under a lesser angle less, while those under equal angles appear equal". In the 36 propositions that follow, Euclid relates the apparent size of an object to its distance from the eye and investigates the apparent shapes of cylinders and cones when viewed from different angles. Pappus believed these results to be important in astronomy and included Euclid's *Optics*, along with his *Phaenomena*, in the *Little Astronomy*, a compendium of smaller works to be studied before the *Syntaxis* (*Almagest*) of Ptolemy.

In 55 BC, Lucretius, a Roman who carried on the ideas of earlier Greek atomists, wrote:

The light and heat of the sun; these are composed of minute atoms which, when they are shoved off, lose no time in shooting right across the interspace of air in the direction imparted by the shove.

— Lucretius, On the nature of the Universe

Despite being similar to later particle theories of light, Lucretius's views were not generally accepted and light was still theorized as emanating from the eye.

In his *Catoptrica*, Hero of Alexandria showed by a geometrical method that the actual path taken by a ray of light reflected from a plane mirror is shorter than any other reflected path that might be drawn between the source and point of observation.

In the second century Claudius Ptolemy, an Alexandrian Greek or Hellenized Egyptian, undertook studies of reflection and refraction. He measured the angles of refraction between air, water, and glass, and his published results indicate that he adjusted his measurements to fit his (incorrect) assumption that the angle of refraction is proportional to the angle of incidence.

The Indian Buddhists, such as Dignāga in the 5th century and Dharmakirti in the 7th century, developed a type of atomism that is a philosophy about reality being composed of atomic entities that are momentary flashes of light or energy. They viewed light as being an atomic entity equivalent to energy, similar to the modern concept of photons, though they also viewed all matter as being composed of these light/energy particles.

The Beginnings of Geometrical Optics

The early writers discussed here treated vision more as a geometrical than as a physical, physiological, or psychological problem. The first known author of a treatise on geometrical optics was the geometer Euclid (c. 325 BC–265 BC). Euclid began his study of optics as he began his study of geometry, with a set of self-evident axioms.

1. Lines (or visual rays) can be drawn in a straight line to the object.

2. Those lines falling upon an object form a cone.

3. Those things upon which the lines fall are seen.

4. Those things seen under a larger angle appear larger.

5. Those things seen by a higher ray, appear higher.

6. Right and left rays appear right and left.

7. Things seen within several angles appear clearer.

Euclid did not define the physical nature of these visual rays but, using the principles of geometry, he discussed the effects of perspective and the rounding of things seen at a distance.

Where Euclid had limited his analysis to simple direct vision, Hero of Alexandria (c. AD 10–70) extended the principles of geometrical optics to consider problems of reflection (catoptrics). Unlike Euclid, Hero occasionally commented on the physical nature of visual rays, indicating that they proceeded at great speed from the eye to the object seen and were reflected from smooth surfaces but could become trapped in the porosities of unpolished surfaces. This has come to be known as *emission theory*.

Hero demonstrated the equality of the angle of incidence and reflection on the grounds that this is the shortest path from the object to the observer. On this basis, he was able to define the fixed relation between an object and its image in a plane mirror. Specifically, the image appears to be as far behind the mirror as the object really is in front of the mirror.

Like Hero, Ptolemy (c. 90–c. 168) considered the visual rays as proceeding from the eye to the object seen, but, unlike Hero, considered that the visual rays were not discrete lines, but formed a continuous cone. Ptolemy extended the study of vision beyond direct and reflected vision; he also studied vision by refracted rays (dioptrics), when we see objects through the interface between two media of different density. He conducted experiments to measure the path of vision when we look from air to water, from air to glass, and from water to glass and tabulated the relationship between the incident and refracted rays.

His tabulated results have been studied for the air water interface, and in general the values he obtained reflect the theoretical refraction given by modern theory, but the outliers are distorted to represent Ptolemy's *a priori* model of the nature of refraction

Optics and Vision in the Islamic World

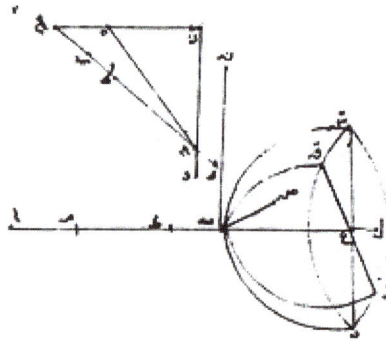

Reproduction of a page of Ibn Sahl's manuscript showing his discovery of the law of refraction, now known as Snell's law.

Al-Kindi (c. 801–873) was one of the earliest important optical writers in the Islamic world. In a work known in the west as *De radiis stellarum*, al-Kindi developed a theory "that everything in the world ... emits rays in every direction, which fill the whole world."

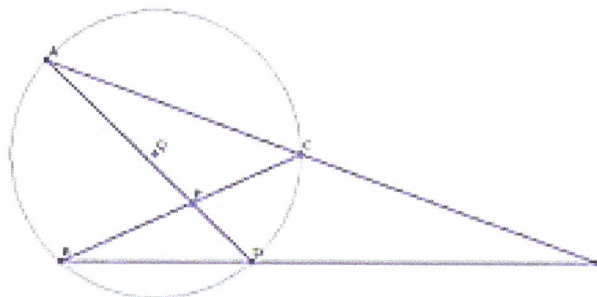

The theorem of Ibn Haytham

This theory of the active power of rays had an influence on later scholars such as Ibn al-Haytham, Robert Grosseteste and Roger Bacon.

Ibn Sahl (c. 940-1000) was an Iraqi mathematician associated with the court of Baghdad. About 984 he wrote a treatise *On Burning Mirrors and Lenses* in which he set out his understanding of how curved mirrors and lenses bend and focus light. In his work he discovered a law of refraction mathematically equivalent to Snell's law. He used his law of refraction to compute the shapes of lenses and mirrors that focus light at a single point on the axis.

Ibn al-Haytham (known in as *Alhacen* or *Alhazen* in Western Europe) (965–1040) produced a comprehensive and systematic analysis of Greek optical theories. Ibn al-Haytham's key achievement was twofold: first, to insist that vision occurred because of rays entering the eye; the second was to define the physical nature of the rays discussed by earlier geometrical optical writers, considering them as the forms of light and color. He then analyzed these physical rays according to the principles of geometrical optics. He wrote many books on optics, most significantly the *Book of Optics* (*Kitab al Manazir* in Arabic), translated into Latin as the *De aspectibus* or *Perspectiva*, which disseminated his ideas to Western Europe and had great influence on the later developments of optics.

Avicenna (980-1037) agreed with Alhazen that the speed of light is finite, as he "observed that if the perception of light is due to the emission of some sort of particles by a luminous source, the speed of light must be finite." Abū Rayhān al-Bīrūnī (973-1048) also agreed that light has a finite speed, and he was the first to discover that the speed of light is much faster than the speed of sound.

Abu 'Abd Allah Muhammad ibn Ma'udh, who lived in Al-Andalus during the second half of the 11th century, wrote a work on optics later translated into Latin as *Liber de crepisculis*, which was mistakenly attributed to Alhazen. This was a "short work containing an estimation of the angle of depression of the sun at the beginning of the morning twilight and at the end of the evening twilight, and an attempt to calculate on the basis of this and other data the height of the atmospheric moisture responsible for the refraction of the sun's rays." Through his experiments, he obtained the value of 18°, which comes close to the modern value.

In the late 13th and early 14th centuries, Qutb al-Din al-Shirazi (1236–1311) and his student Kamāl al-Dīn al-Fārisī (1260–1320) continued the work of Ibn al-Haytham, and they were the first to give the correct explanations for the rainbow phenomenon. Al-Fārisī published his findings in his *Kitab Tanqih al-Manazir* (*The Revision of* [Ibn al-Haytham's] *Optics*).

Optics in Medieval Europe

The English bishop, Robert Grosseteste (c. 1175–1253), wrote on a wide range of scientific topics at the time of the origin of the medieval university and the recovery of the works of Aristotle. Grosseteste reflected a period of transition between the Platonism of early medieval learning and the new Aristotelianism, hence he tended to apply mathematics and the Platonic metaphor of light in many of his writings. He has been credited with discussing light from four different perspectives: an epistemology of light, a metaphysics or cosmogony of light, an etiology or physics of light, and a theology of light.

Setting aside the issues of epistemology and theology, Grosseteste's cosmogony of light describes the origin of the universe in what may loosely be described as a medieval "big bang" theory. Both his biblical commentary, the *Hexaemeron* (1230 x 35), and his scientific *On Light* (1235 x 40), took their inspiration from Genesis 1:3, "God said, let there be light", and described the subsequent process of creation as a natural physical process arising from the generative power of an expanding (and contracting) sphere of light.

His more general consideration of light as a primary agent of physical causation appears in his *On Lines, Angles, and Figures* where he asserts that "a natural agent propagates its power from itself to the recipient" and in *On the Nature of Places* where he notes that "every natural action is varied in strength and weakness through variation of lines, angles and figures."

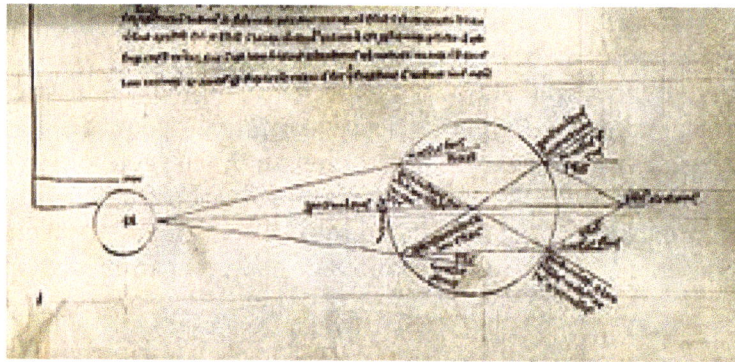

Optical diagram showing light being refracted by a spherical glass container full of water. (from Roger Bacon, *De multiplicatione specierum*)

The English Franciscan, Roger Bacon (c. 1214–1294) was strongly influenced by Grosseteste's writings on the importance of light. In his optical writings (the *Perspectiva*, the *De multiplicatione specierum*, and the *De speculis comburentibus*) he cited a wide range of recently translated optical and philosophical works, including those of Alhacen, Aristotle, Avicenna, Averroes, Euclid, al-Kindi, Ptolemy, Tideus, and Constantine the African. Although he was not a slavish imitator, he drew his mathematical analysis of light and vision from the writings of the Arabic writer, Alhacen. But he added to this the Neoplatonic concept, perhaps drawn from Grosseteste, that every object radiates a power (*species*) by which it acts upon nearby objects suited to receive those *species*. Note that Bacon's optical use of the term "*species*" differs significantly from the genus / species categories found in Aristotelian philosophy.

Another English Franciscan, John Pecham (died 1292) built on the work of Bacon, Grosseteste, and a diverse range of earlier writers to produce what became the most widely used textbook on Optics of the Middle Ages, the *Perspectiva communis*. His book centered on the question of vision, on how we see, rather than on the nature of light and color. Pecham followed the model set forth by Alhacen, but interpreted Alhacen's ideas in the manner of Roger Bacon.

Like his predecessors, Witelo (c. 1230–1280 x 1314) drew on the extensive body of optical works recently translated from Greek and Arabic to produce a massive presentation of the subject entitled the *Perspectiva*. His theory of vision follows Alhacen and he does not consider Bacon's concept of *species*, although passages in his work demonstrate that he was influenced by Bacon's ideas. Judging from the number of surviving manuscripts, his work was not as influential as those of Pecham and Bacon, yet his importance, and that of Pecham, grew with the invention of printing.

- Peter of Limoges (1240–1306)

- Theodoric of Freiberg (ca. 1250–ca. 1310)

Renaissance and Early Modern Optics

Johannes Kepler (1571–1630) picked up the investigation of the laws of optics from his lunar essay of 1600. Both lunar and solar eclipses presented unexplained phenomena, such as unexpected shadow sizes, the red color of a total lunar eclipse, and the reportedly unusual light surrounding a total solar eclipse. Related issues of atmospheric refraction applied to all astronomical observations. Through most of 1603, Kepler paused his other work to focus on optical theory; the resulting manuscript, presented to the emperor on January 1, 1604, was published as *Astronomiae Pars Optica* (*The Optical Part of Astronomy*). In it, Kepler described the inverse-square law governing the intensity of light, reflection by flat and curved mirrors, and principles of pinhole cameras, as well as the astronomical implications of optics such as parallax and the apparent sizes of heavenly bodies. *Astronomiae Pars Optica* is generally recognized as the foundation of modern optics (though the law of refraction is conspicuously absent).

Willebrord Snellius (1580–1626) found the mathematical law of refraction, now known as Snell's law, in 1621. Subsequently René Descartes (1596–1650) showed, by using geometric construction and the law of refraction (also known as Descartes' law), that the angular radius of a rainbow is 42° (i.e. the angle subtended at the eye by the edge of the rainbow and the rainbow's centre is 42°). He also independently discovered the law of reflection, and his essay on optics was the first published mention of this law.

Christiaan Huygens (1629–1695) wrote several works in the area of optics. These included the *Opera reliqua* (also known as *Christiani Hugenii Zuilichemii, dum viveret Zelhemii toparchae, opuscula posthuma*) and the *Traité de la lumière*.

Isaac Newton (1643–1727) investigated the refraction of light, demonstrating that a prism could decompose white light into a spectrum of colours, and that a lens and a second prism could recompose the multicoloured spectrum into white light. He also showed that the coloured light does not change its properties by separating out a coloured beam and shining it on various objects. Newton noted that regardless of whether it was reflected or scattered or transmitted, it stayed the same colour. Thus, he observed that colour is the result of objects interacting with already-coloured light rather than objects generating the colour themselves. This is known as Newton's theory of colour. From this work he concluded that any refracting telescope would suffer from the dispersion of light into colours, and invented a reflecting telescope (today known as a Newtonian telescope) to bypass that problem. By grinding his own mirrors, using Newton's rings to judge the quality of the optics for his telescopes, he was able to produce a superior instrument to the refracting telescope, due primarily to the wider diameter of the mirror. In 1671 the Royal Society asked for a demonstration of his reflecting telescope. Their interest encouraged him to publish his notes *On Colour*, which he later expanded into his *Opticks*. Newton argued that light is composed of particles or *corpuscles* and were refracted by accelerating toward the denser medium, but he had to associate them with waves to explain the diffraction of light (*Opticks* Bk. II, Props. XII-L). Later physicists instead favoured a purely wavelike explanation of light to account for diffraction. Today's quantum mechanics, photons and the idea of wave-particle duality bear only a minor resemblance to Newton's understanding of light.

In his *Hypothesis of Light* of 1675, Newton posited the existence of the ether to transmit forces between particles. In 1704, Newton published *Opticks*, in which he expounded his corpuscular theory of light. He considered light to be made up of extremely subtle corpuscles, that ordinary matter was made of grosser corpuscles and speculated that through a kind of alchemical transmutation "Are not gross Bodies and Light convertible into one another, ...and may not Bodies receive much of their Activity from the Particles of Light which enter their Composition?"

The Beginnings of Diffractive Optics

The effects of diffraction of light were first carefully observed and characterized by Francesco Maria Grimaldi, who also coined the term *diffraction*, from the Latin *diffringere*, 'to break into pieces', referring to light breaking up into different directions. The results of Grimaldi's observations were published posthumously in 1665. Isaac Newton studied these effects and attributed them to *inflexion* of light rays. James Gregory (1638–1675) observed the diffraction patterns caused by a bird feather, which was effectively the first diffraction grating. In 1803 Thomas Young did his famous experiment observing interference from two closely spaced slits in his double slit interferometer. Explaining his results by interference of the waves emanating from the two different slits, he deduced that light must propagate as waves. Augustin-Jean Fresnel did more definitive studies and calculations of diffraction, published in 1815 and 1818, and thereby gave great support to the wave theory of light that had been advanced by Christiaan Huygens and reinvigorated by Young, against Newton's particle theory.

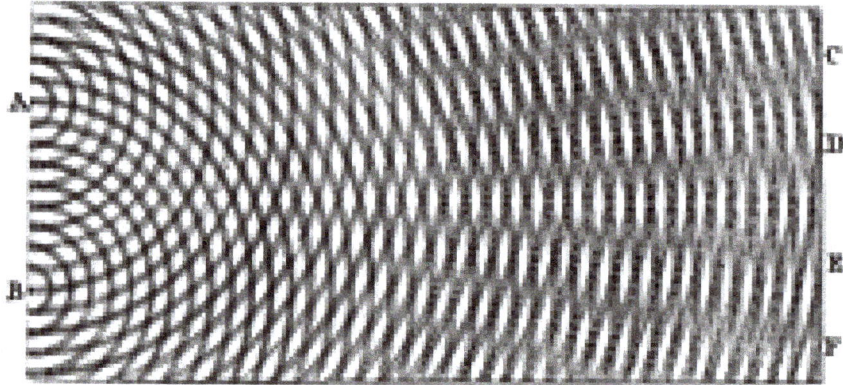

Thomas Young's sketch of two-slit diffraction, which he presented to the Royal Society in 1803

Lenses and Lensmaking

The earliest known lenses were made from polished crystal, often quartz, and have been dated as early as 750 BC for Assyrian lenses such as the Layard / Nimrud lens. There are many similar lenses from ancient Egypt, Greece and Babylon. The ancient Romans and Greeks filled glass spheres with water to make lenses.

The earliest historical reference to magnification dates back to ancient Egyptian hieroglyphs in the 5th century BC, which depict "simple glass meniscal lenses". The earliest written record of magnification dates back to the 1st century AD, when Seneca the Younger, a tutor of Emperor Nero, wrote: "Letters, however small and indistinct, are seen enlarged and more clearly through a globe or glass filled with water". Emperor Nero is also said to have watched the gladiatorial games using an emerald as a corrective lens.

Ibn al-Haytham (Alhacen) wrote about the effects of pinhole, concave lenses, and magnifying glasses in his *Book of Optics*. Roger Bacon used parts of glass spheres as magnifying glasses and recommended them to be used to help people read. Roger Bacon got his inspiration from Alhacen in the 11th century. He discovered that light reflects from objects and does not get released from them. Around 1284 in Italy, Salvino D'Armate is credited with inventing the first wearable eye glasses.

Between the 11th and 13th century "reading stones" were invented. Often used by monks to assist in illuminating manuscripts, these were primitive plano-convex lenses initially made by cutting a glass sphere in half. As the stones were experimented with, it was slowly understood that shallower lenses magnified more effectively.

The earliest known working telescopes were the refracting telescopes that appeared in the Netherlands in 1608. Their development is credited to three individuals: Hans Lippershey and Zacharias Janssen, who were spectacle makers in Middelburg, and Jacob Metius of Alkmaar. Galileo greatly improved upon these designs the following year. Isaac Newton is credited with constructing the first functional reflecting telescope in 1668, his Newtonian reflector.

The first microscope was made around 1595 in Middelburg in the Dutch Republic. Three different eyeglass makers have been given credit for the invention: Hans Lippershey (who also developed the first real telescope); Hans Janssen; and his son, Zacharias. The coining of the name "microscope" has been credited to Giovanni Faber, who gave that name to Galileo Galilei's compound microscope in 1625.

Quantum Optics

Light is made up of particles called photons and hence inherently is quantized. Quantum optics is the study of the nature and effects of light as quantized photons. The first indication that light might be quantized came from Max Planck in 1899 when he correctly modelled blackbody radiation by assuming that the exchange of energy between light and matter only occurred in discrete amounts he called quanta. It was unknown whether the source of this discreteness was the matter or the light. In 1905, Albert Einstein published the theory of the photoelectric effect. It appeared that the only possible explanation for the effect was the quantization of light itself. Later, Niels Bohr showed that atoms could only emit discrete amounts of energy. The understanding of the interaction between light and matter following from these developments not only formed the basis of quantum optics but also were crucial for the development of quantum mechanics as a whole. However, the subfields of quantum mechanics dealing with matter-light interaction were principally regarded as research into matter rather than into light and hence, one rather spoke of atom physics and quantum electronics.

This changed with the invention of the maser in 1953 and the laser in 1960. Laser science—research into principles, design and application of these devices—became an important field, and the quantum mechanics underlying the laser's principles was studied now with more emphasis on the properties of light, and the name *quantum optics* became customary.

As laser science needed good theoretical foundations, and also because research into these soon proved very fruitful, interest in quantum optics rose. Following the work of Dirac in quantum field theory, George Sudarshan, Roy J. Glauber, and Leonard Mandel applied quantum theory to the electromagnetic field in the 1950s and 1960s to gain a more detailed understanding of photodetection and the statistics of light. This led to the introduction of the coherent state as a quantum description of laser light and the realization that some states of light could not be described with classical waves. In 1977, Kimble et al. demonstrated the first source of light which required a quantum description: a single atom that emitted one photon at a time. Another quantum state of light with certain advantages over any classical state, squeezed light, was soon proposed. At the same time, development of short and ultrashort laser pulses—created by Q-switching and mode-locking techniques—opened the way to the study of unimaginably fast ("ultrafast") processes. Applications for solid state research (e.g. Raman spectroscopy) were found, and mechanical forces of light on matter were studied. The latter led to levitating and positioning clouds of atoms or even small biological samples in an optical trap or optical tweezers by laser beam. This, along with Doppler cooling was the crucial technology needed to achieve the celebrated Bose–Einstein condensation.

Other remarkable results are the demonstration of quantum entanglement, quantum teleportation, and (recently, in 1995) quantum logic gates. The latter are of much interest in quantum information theory, a subject which partly emerged from quantum optics, partly from theoretical computer science.

Today's fields of interest among quantum optics researchers include parametric down-conversion, parametric oscillation, even shorter (attosecond) light pulses, use of quantum optics for quantum information, manipulation of single atoms, Bose–Einstein condensates, their application, and how to manipulate them (a sub-field often called atom optics), and much more.

References

- Morelon, Régis; Rashed, Roshdi (1996), Encyclopedia of the History of Arabic Science, 2, Routledge, ISBN 0-415-12410-7, OCLC 34731151 .

- Wade, Nicholas J. (1998), A Natural History of Vision, Cambridge, Massachusetts: MIT Press, ISBN 0-262-23194-8, OCLC 37246567 .

3

Branches of Optics

The branches of optics discussed in this section are geometrical optics, physical optics, quantum optics, nonimaging optics and nonlinear optics. Geometrical optics defines light propagation in terms of rays whereas physics optics is a branch of optics that concerns itself with diffraction and polarization. This chapter is a compilation of the various branches of optics that form an integral part of the broader subject matter.

Geometrical Optics

Geometrical optics, or ray optics, describes light propagation in terms of rays. The ray in geometric optics is an abstraction, or instrument, useful in approximating the paths along which light propagates in certain classes of circumstances.

The simplifying assumptions of geometrical optics include that light rays:

- propagate in rectilinear paths as they travel in a homogeneous medium
- bend, and in particular circumstances may split in two, at the interface between two dissimilar media
- follow curved paths in a medium in which the refractive index changes
- may be absorbed or reflected.

Geometrical optics does not account for certain optical effects such as diffraction and interference. This simplification is useful in practice; it is an excellent approximation when the wavelength is small compared to the size of structures with which the light interacts. The techniques are particularly useful in describing geometrical aspects of imaging, including optical aberrations.

Explanation

A light ray is a line or curve that is perpendicular to the light's wavefronts (and is therefore collinear with the wave vector).

A slightly more rigorous definition of a light ray follows from Fermat's principle, which states that the path taken between two points by a ray of light is the path that can be traversed in the least time.

Geometrical optics is often simplified by making the paraxial approximation, or "small angle approximation." The mathematical behavior then becomes linear, allowing optical components and systems to be described by simple matrices. This leads to the techniques of Gaussian optics and *paraxial ray tracing*, which are used to find basic properties of optical systems, such as approximate image and object positions and magnifications.

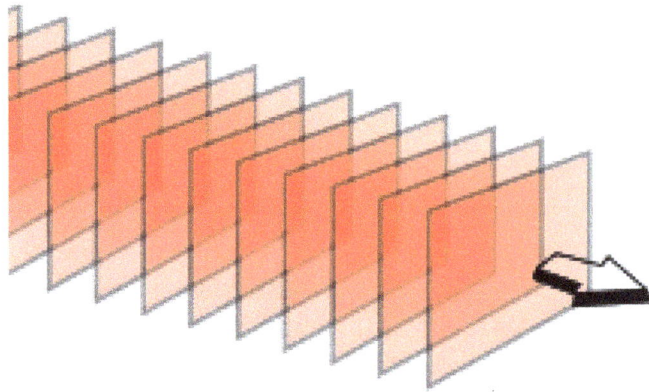

As light travels through space, it oscillates in amplitude. In this image, each maximum amplitude crest is marked with a plane to illustrate the wavefront. The ray is the arrow perpendicular to these parallel surfaces.

Reflection

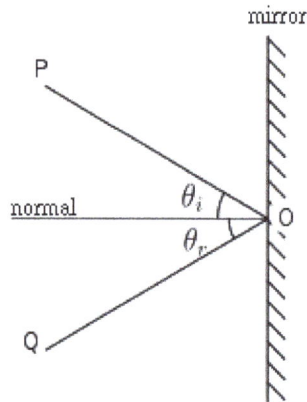

Diagram of specular reflection

Glossy surfaces such as mirrors reflect light in a simple, predictable way. This allows for production of reflected images that can be associated with an actual (real) or extrapolated (virtual) location in space.

With such surfaces, the direction of the reflected ray is determined by the angle the incident ray makes with the surface normal, a line perpendicular to the surface at the point where the ray hits. The incident and reflected rays lie in a single plane, and the angle between the reflected ray and the surface normal is the same as that between the incident ray and the normal. This is known as the Law of Reflection.

For flat mirrors, the law of reflection implies that images of objects are upright and the same distance behind the mirror as the objects are in front of the mirror. The image size is the same as the object size. (The magnification of a flat mirror is equal to one.) The law also implies that mirror images are parity inverted, which is perceived as a left-right inversion.

Mirrors with curved surfaces can be modeled by ray tracing and using the law of reflection at each point on the surface. For mirrors with parabolic surfaces, parallel rays incident on the mirror produce reflected rays that converge at a common focus. Other curved surfaces may also focus light, but with aberrations due to the diverging shape causing the focus to be smeared out in space. In particular, spherical mirrors exhibit spherical aberration. Curved mirrors can form images with magnification greater than or less than one, and the image can be upright or inverted. An upright image formed by reflection in a mirror is always virtual, while an inverted image is real and can be projected onto a screen.

Refraction

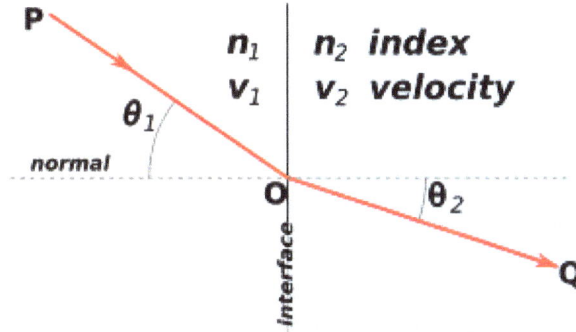

Illustration of Snell's Law

Refraction occurs when light travels through an area of space that has a changing index of refraction. The simplest case of refraction occurs when there is an interface between a uniform medium with index of refraction n_1 and another medium with index of refraction n_2. In such situations, Snell's Law describes the resulting deflection of the light ray:

$$n_1 \sin \theta_1 = n_2 \sin \theta_2$$

where θ_1 and θ_2 are the angles between the normal (to the interface) and the incident and refracted waves, respectively. This phenomenon is also associated with a changing speed of light as seen from the definition of index of refraction provided above which implies:

$$v_1 \sin \theta_2 = v_2 \sin \theta_1$$

where v_1 and v_2 Are the wave velocities through the respective media.

Various consequences of Snell's Law include the fact that for light rays traveling from a material with a high index of refraction to a material with a low index of refraction, it is possible for the interaction with the interface to result in zero transmission. This phenomenon is called total internal reflection and allows for fiber optics technology. As light signals travel down a fiber optic cable, it undergoes total internal reflection allowing for essentially no light lost over the length of the cable. It is also possible to produce polarized light rays using a combination of reflection and refraction: When a refracted ray and the reflected ray form a right angle, the reflected ray has the property of "plane polarization". The angle of incidence required for such a scenario is known as Brewster's angle.

Snell's Law can be used to predict the deflection of light rays as they pass through "linear media" as long as the indexes of refraction and the geometry of the media are known. For example, the propagation of light through a prism results in the light ray being deflected depending on the shape and orientation of the prism. Additionally, since different frequencies of light have slightly different indexes of refraction in most materials, refraction can be used to produce dispersion spectra that appear as rainbows. The discovery of this phenomenon when passing light through a prism is famously attributed to Isaac Newton.

Some media have an index of refraction which varies gradually with position and, thus, light rays curve through the medium rather than travel in straight lines. This effect is what is responsible for mirages seen on hot days where the changing index of refraction of the air causes the light rays to bend creating the appearance of specular reflections in the distance (as if on the surface of a pool of water). Material that has a varying index of refraction is called a gradient-index (GRIN) material and has many useful properties used in modern optical scanning technologies including photocopiers and scanners. The phenomenon is studied in the field of gradient-index optics.

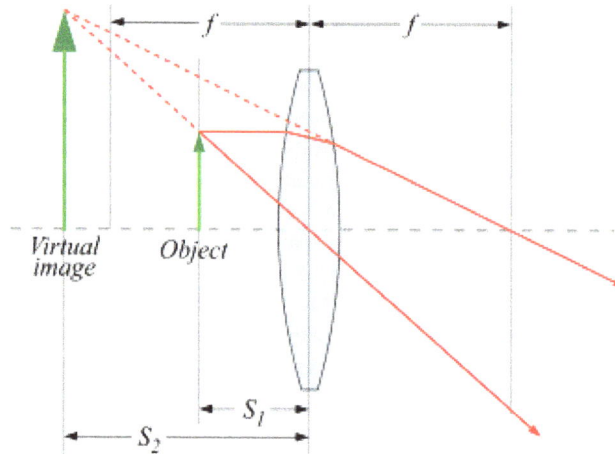

A ray tracing diagram for a simple converging lens.

A device which produces converging or diverging light rays due to refraction is known as a lens. Thin lenses produce focal points on either side that can be modeled using the lensmaker's equation. In general, two types of lenses exist: convex lenses, which cause parallel light rays to converge, and concave lenses, which cause parallel light rays to diverge. The detailed prediction of how images are produced by these lenses can be made using ray-tracing similar to curved mirrors. Similarly to curved mirrors, thin lenses follow a simple equation that determines the location of the images given a particular focal length (f) and object distance (S_1):

$$\frac{1}{S_1} + \frac{1}{S_2} = \frac{1}{f}$$

where S_2 is the distance associated with the image and is considered by convention to be negative if on the same side of the lens as the object and positive if on the opposite side of the lens. The focal length f is considered negative for concave lenses.

Incoming parallel rays are focused by a convex lens into an inverted real image one focal length from the lens, on the far side of the lens.

Incoming parallel rays are focused by a convex lens into an inverted real image one focal length from the lens, on the far side of the lens

Rays from an object at finite distance are focused further from the lens than the focal distance;

the closer the object is to the lens, the further the image is from the lens. With concave lenses, incoming parallel rays diverge after going through the lens, in such a way that they seem to have originated at an upright virtual image one focal length from the lens, on the same side of the lens that the parallel rays are approaching on.

With concave lenses, incoming parallel rays diverge after going through the lens, in such a way that they seem to have originated at an upright virtual image one focal length from the lens, on the same side of the lens that the parallel rays are approaching on.

Rays from an object at finite distance are associated with a virtual image that is closer to the lens than the focal length, and on the same side of the lens as the object. The closer the object is to the lens, the closer the virtual image is to the lens.

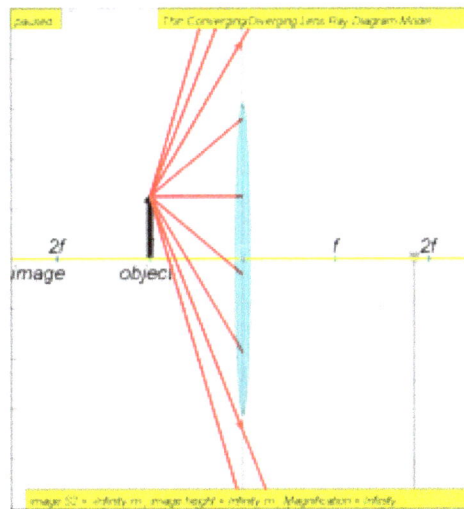

Rays from an object at finite distance are associated with a virtual image that is closer to the lens than the focal length, and on the same side of the lens as the object.

Likewise, the magnification of a lens is given by

$$M = -\frac{S_2}{S_1} = \frac{f}{f - S_1}$$

where the negative sign is given, by convention, to indicate an upright object for positive values and an inverted object for negative values. Similar to mirrors, upright images produced by single lenses are virtual while inverted images are real.

Lenses suffer from aberrations that distort images and focal points. These are due to both to geometrical imperfections and due to the changing index of refraction for different wavelengths of light (chromatic aberration).

Underlying Mathematics

As a mathematical study, geometrical optics emerges as a short-wavelength limit for solutions to hyperbolic partial differential equations. In this short-wavelength limit, it is possible to approximate the solution locally by

$$u(t, x) \approx a(t, x)e^{i(k \cdot x - \omega t)}$$

where k, ω satisfy a dispersion relation, and the amplitude $a(t,x)$ varies slowly. More precisely, the leading order solution takes the form

$$a_0(t,x)e^{i\varphi(t,x)/\varepsilon}.$$

The phase $\varphi(t,x)/\varepsilon$ can be linearized to recover large wavenumber $k := \nabla_x \varphi$, and frequency $\omega := -\partial_t \varphi$. The amplitude a_0 satisfies a transport equation. The small parameter ε enters the scene due to highly oscillatory initial conditions. Thus, when initial conditions oscillate much faster than the coefficients of the differential equation, solutions will be highly oscillatory, and transported along rays. Assuming coefficients in the differential equation are smooth, the rays will be too. In other words, refraction does not take place. The motivation for this technique comes from studying the typical scenario of light propagation where short wavelength light travels along rays that minimize (more or less) its travel time. Its full application requires tools from microlocal analysis.

A Simple Example

Starting with the wave equation for $(t,x) \in \mathbb{R} \times \mathbb{R}^n$

$$L(\partial_t, \nabla_x)u := \left(\frac{\partial^2}{\partial t^2} - c(x)^2 \Delta \right) u(t,x) = 0, \quad u(0,x) = u_0(x), \quad u_t(0,x) = 0$$

assume an asymptotic series solution of the form

$$u(t,x) \sim a_\varepsilon(t,x)e^{i\varphi(t,x)/\varepsilon} = \sum_{j=0}^{\infty} i^j \varepsilon^j a_j(t,x)e^{i\varphi(t,x)/\varepsilon}.$$

Check that

$$L(\partial_t, \nabla_x)(e^{i\varphi(t,x)/\varepsilon})a_\varepsilon(t,x) = e^{i\varphi(t,x)/\varepsilon} \left(\left(\frac{i}{\varepsilon} \right)^2 L(\varphi_t, \nabla_x \varphi)a_\varepsilon + \frac{2i}{\varepsilon} V(\partial_t, \nabla_x)a_\varepsilon + \right.$$

$$\left. \frac{i}{\varepsilon}(a_\varepsilon L(\partial_t, \nabla_x)\varphi) + L(\partial_t, \nabla_x)a_\varepsilon \right).$$

with

$$V(\partial_t, \nabla_x) := \frac{\partial \varphi}{\partial t}\frac{\partial}{\partial t} - c^2(x)\sum_j \frac{\partial \varphi}{\partial x_j}\frac{\partial}{\partial x_j}$$

Plugging the series into this equation, and equating powers of ε, the most singular term $O(\varepsilon^{-2})$ satisfies the eikonal equation (in this case called a dispersion relation),

$$0 = L(\varphi_t, \nabla_x \varphi) = (\varphi_t)^2 - c(x)^2(\nabla_x \varphi)^2.$$

To order ε^{-1}, the leading-order amplitude must satisfy a transport equation

$$2Va_0 + (L\varphi)a_0 = 0$$

With the definition $k := \nabla_x \varphi$, $\omega := -\varphi_t$, the eikonal equation is precisely the dispersion relation that results by plugging the plane wave solution $e^{i(k \cdot x - \omega t)}$ into the wave equation. The value of this more complicated expansion is that plane waves cannot be solutions when the wavespeed c is non-constant. However, it can be shown that the amplitude a_0 and phase φ are smooth, so that on a local scale there are plane waves.

To justify this technique, the remaining terms must be shown to be small in some sense. This can be done using energy estimates, and an assumption of rapidly oscillating initial conditions. It also must be shown that the series converges in some sense.

Physical Optics

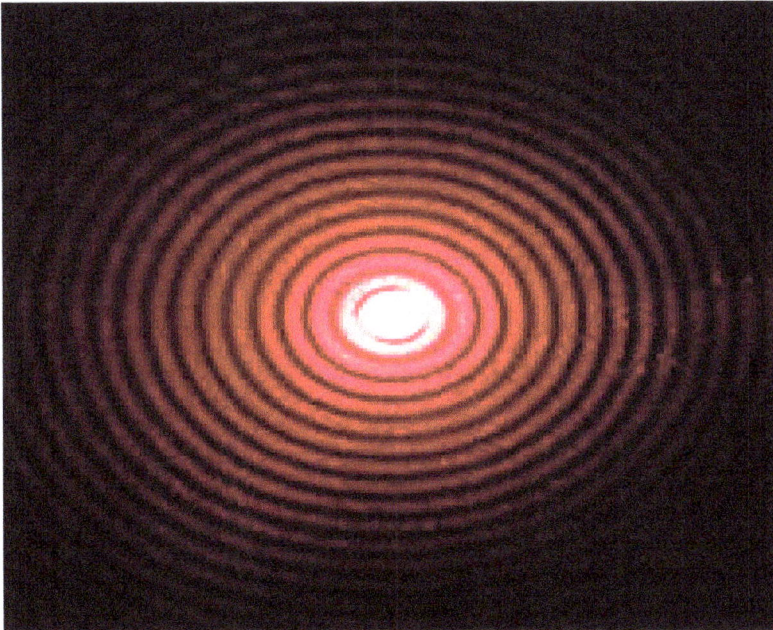

Physical optics is used to explain effects such as diffraction.

In physics, physical optics, or wave optics, is the branch of optics which studies interference, diffraction, polarization, and other phenomena for which the ray approximation of geometric optics is not valid. This usage tends not to include effects such as quantum noise in optical communication, which is studied in the sub-branch of coherence theory.

The Physical Optics Approximation

Physical optics is also the name of an approximation commonly used in optics, electrical engineering and applied physics. In this context, it is an intermediate method between geometric optics, which ignores wave effects, and full wave electromagnetism, which is a precise theory. The word "physical" means that it is more physical than geometric or ray optics and not that it is an exact physical theory.

This approximation consists of using ray optics to estimate the field on a surface and then integrating that field over the surface to calculate the transmitted or scattered field. This resembles the Born approximation, in that the details of the problem are treated as a perturbation.

In optics, it is a standard way of estimating diffraction effects. In radio, this approximation is used to estimate some effects that resemble optical effects. It models several interference, diffraction and polarization effects but not the dependence of diffraction on polarization. Since it is a high frequency approximation, it is often more accurate in optics than for radio.

In optics, it typically consists of integrating ray estimated field over a lens, mirror or aperture to calculate the transmitted or scattered field.

In radar scattering it usually means taking the current that would be found on a tangent plane of similar material as the current at each point on the front, i. e. the geometrically illuminated part, of a scatterer. Current on the shadowed parts is taken as zero. The approximate scattered field is then obtained by an integral over these approximate currents. This is useful for bodies with large smooth convex shapes and for lossy (low reflection) surfaces.

The ray optics field or current is generally not accurate near edges or shadow boundaries, unless supplemented by diffraction and creeping wave calculations.

The standard theory of physical optics has some defects in the evaluation of scattered fields, leading to decreased accuracy away from the specular direction. An improved theory introduced in 2004 gives exact solutions to problems involving wave diffraction by conducting scatterers.

Quantum Optics

Quantum optics is a field of research that uses semi-classical and quantum-mechanical physics to investigate phenomena involving light and its interactions with matter at submicroscopic levels.

History of Quantum Optics

Light propagating in a vacuum has its energy and momentum quantized according to an integer number of particles known as photons. Quantum optics studies the nature and effects of light as quantized photons. The first major development leading to that understanding was the correct modeling of the blackbody radiation spectrum by Max Planck in 1899 under the hypothesis of light being emitted in discrete units of energy. The photoelectric effect was further evidence of this quantization as explained by Einstein in a 1905 paper, a discovery for which he was to be awarded the Nobel Prize in 1921. Niels Bohr showed that the hypothesis of optical radiation being quantized corresponded to his theory of the quantized energy levels of atoms, and the spectrum of discharge emission from hydrogen in particular. The understanding of the interaction between light and matter following these developments was crucial for the development of quantum mechanics as a whole. However, the subfields of quantum mechanics dealing with matter-light interaction were principally regarded as research into matter rather than into light; hence one rather spoke of atom physics and quantum electronics in 1960. Laser science—i.e., research into principles, design and application of these devices—became an important field, and the quantum mechanics underlying the laser's principles was studied now with more emphasis on the properties of light, and the name *quantum optics* became customary.

As laser science needed good theoretical foundations, and also because research into these soon proved very fruitful, interest in quantum optics rose. Following the work of Dirac in quantum field theory, George Sudarshan, Roy J. Glauber, and Leonard Mandel applied quantum theory to the electromagnetic field in the 1950s and 1960s to gain a more detailed understanding of photodetec-tion and the statistics of light. This led to the introduction of the coherent state as a concept which addressed variations between laser light, thermal light, exotic squeezed states, etc. as it became understood that light cannot be fully described just referring to the electro-magnetic fields describing the waves in the classical picture. In 1977, Kimble et al. demonstrated a single atom emitting one photon at a time, further compelling evidence that light consists of pho-tons. Previously unknown quantum states of light with characteristics unlike classical states, such as squeezed light were subsequently discovered.

Development of short and ultrashort laser pulses—created by Q switching and modelocking techniques—opened the way to the study of what became known as ultrafast processes. Appli-cations for solid state research (e.g. Raman spectroscopy) were found, and mechanical forces of light on matter were studied. The latter led to levitating and positioning clouds of atoms or even small biological samples in an optical trap or optical tweezers by laser beam. This, along with Doppler cooling, was the crucial technology needed to achieve the celebrated Bose–Einstein condensation.

Other remarkable results are the demonstration of quantum entanglement, quantum teleporta-tion, and quantum logic gates. The latter are of much interest in quantum information theory, a subject which partly emerged from quantum optics, partly from theoretical computer science.

Today's fields of interest among quantum optics researchers include parametric down-conversion, parametric oscillation, even shorter (attosecond) light pulses, use of quantum optics for quantum information, manipulation of single atoms, Bose–Einstein condensates, their application, and how to manipulate them (a sub-field often called atom optics), coherent perfect absorbers, and much more. Topics classified under the term of quantum optics, especially as applied to engineering and technological innovation, often go under the modern term photonics.

Several Nobel prizes have been awarded for work in quantum optics. These were awarded:

- in 2012, Serge Haroche and David J. Wineland "for ground-breaking experimental meth-ods that enable measuring & manipulation of individual quantum systems".

- in 2005, Theodor W. Hänsch, Roy J. Glauber and John L. Hall

- in 2001, Wolfgang Ketterle, Eric Allin Cornell and Carl Wieman

- in 1997, Steven Chu, Claude Cohen-Tannoudji and William Daniel Phillips

Concepts of Quantum Optics

According to quantum theory, light may be considered not only as an electro-magnetic wave but also as a "stream" of particles called photons which travel with c, the vacuum speed of light. These particles should not be considered to be classical billiard balls, but as quantum mechanical parti-cles described by a wavefunction spread over a finite region.

Each particle carries one quantum of energy, equal to hf, where h is Planck's constant and f is the frequency of the light. That energy possessed by a single photon corresponds exactly to the transi-tion between discrete energy levels in an atom (or other system) that emitted the photon; material absorption of a photon is the reverse process. Einstein's explanation of spontaneous emission also predicted the existence of stimulated emission, the principle upon which the laser rests. However, the actual invention of the maser (and laser) many years later was dependent on a method to produce a population inversion.

The use of statistical mechanics is fundamental to the concepts of quantum optics: Light is described in terms of field operators for creation and annihilation of photons—i.e. in the language of quantum electrodynamics.

A frequently encountered state of the light field is the coherent state, as introduced by Roy J. Glauber in 1963. This state, which can be used to approximately describe the output of a single-frequency laser well above the laser threshold, exhibits Poissonian photon number statistics. Via certain nonlinear interactions, a coherent state can be transformed into a squeezed coherent state, by applying a squeezing operator which can exhibit super- or sub-Poissonian photon statistics. Such light is called squeezed light. Other important quantum aspects are related to correlations of photon statistics between different beams. For example, spontaneous parametric down-conversion can generate so-called 'twin beams', where (ideally) each photon of one beam is associated with a photon in the other beam.

Atoms are considered as quantum mechanical oscillators with a discrete energy spectrum, with the transitions between the energy eigenstates being driven by the absorption or emission of light according to Einstein's theory.

For solid state matter, one uses the energy band models of solid state physics. This is important for understanding how light is detected by solid-state devices, commonly used in experiments.

Quantum Electronics

Quantum electronics is a term that was used mainly between the 1950s and 1970s to denote the area of physics dealing with the effects of quantum mechanics on the behavior of electrons in matter, together with their interactions with photons. Today, it is rarely considered a sub-field in its own right, and it has been absorbed by other fields. Solid state physics regularly takes quantum mechanics into account, and is usually concerned with electrons. Specific applications of quantum mechanics in electronics is researched within semiconductor physics . The term also encompassed the basic processes of laser operation, which is today studied as a topic in quantum optics. Usage of the term overlapped early work on the quantum Hall effect and quantum cellular automata.

Nonimaging Optics

Nonimaging optics (also called anidolic optics) is the branch of optics concerned with the optimal transfer of light radiation between a source and a target. Unlike traditional imaging optics, the techniques involved do not attempt to form an image of the source; instead an optimized optical system for optical radiative transfer from a source to a target is desired.

Applications

The two design problems that nonimaging optics solves better than imaging optics are:

- solar energy concentration: maximizing the amount of energy applied to a receiver, typically a solar cell or a thermal receiver

- illumination: controlling the distribution of light, typically so it is "evenly" spread over some areas and completely blocked from other areas

Typical variables to be optimized at the target include the total radiant flux, the angular distribution of optical radiation, and the spatial distribution of optical radiation. These variables on the target side of the optical system often must be optimized while simultaneously considering the collection efficiency of the optical system at the source.

Solar Energy Concentration

For a given concentration, nonimaging optics provide the widest possible acceptance angles and, therefore, are the most appropriate for use in solar concentration as, for example, in concentrated photovoltaics. When compared to "traditional" imaging optics (such as parabolic reflectors or fresnel lenses), the main advantages of nonimaging optics for concentrating solar energy are:

- wider acceptance angles resulting in higher tolerances (and therefore higher efficiencies) for:
  - o less precise tracking
  - o imperfectly manufactured optics
  - o imperfectly assembled components
  - o movements of the system due to wind
  - o finite stiffness of the supporting structure
  - o deformation due to aging
  - o capture of circumsolar radiation
  - o other imperfections in the system

- higher solar concentrations
  - o smaller solar cells (in concentrated photovoltaics)
  - o higher temperatures (in concentrated solar thermal)
  - o lower thermal losses (in concentrated solar thermal)
  - o widen the applications of concentrated solar power, for example to solar lasers

- possibility of a uniform illumination of the receiver
  - o improve reliability and efficiency of the solar cells (in concentrated photovoltaics)
  - o improve heat transfer (in concentrated solar thermal)

- design flexibility: different kinds of optics with different geometries can be tailored for different applications

Also, for low concentrations, the very wide acceptance angles of nonimaging optics can avoid solar tracking altogether or limit it to a few positions a year.

The main disadvantage of nonimaging optics when compared to parabolic reflectors or fresnel lenses is that, for high concentrations, they typically have one more optical surface, slightly decreasing efficiency. That, however, is only noticeable when the optics are aiming perfectly towards the sun, which is typically not the case because of imperfections in practical systems.

Illumination Optics

Examples of nonimaging optical devices include optical light guides, nonimaging reflectors, nonimaging lenses or a combination of these devices. Common applications of nonimaging optics include many areas of illumination engineering (lighting). Examples of modern implementations of nonimaging optical designs include automotive headlamps, LCD backlights, illuminated instrument panel displays, fiber optic illumination devices, LED lights, projection display systems and luminaires.

When compared to "traditional" design techniques, nonimaging optics has the following advantages for illumination:

- better handling of extended sources

- more compact optics

- color mixing capabilities

- combination of light sources and light distribution to different places

- well suited to be used with increasingly popular LED light sources

- tolerance to variations in the relative position of light source and optic

Examples of nonimaging illumination optics using solar energy are anidolic lighting or solar pipes.

Other Applications

Collecting radiation emitted by high-energy particle collisions using the fewest number of photomultiplier tubes.

Some of the design methods for nonimaging optics are also finding application in imaging devices, for example some with ultra-high numerical aperture.

Theory

Early academic research in nonimaging optical mathematics seeking closed form solutions was first published in textbook form in a 1978 book. A modern textbook illustrating the depth and breadth of research and engineering in this area was published in 2004. A thorough introduction to this field was published in 2008.

Special applications of nonimaging optics such as Fresnel lenses for solar concentration or solar concentration in general have also been published, although this last reference by O'Gallagher describes mostly the work developed some decades ago.

Imaging optics can concentrate sunlight to, at most, the same flux found at the surface of the sun. Nonimaging optics have been demonstrated to concentrate sunlight to 84,000 times the ambient intensity of sunlight, exceeding the flux found at the surface of the sun, and approaching the theoretical (2nd law of thermodynamics) limit of heating objects up to the temperature of the sun's surface.

The simplest way to design nonimaging optics is called "the method of strings", based on the edge ray principle. Other more advanced methods were developed starting in the early 1990s that can better handle extended light sources than the edge-ray method. These were developed primarily to solve the design problems related to solid state automobile headlamps and complex illumination systems. One of these advanced design methods is the Simultaneous Multiple Surface design method (SMS). The 2D SMS design method (U.S. Patent 6,639,733) is described in detail in the aforementioned textbooks. The 3D SMS design method (U.S. Patent 7,460,985) was developed in 2003 by a team of optical scientists at Light Prescriptions Innovators.

Edge Ray Principle

In simple terms, the edge ray principle states that if the light rays coming from the edges of the source are redirected towards the edges of the receiver, this will ensure that all light rays coming from the inner points in the source will end up on the receiver. There is no condition on image formation, the only goal is to transfer the light from the source to the target.

Figure "edge ray principle" on the right illustrates this principle. A lens collects light from a source S_1S_2 and redirects it towards a receiver R_1R_2.

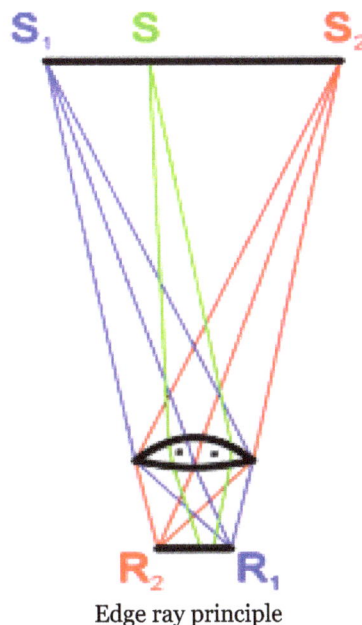

Edge ray principle

The lens has two optical surfaces and, therefore, it is possible to design it (using the SMS design method) so that the light rays coming from the edge S_1 of the source are redirected towards edge R_1 of the receiver, as indicated by the blue rays. By symmetry, the rays coming from edge S_2 of the source are redirected towards edge R_2 of the receiver, as indicated by the red rays. The rays coming from an inner point S in the source are redirected towards the target, but they are not concentrated onto a point and, therefore, no image is formed.

Actually, if we consider a point P on the top surface of the lens, a ray coming from S_1 through P will be redirected towards R_1. Also a ray coming from S_2 through P will be redirected towards R_2. A ray coming through P from an inner point S in the source will be redirected towards an inner point of the receiver. This lens then guarantees that all light from the source crossing it will be redirected towards the receiver. However, no image of the source is formed on the target. Imposing the condition of image formation on the receiver would imply using more optical surfaces, making the optic more complicated, but would not improve light transfer between source and target (since all light is already transferred). For that reason nonimaging optics are simpler and more efficient than imaging optics in transferring radiation from a source to a target.

Design Methods

Nonimaging optics devices are obtained using different methods. The most important are: the flow-line or Winston-Welford design method, the SMS or Miñano-Benitez design method and the Miñano design method using Poisson brackets. The first (flow-line) is probably the most used, although the second (SMS) has proven very versatile, resulting in a wide variety of optics. The third has remained in the realm of theoretical optics and has not found real world application to date. Often optimization is also used

Typically optics have refractive and reflective surfaces and light travels though media of different refractive indices as it crosses the optic. In those cases a quantity called optical path length (OPL) may be defined as $S = \sum_i n_i d_i$ where index i indicates different ray sections between successive deflections (refractions or reflections), n_i is the refractive index and d_i the distance in each section i of the ray path.

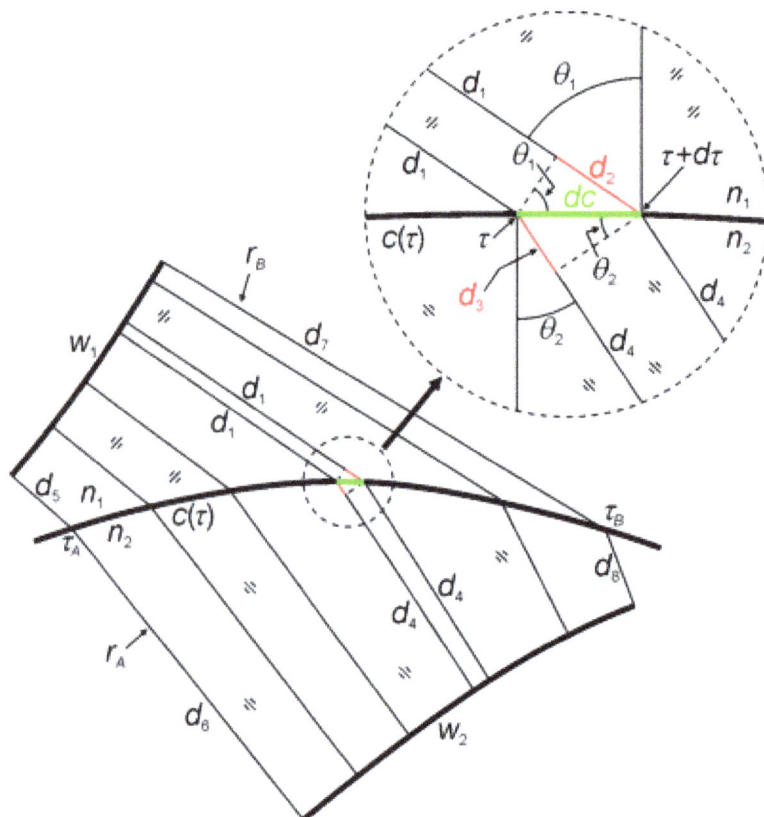

Constant OPL

The OPL is constant between wavefronts. This can be seen for refraction in the figure "constant OPL" to the right. It shows a separation $c(\tau)$ between two media of refractive indices n_1 and n_2, where $c(\tau)$ is described by a parametric equation with parameter τ. Also shown are a set of rays perpendicular to wavefront w_1 and traveling in the medium of refractive index n_1. These rays refract at $c(\tau)$ into the medium of refractive index n_2 in directions perpendicular to wavefront w_2. Ray r_A crosses c at point $c(\tau_A)$ and, therefore, ray r_A is identified by parameter τ_A on c. Likewise, ray r_B is identified by parameter τ_B on c. Ray r_A has optical path length $S(\tau_A) = n_1 d_5 + n_2 d_6$. Also, ray r_B has optical path length $S(\tau_B) = n_1 d_7 + n_2 d_8$. The difference in optical path length for rays r_A and r_B is given by:

$$S(\tau_B) - S(\tau_A) = \int_A^B dS = \int_{\tau_A}^{\tau_B} \frac{dS}{d\tau} d\tau = \int_{\tau_A}^{\tau_B} \frac{S(\tau + d\tau) - S(\tau)}{(\tau + d\tau) - \tau} d\tau$$

In order to calculate the value of this integral, we evaluate $S(\tau + d\tau) - S(\tau)$, again with the help of the same figure. We have $S(\tau) = n_1 d_1 + n_2 (d_3 + d_4)$ and $S(\tau + d\tau) = n_1 (d_1 + d_2) + n_2 d_4$. These expressions can be rewritten as $S(\tau) = n_1 d_1 + n_2 dc\ \sin\theta_2 + n_2 d_4$ and $S(\tau + d\tau) = n_1 d_1 + n_1 dc\ \sin\theta_1 + n_2 d_4$. From the law of refraction $n_1 \sin\theta_1 = n_2 \sin\theta_2$ and therefore $S(\tau + d\tau) = S(\tau)$, leading to $S(\tau_A) = S(\tau_B)$. Since these may be arbitrary rays crossing c, it may be concluded that the optical path length between w_1 and w_2 is the same for all rays perpendicular to incoming wavefront w_1 and outgoing wavefront w_2.

Similar conclusions may be drawn for the case of reflection, only in this case $n_1 = n_2$. This relationship between rays and wavefronts is valid in general.

Flow-line Design Method

The flow-line (or Winston-Welford) design method typically leads to optics which guide the light confining it between two reflective surfaces. The best known of these devices is the CPC (Compound Parabolic Concentrator).

These types of optics may be obtained, for example, by applying the edge ray of nonimaging optics to the design of mirrored optics, as show in figure "CEC" on the right. It is composed of two elliptical mirrors e_1 with foci S_1 and R_1 and its symmetrical e_2 with foci S_2 and R_2.

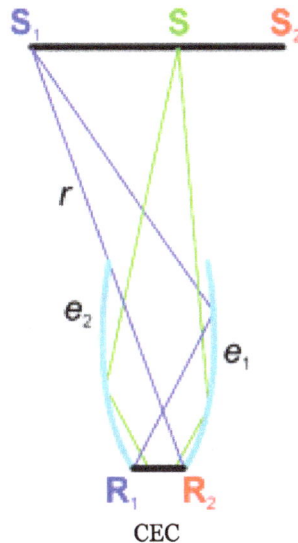

CEC

Mirror e_1 redirects the rays coming from the edge S_1 of the source towards the edge R_1 of the receiver and, by symmetry, mirror e_2 redirects the rays coming from the edge S_2 of the source towards the edge R_2 of the receiver. This device does not form an image of the source S_1S_2 on the receiver R_1R_2 as indicated by the green rays coming from a point S in the source that end up on the receiver but are not focused onto an image point. Mirror e_2 starts at the edge R_1 of the receiver since leaving a gap between mirror and receiver would allow light to escape between the two. Also, mirror e_2 ends at ray r connecting S_1 and R_2 since cutting it short would prevent it from capturing as much light as possible, but extending it above r would shade light coming from S_1 and its neighboring points of the source. The resulting device is called a CEC (Compound Elliptical Concentrator).

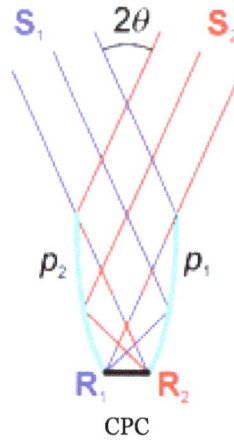

CPC

A particular case of this design happens when the source S_1S_2 becomes infinitely large and moves to an infinite distance. Then the rays coming from S_1 become parallel rays and the same for those coming from S_2 and the elliptical mirrors e_1 and e_2 converge to parabolic mirrors p_1 and p_2. The resulting device is called a CPC (Compound Parabolic Concentrator), and shown in the "CPC" figure on the left. CPCs are the most common seen nonimaging optics. They are often used to demonstrate the difference between Imaging optics and nonimaging optics.

When seen from the CPC, the incoming radiation (emitted from the infinite source at an infinite distance) subtends an angle $\pm\theta$ (total angle 2θ). This is called the acceptance angle of the CPC. The reason for this name can be appreciated in the figure "rays showing the acceptance angle" on the right. An incoming ray r_1 at an angle θ to the vertical (coming from the edge of the infinite source) is redirected by the CPC towards the edge R_1 of the receiver.

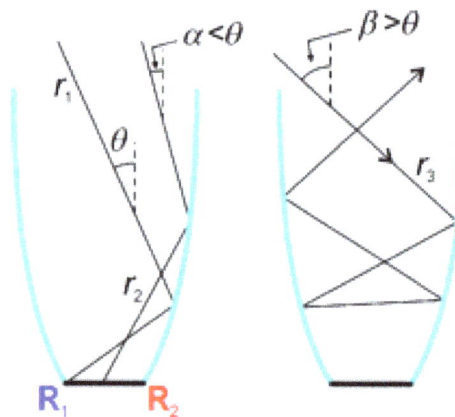

Rays showing the acceptance angle

Another ray r_2 at an angle $\alpha<\theta$ to the vertical (coming from an inner point of the infinite source) is redirected towards an inner point of the receiver. However, a ray r_3 at an angle $\beta>\theta$ to the vertical (coming from a point outside the infinite source) bounces around inside the CPC until it is rejected by it. Therefore, only the light inside the acceptance angle $\pm\theta$ is captured by the optic; light outside it is rejected.

The ellipses of a CEC can be obtained by the (pins and) string method, as shown in the figure "string method" on the left. A string of constant length is attached to edge point S_1 of the source and edge point R_1 of the receiver.

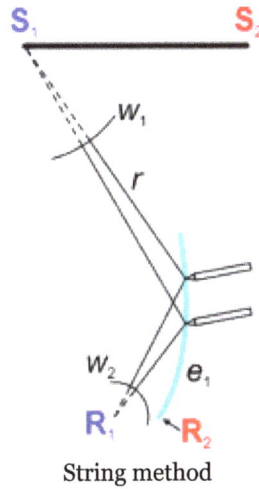
String method

The string is kept stretched while moving a pencil up and down, drawing the elliptical mirror e_1. We can now consider a wavefront w_1 as a circle centered at S_1. This wavefront is perpendicular to all rays coming out of S_1 and the distance from S_1 to w_1 is constant for all its points. The same is valid for wavefront w_2 centered at R_1. The distance from w_1 to w_2 is then constant for all light rays reflected at e_1 and these light rays are perpendicular to both, incoming wavefront w_1 and outgoing wavefront w_2.

Optical path length (OPL) is constant between wavefronts. When applied to nonimaging optics, this result extends the string method to optics with both refractive and reflective surfaces. Figure "DTIRC" (Dielectric Total Internal Reflection Concentrator) on the left shows one such example.

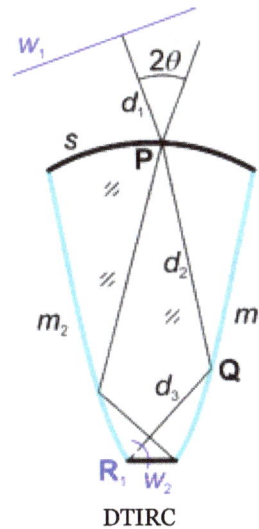
DTIRC

The shape of the top surface s is prescribed, for example, as a circle. Then the lateral wall m_1 is calculated by the condition of constant optical path length $S=d_1+n\,d_2+n\,d_3$ where d_1 is the distance between incoming wavefront w_1 and point P on the top surface s, d_2 is the distance between P and Q and d_3 the distance between Q and outgoing wavefront w_2, which is circular and centered at R_1. Lateral wall m_2 is symmetrical to m_1. The acceptance angle of the device is 2θ.

These optics are called flow-line optics and the reason for that is illustrated in figure "CPC flow-lines" on the right. It shows a CPC with an acceptance angle 2θ, highlighting one of its inner points P.

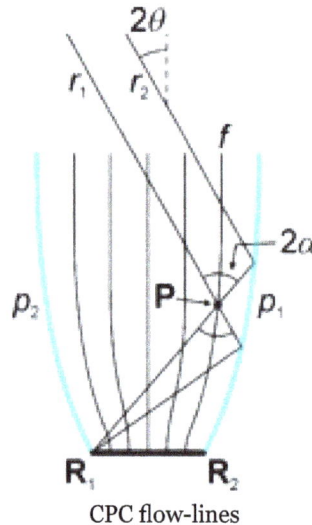

CPC flow-lines

The light crossing this point is confined to a cone of angular aperture 2α. A line f is also shown whose tangent at point P bisects this cone of light and, therefore, points in the direction of the "light flow" at P. Several other such lines are also shown in the figure. They all bisect the edge rays at each point inside the CPC and, for that reason, their tangent at each point points in the direction of the flow of light. These are called flow-lines and the CPC itself is just a combination of flow line p_1 starting at R_2 and p_2 starting at R_1.

Variations to the Flow-line Design Method

There are some variations to the flow-line design method.

A variation are the multichannel or stepped flow-line optics in which light is split into several "channels" and then recombined again into a single output. Aplanatic (a particular case of SMS) versions of these designs have also been developed. The main application of this method is in the design of ultra-compact optics.

Another variation is the confinement of light by caustics. Instead of light being confined by two reflective surfaces, it is confined by a reflective surface and a caustic of the edge rays. This provides the possibility to add lossless non-optical surfaces to the optics.

Simultaneous Multiple Surface (SMS) Design Method

A nonimaging optics design method known in the field as the simultaneous multiple surface (SMS) or the Miñano-Benitez design method. The abbreviation SMS comes from the fact that it enables the simultaneous design of multiple optical surfaces. The original idea came from Miñano. The design method itself was initially developed in 2-D by Miñano and later also by Benítez. The first generalization to 3-D geometry came from Benítez. It was then much further developed by contributions of Miñano and Benítez. Other people have worked initially with Miñano and later with Miñano and Benítez on programming the method.

The Design Procedure is related to the algorithm used by Schulz in the design of aspheric imaging lenses.

The SMS (or Miñano-Benitez) design method is very versatile and many different types of optics have been designed using it. The 2D version allows the design of two (although more are also possible) aspheric surfaces simultaneously. The 3D version allows the design of optics with freeform surfaces (also called anamorphic) surfaces which may not have any kind of symmetry.

SMS optics are also calculated by applying a constant optical path length between wavefronts. Figure "SMS chain" on the right illustrates how these optics are calculated. In general, the rays perpendicular to incoming wavefront w_1 will be coupled to outgoing wavefront w_4 and the rays perpendicular to incoming wavefront w_2 will be coupled to outgoing wavefront w_3 and these wavefronts may be any shape. However, for the sake of simplicity, this figure shows a particular case or circular wavefronts. This example shows a lens of a given refractive index n designed for a source S_1S_2 and a receiver R_1R_2.

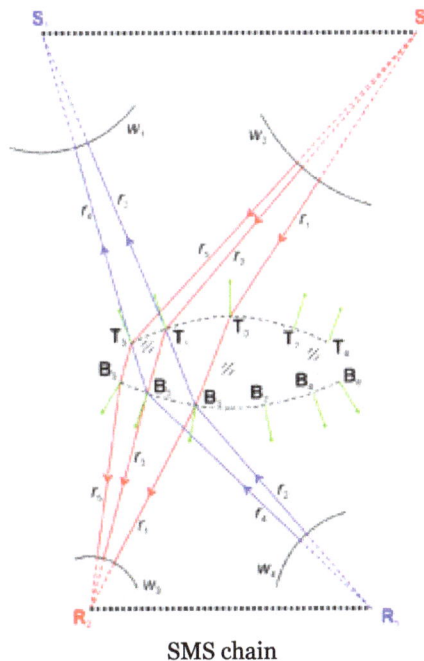

SMS chain

The rays emitted from edge S_1 of the source are focused onto edge R_1 of the receiver and those emitted from edge S_2 of the source are focused onto edge R_2 of the receiver. We first choose a point T_0 and its normal on the top surface of the lens. We can now take a ray r_1 coming from S_2 and refract it at T_0. Choosing now the optical path length S_{22} between S_2 and R_2 we have one condition that allows us to calculate point B_1 on the bottom surface of the lens. The normal at B_1 can also be calculated from the directions of the incoming and outgoing rays at this point and the refractive index of the lens. Now we can repeat the process taking a ray r_2 coming from R_1 and refracting it at

B_1. Choosing now the optical path length S_{11} between R_1 and S_1 we have one condition that allows us to calculate point T_1 on the top surface of the lens. The normal at T_1 can also be calculated from the directions of the incoming and outgoing rays at this point and the refractive index of the lens. Now, refracting at T_1 a ray r_3 coming from S_2 we can calculate a new point B_3 and corresponding normal on the bottom surface using the same optical path length S_{22} between S_2 and R_2. Refracting at B_3 a ray r_4 coming from R_1 we can calculate a new point T_3 and corresponding normal on the top surface using the same optical path length S_{11} between R_1 and S_1. The process continues by calculating another point B_5 on the bottom surface using another edge ray r_5, and so on. The sequence of points T_0 B_1 T_1 B_3 T_3 B_5 is called an SMS chain.

Another SMS chain can be constructed towards the right starting at point T_0. A ray from S_1 refracted at T_0 defines a point and normal B_2 on the bottom surface, by using constant optical path length S_{11} between S_1 and R_1. Now a ray from R_2 refracted at B_2 defines a new point and normal T_2 on the top surface, by using constant optical path length S_{22} between S_2 and R_2. The process continues as more points are added to the SMS chain. In this example shown in the figure, the optic has a left-right symmetry and, therefore, points B_2 T_2 B_4 T_4 B_6 can also be obtained by symmetry about the vertical axis of the lens.

Now we have a sequence of spaced points on the plane. Figure "SMS skinning" on the left illustrates the process used to fill the gaps between points, completely defining both optical surfaces.

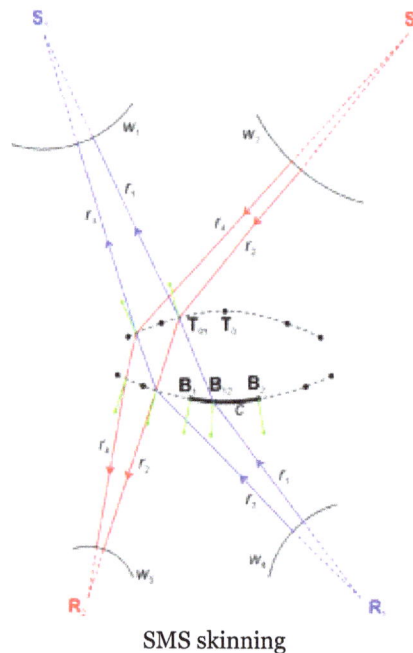

SMS skinning

We pick two points, say B_1 and B_2, with their corresponding normals and interpolate a curve c between them. Now we pick a point B_{12} and its normal on c. A ray r_1 coming from R_1 and refracted at B_{12} defines a new point T_{01} and its normal between T_0 and T_1 on the top surface, by applying the same constant optical path length S_{11} between S_1 and R_1. Now a ray r_2 coming from S_2 and refracted at T_{01} defines a new point and normal on the bottom surface, by applying the same constant optical path length S_{22} between S_2 and R_2. The process continues with rays r_3 and r_4 building a new SMS chain filling the gaps between points. Picking other points and corresponding normals on curve c gives us more points in between the other SMS points calculated originally.

In general, the two SMS optical surfaces do not need to be refractive. Refractive surfaces are noted R (from Refraction) while reflective surfaces are noted X (from the Spanish word refleXión). Total Internal Reflection (TIR) is noted I. Therefore, a lens with two refractive surfaces is an RR optic, while another configuration with a reflective and a refractive surface is an XR optic. Configurations with more optical surfaces are also possible and, for example, if light is first refracted (R), then reflected (X) then reflected again by TIR (I), the optic is called an RXI.

The SMS 3D is similar to the SMS 2D, only now all calculations are done in 3D space. Figure "SMS 3D chain" on the right illustrates the algorithm of an SMS 3D calculation.

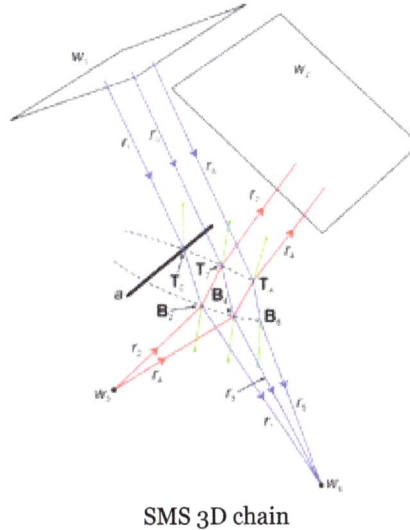

SMS 3D chain

The first step is to choose the incoming wavefronts w_1 and w_2 and outgoing wavefronts w_3 and w_4 and the optical path length S_{14} between w_1 and w_4 and the optical path length S_{23} between w_2 and w_3. In this example the optic is a lens (an RR optic) with two refractive surfaces, so its refractive index must also be specified. One difference between the SMS 2D and the SMS 3D is on how to choose initial point T_0, which is now on a chosen 3D curve a. The normal chosen for point T_0 must be perpendicular to curve a. The process now evolves similarly to the SMS 2D. A ray r_1 coming from w_1 is refracted at T_0 and, with the optical path length S_{14}, a new point B_2 and its normal is obtained on the bottom surface. Now ray r_2 coming from w_3 is refracted at B_2 and, with the optical path length S_{23}, a new point T_2 and its normal is obtained on the top surface. With ray r_3 a new point B_2 and its normal are obtained, with ray r_4 a new point T_4 and its normal are obtained, and so on. This process is performed in 3D space and the result is a 3D SMS chain. As with the SMS 2D, a set of points and normals to the left of T_0 can also be obtained using the same method. Now, choosing another point T_0 on curve a the process can be repeated and more points obtained on the top and bottom surfaces of the lens.

The power of the SMS method lies in the fact that the incoming and outgoing wavefronts can themselves be free-form, giving the method great flexibility. Also, by designing optics with reflective surfaces or combinations of reflective and refractive surfaces, different configurations are possible.

Miñano Design Method Using Poisson Brackets

This design method was developed by Miñano and is based on Hamiltonian optics, the Hamiltonian formulation of geometrical optics which shares much of the mathematical formulation with

Hamiltonian mechanics. It allows the design of optics with variable refractive index, and therefore solves some nonimaging problems that are not solvable using other methods. However, manufacturing of variable refractive index optics is still not possible and this method, although potentially powerful, did not yet find a practical application.

Conservation of Etendue

Conservation of etendue is a central concept in nonimaging optics. In concentration optics, it relates the acceptance angle with the maximum concentration possible. Conservation of etendue may be seen as constant a volume moving in phase space.

Kohler Integration

In some applications it is important to achieve a given irradiance (or illuminance) pattern on a target, while allowing for movements or inhomogeneities of the source. Figure "Kohler integrator" on the right illustrates this for the particular case of solar concentration. Here the light source is the sun moving in the sky. On the left this figure shows a lens $L_1 L_2$ capturing sunlight incident at an angle a to the optical axis and concentrating it onto a receiver $L_3 L_4$. As seen, this light is concentrated onto a hotspot on the receiver. This may be a problem in some applications. One way around this is to add a new lens extending from L_3 to L_4 that captures the light from $L_1 L_2$ and redirects it onto a receiver $R_1 R_2$, as shown in the middle of the figure.

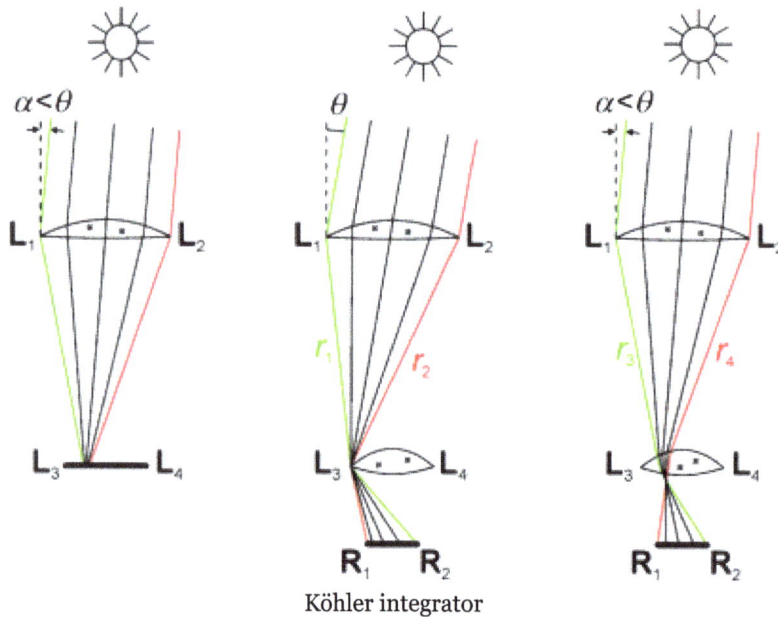

Köhler integrator

The situation in the middle of the figure shows a nonimaging lens $L_1 L_2$ is designed in such a way that sunlight (here considered as a set of parallel rays) incident at an angle θ to the optical axis will be concentrated to point L_3. On the other hand, nonimaging lens $L_3 L_4$ is designed in such a way that light rays coming from L_1 are focused on R_2 and light rays coming from L_2 are focused on R_1. Therefore, ray r_1 incident on the first lens at an angle θ will be redirected towards L_3. When it hits the second lens, it is coming from point L_1 and it is redirected by the second lens to R_2. On the other hand, ray r_2 also incident on the first lens at an angle θ will also be redirected towards L_3. However, when it hits the second lens, it is coming from point L_2 and it is redirected by the second lens to R_1.

Intermediate rays incident on the first lens at an angle θ will be redirected to points between R_1 and R_2, fully illuminating the receiver.

Something similar happens in the situation shown in the same figure, on the right. Ray r_3 incident on the first lens at an angle $\alpha<\theta$ will be redirected towards a point between L_3 and L_4. When it hits the second lens, it is coming from point L_1 and it is redirected by the second lens to R_2. Also, Ray r_4 incident on the first lens at an angle $\alpha<\theta$ will be redirected towards a point between L_3 and L_4. When it hits the second lens, it is coming from point L_2 and it is redirected by the second lens to R_1. Intermediate rays incident on the first lens at an angle $\alpha<\theta$ will be redirected to points between R_1 and R_2, also fully illuminating the receiver.

This combination of optical elements is called Köhler illumination. Although the example given here was for solar energy concentration, the same principles apply for illumination in general. In practice, Köhler optics are typically not designed as a combination of nonimaging optics, but they are simplified versions with a lower number of active optical surfaces. This decreases the effectiveness of the method, but allows for simpler optics. Also, Köhler optics are often divided into several sectors, each one of them channeling light separately and then combining all the light on the target.

An example of one of these optics used for solar concentration is the Fresnel-R Köhler.

Compound Parabolic Concentrator

In the drawing opposite there are two parabolic mirrors CC' (red) and DD' (blue). Both parabolas are cut at B and A respectively. A is the focal point of parabola CC' and B is the focal point of the parabola DD' The area DC is the entrance aperture and the flat absorber is AB. This CPC has an acceptance angle of θ.

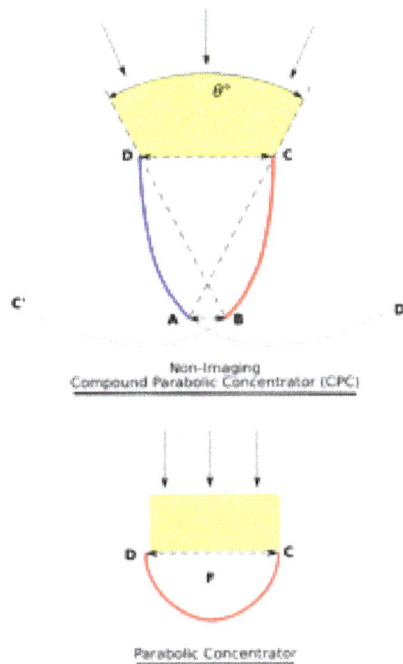

Comparison between Non-imaging Compound Parabolic Concentrator and Parabolic Concentrator

The Parabolic Concentrator has an entrance aperture of DC and a focal point F.

The Parabolic concentrator only accepts rays of light that are perpendicular to the entrance aperture DC. The tracking of this type of concentrator must be more exact and requires expensive equipment .

The Compound Parabolic Concentrator accepts a greater amount of light and needs less accurate tracking

For a 3-dimensional "nonimaging compound parabolic concentrator", the maximum concentration possible in air or in vacuum (equal to the ratio of input and output aperture areas), is:

$$C = \frac{1}{\sin^2 \theta},$$

where θ is the half-angle of the acceptance angle (of the larger aperture).

History

The development started in the mid-1960s at three different locations by V. K. Baranov (USSR) with the study of the focons (focusing cones) Martin Ploke (Germany), and Roland Winston (United States), and led to the independent origin of the first nonimaging concentrators, later applied to solar energy concentration. Among these three earliest works, the one most developed was the American one, resulting in what nonimaging optics is today.

There are different commercial companies and universities working on nonimaging optics. Currently the largest research group in this subject is the Advanced Optics group at the CeDInt, part of the Technical University of Madrid (UPM)

Nonlinear Optics

Nonlinear optics (NLO) is the branch of optics that describes the behavior of light in *nonlinear media*, that is, media in which the dielectric polarization P responds nonlinearly to the electric field E of the light. This nonlinearity is typically only observed at very high light intensities (values of the electric field comparable to interatomic electric fields, typically 10^8 V/m) such as those provided by lasers. Above the Schwinger limit, the vacuum itself is expected to become nonlinear. In nonlinear optics, the superposition principle no longer holds.

Nonlinear optics remained unexplored until the discovery in 1961 of second-harmonic generation by Peter Franken *et al.* at University of Michigan, shortly after the construction of the first laser by Theodore Harold Maiman. However, some nonlinear effects were discovered before the development of the laser. The theoretical basis for many nonlinear processes were first described in Bloembergen's monograph "Nonlinear Optics".

Nonlinear Optical Processes

Nonlinear optics explains nonlinear response of properties such as frequency, polarization, phase or path of incident light. These nonlinear interactions give rise to a host of optical phenomena:

Frequency-mixing Processes

- Second-harmonic generation (SHG), or *frequency doubling*, generation of light with a doubled frequency (half the wavelength), two photons are destroyed, creating a single photon at two times the frequency.

- Third-harmonic generation (THG), generation of light with a tripled frequency (one-third the wavelength), three photons are destroyed, creating a single photon at three times the frequency.

- High-harmonic generation (HHG), generation of light with frequencies much greater than the original (typically 100 to 1000 times greater).

- Sum-frequency generation (SFG), generation of light with a frequency that is the sum of two other frequencies (SHG is a special case of this).

- Difference-frequency generation (DFG), generation of light with a frequency that is the difference between two other frequencies.

- Optical parametric amplification (OPA), amplification of a signal input in the presence of a higher-frequency pump wave, at the same time generating an *idler* wave (can be considered as DFG).

- Optical parametric oscillation (OPO), generation of a signal and idler wave using a parametric amplifier in a resonator (with no signal input).

- Optical parametric generation (OPG), like parametric oscillation but without a resonator, using a very high gain instead.

- Spontaneous parametric down-conversion (SPDC), the amplification of the vacuum fluctuations in the low-gain regime.

- Optical rectification (OR), generation of quasi-static electric fields.

- Nonlinear light-matter interaction with free electrons and plasmas.

Other Nonlinear Processes

- Optical Kerr effect, intensity-dependent refractive index (a $\chi^{(3)}$ effect).

  o Self-focusing, an effect due to the optical Kerr effect (and possibly higher-order nonlinearities) caused by the spatial variation in the intensity creating a spatial variation in the refractive index.

  o Kerr-lens modelocking (KLM), the use of self-focusing as a mechanism to mode-lock laser.

  o Self-phase modulation (SPM), an effect due to the optical Kerr effect (and possibly higher-order nonlinearities) caused by the temporal variation in the intensity creating a temporal variation in the refractive index.

    o   Optical solitons, an equilibrium solution for either an optical pulse (temporal soliton) or spatial mode (spatial soliton) that does not change during propagation due to a balance between dispersion and the Kerr effect (e.g. self-phase modulation for temporal and self-focusing for spatial solitons).

- Cross-phase modulation (XPM), where one wavelength of light can affect the phase of another wavelength of light through the optical Kerr effect.

- Four-wave mixing (FWM), can also arise from other nonlinearities.

- Cross-polarized wave generation (XPW), a $\chi^{(3)}$ effect in which a wave with polarization vector perpendicular to the input one is generated.

- Modulational instability.

- Raman amplification

- Optical phase conjugation.

- Stimulated Brillouin scattering, interaction of photons with acoustic phonons

- Multi-photon absorption, simultaneous absorption of two or more photons, transferring the energy to a single electron.

- Multiple photoionisation, near-simultaneous removal of many bound electrons by one photon.

- Chaos in optical systems.

Related Processes

In these processes, the medium has a linear response to the light, but the properties of the medium are affected by other causes:

- Pockels effect, the refractive index is affected by a static electric field; used in electro-optic modulators.

- Acousto-optics, the refractive index is affected by acoustic waves (ultrasound); used in acousto-optic modulators.

- Raman scattering, interaction of photons with optical phonons.

Parametric Processes

Nonlinear effects fall into two qualitatively different categories, parametric and non-parametric effects. A parametric non-linearity is an interaction in which the quantum state of the nonlinear material is not changed by the interaction with the optical field. As a consequence of this, the process is "instantaneous". Energy and momentum are conserved in the optical field, making phase matching important and polarization-dependent.

Theory

Parametric and "instantaneous" (i.e. material must be lossless and dispersionless through the Kramers–Kronig relations) nonlinear optical phenomena, in which the optical fields are not too large, can be described by a Taylor series expansion of the dielectric polarization density (dipole moment per unit volume) P(t) at time t in terms of the electrical field E(t):

$$\mathbf{P}(t) = \varepsilon_0 (\chi^{(1)}\mathbf{E}(t) + \chi^{(2)}\mathbf{E}^2(t) + \chi^{(3)}\mathbf{E}^3(t) + \ldots),$$

where the coefficients $\chi^{(n)}$ are the n-th-order susceptibilities of the medium, and the presence of such a term is generally referred to as an n-th-order nonlinearity. Note that the polarization density P(t) and electrical field E(t) are considered as scalar for simplicity. In general, $\chi^{(n)}$ is an $(n + 1)$-th-order tensor representing both the polarization-dependent nature of the parametric interaction and the symmetries (or lack of) of the nonlinear material.

Wave Equation in a Nonlinear Material

Central to the study of electromagnetic waves is the wave equation. Starting with Maxwell's equations in an isotropic space, containing no free charge, it can be shown that

$$\nabla \times \nabla \times \mathbf{E} + \frac{n^2}{c^2}\frac{\partial^2}{\partial t^2}\mathbf{E} = -\frac{1}{\varepsilon_0 c^2}\frac{\partial^2}{\partial t^2}\mathbf{P}^{\mathrm{NL}},$$

where $\mathbf{P}^{\mathrm{NL}}$ is the nonlinear part of the polarization density, and n is the refractive index, which comes from the linear term in P.

Note that one can normally use the vector identity

$$\nabla \times (\nabla \times \mathbf{V}) = \nabla(\nabla \cdot \mathbf{V}) - \nabla^2 \mathbf{V}$$

and Gauss's law (assuming no free charges, $\rho_{\mathrm{free}} = 0$),

$$\nabla \cdot \mathbf{D} = 0,$$

to obtain the more familiar wave equation

$$\nabla^2 \mathbf{E} - \frac{n^2}{c^2}\frac{\partial^2}{\partial t^2}\mathbf{E} = 0.$$

For nonlinear medium, Gauss's law does not imply that the identity

$$\nabla \cdot \mathbf{E} = 0$$

is true in general, even for an isotropic medium. However, even when this term is not identically 0, it is often negligibly small and thus in practice is usually ignored, giving us the standard nonlinear wave equation:

$$\nabla^2 \mathbf{E} - \frac{n^2}{c^2}\frac{\partial^2}{\partial t^2}\mathbf{E} = \frac{1}{\varepsilon_0 c^2}\frac{\partial^2}{\partial t^2}\mathbf{P}^{\mathrm{NL}}.$$

Nonlinearities as a wave-mixing process

The nonlinear wave equation is an inhomogeneous differential equation. The general solution comes from the study of ordinary differential equations and can be obtained by the use of a Green's function. Physically one gets the normal electromagnetic wave solutions to the homogeneous part of the wave equation:

$$\nabla^2 \mathbf{E} - \frac{n^2}{c^2}\frac{\partial^2}{\partial t^2}\mathbf{E} = 0,$$

and the inhomogeneous term

$$\frac{1}{\varepsilon_0 c^2}\frac{\partial^2}{\partial t^2}\mathbf{P}^{NL}$$

acts as a driver/source of the electromagnetic waves. One of the consequences of this is a nonlinear interaction that results in energy being mixed or coupled between different frequencies, which is often called a "wave mixing".

In general, an n-th order will lead to $(n + 1)$-wave mixing. As an example, if we consider only a second-order nonlinearity (three-wave mixing), then the polarization P takes the form

$$\mathbf{P}^{NL} = \varepsilon_0 \chi^{(2)}\mathbf{E}^2(t).$$

If we assume that $E(t)$ is made up of two components at frequencies ω_1 and ω_2, we can write $E(t)$ as

$$\mathbf{E}(t) = E_1 \cos(\omega_1 t) + E_2 \cos(\omega_2 t),$$

and using Euler's formula to convert to exponentials,

$$\mathbf{E}(t) = \frac{1}{2}E_1 e^{-i\omega_1 t} + \frac{1}{2}E_2 e^{-i\omega_2 t} + \text{c.c.},$$

where "c.c." stands for complex conjugate. Plugging this into the expression for P gives

$$\mathbf{P}^{NL} = \varepsilon_0 \chi^{(2)}\mathbf{E}^2(t) = \frac{\varepsilon_0}{4}\chi^{(2)}\Big[|E_1|^2\,e^{-i2\omega_1 t} + |E_2|^2\,e^{-i2\omega_2 t}$$

$$+2E_1 E_2 e^{-i(\omega_1+\omega_2)t}\,.$$

$$+2E_1 E_2^* e^{-i(\omega_1-\omega_2)t}$$

$$+\Big(|E_1|^2 + |E_2|^2\Big)e^0 + \text{c.c.}\Big],$$

which has frequency components at $2\omega_1$, $2\omega_2$, $\omega_1 + \omega_2$, $\omega_1 - \omega_2$, and 0. These three-wave mixing processes correspond to the nonlinear effects known as second-harmonic generation, sum-frequency generation, difference-frequency generation and optical rectification respectively.

Note: Parametric generation and amplification is a variation of difference-frequency generation, where the lower frequency of one of the two generating fields is much weaker (parametric amplification) or completely absent (parametric generation). In the latter case, the fundamental quantum-mechanical uncertainty in the electric field initiates the process.

Phase Matching

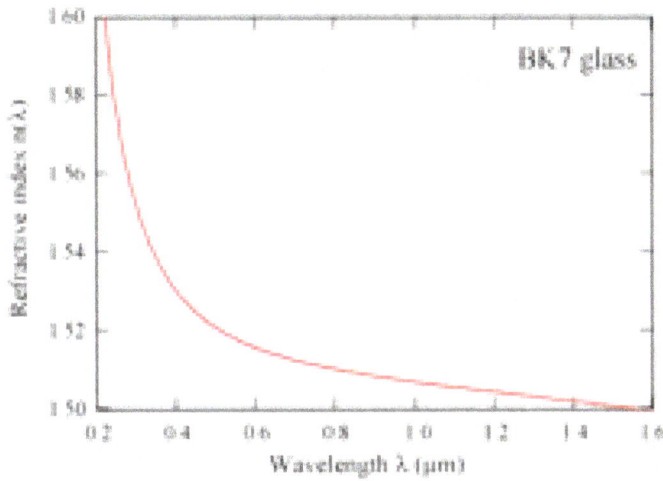

Most transparent materials, like the BK7 glass shown here, have normal dispersion: the index of refraction decreases monotonically as a function of wavelength (or increases as a function of frequency). This makes phase matching impossible in most frequency-mixing processes. For example, in SHG, there is no simultaneous solution to $\omega' = 2\omega$ and $\mathbf{k}' = 2\mathbf{k}$ in these materials. Birefringent materials avoid this problem by having two indices of refraction at once.

The above ignores the position dependence of the electrical fields. In a typical situation, the electrical fields are traveling waves described by

$$E_j(\mathbf{x},t) = e^{i(\mathbf{k}_j \cdot \mathbf{x} - \omega_j t)} + \text{c.c.}$$

at position $\mathbf{x}$, with the wave vector $\|\mathbf{k}_j\| = n(\omega_j)\omega_j / c$, where c is the velocity of light in vacuum, and $n(\omega_j)$ is the index of refraction of the medium at angular frequency ω_j Thus, the second-order polarization at angular frequency $\omega_3 = \omega_1 + \omega_2$ is

$$P^{(2)}(\mathbf{x},t) \propto E_1^{n_1} E_2^{n_2} e^{i[(\mathbf{k}_1 + \mathbf{k}_2)\cdot\mathbf{x} - \omega_3 t]} + \text{c.c.}$$

At each position $\mathbf{x}$ within the nonlinear medium, the oscillating second-order polarization radiates at angular frequency ω_3 and a corresponding wave vector $\|\mathbf{k}_3\| = n(\omega_3)\omega_3 / c$. Constructive interference, and therefore a high-intensity ω_3 field, will occur only if

$$\vec{k}_3 = \vec{k}_1 + \vec{k}_2.$$

The above equation is known as the *phase-matching condition*. Typically, three-wave mixing is done in a birefringent crystalline material, where the refractive index depends on the polarization and direction of the light that passes through. The polarizations of the fields and the orientation of the crystal are chosen such that the phase-matching condition is fulfilled. This phase-matching technique is called angle tuning. Typically a crystal has three axes, one or two of which have a different refractive index than the other one(s). Uniaxial crystals, for example, have a single preferred axis, called the extraordinary (e) axis, while the other two are ordinary axes (o). There are several schemes of choosing the polarizations for this crystal type. If the signal and idler have the same polarization, it is called "type-I phase matching", and if their polarizations are perpendicular, it is called "type-II phase matching". However, other conventions exist that specify further which frequency has what polarization relative to the crystal axis. These types are listed below, with the convention that the signal wavelength is shorter than the idler wavelength.

$$(\lambda_p \leq \lambda_s \leq \lambda_i)$$

Phase-matching types

Polarizations Scheme			
Pump	**Signal**	**Idler**	
e	o	o	Type I
e	o	e	Type II (or IIA)
e	e	o	Type III (or IIB)
e	e	e	Type IV
o	o	o	Type V (or type o, or "zero")
o	o	e	Type VI (or IIB or IIIA)
o	e	o	Type VII (or IIA or IIIB)
o	e	e	Type VIII (or I)

Most common nonlinear crystals are negative uniaxial, which means that the *e* axis has a smaller refractive index than the *o* axes. In those crystals, type-I and -II phase matching are usually the most suitable schemes. In positive uniaxial crystals, types VII and VIII are more suitable. Types II and III are essentially equivalent, except that the names of signal and idler are swapped when the signal has a longer wavelength than the idler. For this reason, they are sometimes called IIA and IIB. The type numbers V–VIII are less common than I and II and variants.

One undesirable effect of angle tuning is that the optical frequencies involved do not propagate collinearly with each other. This is due to the fact that the extraordinary wave propagating through a birefringent crystal possesses a Poynting vector that is not parallel to the propagation vector. This would lead to beam walk-off, which limits the nonlinear optical conversion efficiency. Two other methods of phase matching avoid beam walk-off by forcing all frequencies to propagate at a 90° with respect to the optical axis of the crystal. These methods are called temperature tuning and quasi-phase-matching.

Temperature tuning is used when the pump (laser) frequency polarization is orthogonal to the signal and idler frequency polarization. The birefringence in some crystals, in particular lithium niobate is highly temperature-dependent. The crystal temperature is controlled to achieve phase-matching conditions.

The other method is quasi-phase-matching. In this method the frequencies involved are not constantly locked in phase with each other, instead the crystal axis is flipped at a regular interval Λ, typically 15 micrometres in length. Hence, these crystals are called periodically poled. This results in the polarization response of the crystal to be shifted back in phase with the pump beam by reversing the nonlinear susceptibility. This allows net positive energy flow from the pump into the signal and idler frequencies. In this case, the crystal itself provides the additional wavevector $k = 2\pi/\lambda$ (and hence momentum) to satisfy the phase-matching condition. Quasi-phase-matching can be expanded to chirped gratings to get more bandwidth and to shape an SHG pulse like it is done in a dazzler. SHG of a pump and self-phase modulation (emulated by second-order processes) of the signal and an optical parametric amplifier can be integrated monolithically.

Higher-order Frequency Mixing

The above holds for $\chi^{(2)}$ processes. It can be extended for processes where $\chi^{(3)}$ is nonzero, something that is generally true in any medium without any symmetry restrictions; in particular resonantly enhanced sum or difference frequency mixing in gasses is frequently used for extreme or "vacuum" Ultra Violet light generation. In common scenarios, such as mixing in dilute gases, the non-linearity is weak and so the light beams are focused which, unlike the plane wave approximation used above, introduces a pi phase shift on each light beam, complicating the phase matching requirements. Conveniently, difference frequency mixing with $\chi^{(3)}$ cancels this focal phase shift and often has a nearly self-canceling overall phase matching condition, which relatively simplifies broad wavelength tuning compared to sum frequency generation. In $\chi^{(3)}$ all four frequencies are mixing simultaneously, as opposed to sequential mixing via two $\chi^{(2)}$ processes.

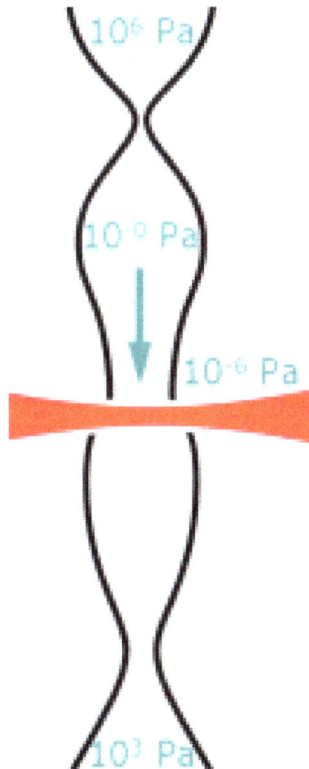

The Kerr effect can be described as a $\chi^{(3)}$ as well. At high peak powers the Kerr effect can cause filamentation of light in air, in which the light travels without dispersion or divergence in a self-generated waveguide. At even high intensities the Taylor series, which led the domination of the lower orders, does not converge anymore and instead a time based model is used. When a noble gas atom is hit by an intense laser pulse, which has an electric field strength comparable to the Coulomb field of the atom, the outermost electron may be ionized from the atom. Once freed, the electron can be accelerated by the electric field of the light, first moving away from the ion, then back toward it as the field changes direction. The electron may then recombine with the ion, releasing its energy in the form of a photon. The light is emitted at every peak of the laser light field which is intense enough, producing a series of attosecond light flashes. The photon energies generated by this process can extend past the 800th harmonic order up to a few KeV. This is called high-order harmonic generation. The laser must be linearly polarized, so that the electron returns to the vicinity of the parent ion. High-order harmonic generation has been observed in noble gas jets, cells, and gas-filled capillary waveguides.

Example Uses of Nonlinear Optics

Frequency Doubling

One of the most commonly used frequency-mixing processes is frequency doubling, or second-harmonic generation. With this technique, the 1064 nm output from Nd:YAG lasers or the 800 nm output from Ti:sapphire lasers can be converted to visible light, with wavelengths of 532 nm (green) or 400 nm (violet) respectively.

Practically, frequency doubling is carried out by placing a nonlinear medium in a laser beam. While there are many types of nonlinear media, the most common media are crystals. Commonly used crystals are BBO (β-barium borate), KDP (potassium dihydrogen phosphate), KTP (potassium titanyl phosphate), and lithium niobate. These crystals have the necessary properties of being strongly birefringent (necessary to obtain phase matching), having a specific crystal symmetry, being transparent for both the impinging laser light and the frequency-doubled wave-length, and having high damage thresholds, which makes them resistant against the high-intensity laser light.

Optical Phase Conjugation

It is possible, using nonlinear optical processes, to exactly reverse the propagation direction and phase variation of a beam of light. The reversed beam is called a *conjugate* beam, and thus the technique is known as optical phase conjugation (also called *time reversal, wavefront reversal* and *retroreflection*).

One can interpret this nonlinear optical interaction as being analogous to a real-time holographic process. In this case, the interacting beams simultaneously interact in a nonlinear optical material to form a dynamic hologram (two of the three input beams), or real-time diffraction pattern, in the material. The third incident beam diffracts at this dynamic hologram, and, in the process, reads out the *phase-conjugate* wave. In effect, all three incident beams interact (essentially) simultaneously to form several real-time holograms, resulting in a set of diffracted output waves that phase up as the "time-reversed" beam. In the language of nonlinear optics, the interacting beams result in a nonlinear polarization within the material, which coherently radiates to form the phase-conjugate wave.

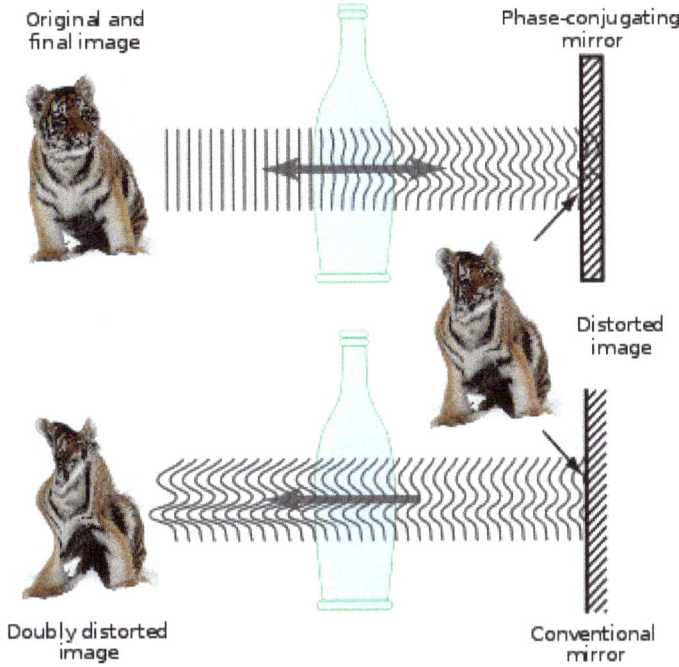

Comparison of a phase-conjugate mirror with a conventional mirror. With the phase-conjugate mirror the image is not deformed when passing through an aberrating element twice.

The most common way of producing optical phase conjugation is to use a four-wave mixing technique, though it is also possible to use processes such as stimulated Brillouin scattering. A device producing the phase-conjugation effect is known as a phase-conjugate mirror (PCM).

For the four-wave mixing technique, we can describe four beams (j = 1, 2, 3, 4) with electric fields:

$$\Xi_j(\mathbf{x},t) = \frac{1}{2}E_j(\mathbf{x})e^{i(\omega_j t - \mathbf{k}\cdot\mathbf{x})} + \text{c.c.},$$

where E_j are the electric field amplitudes. Ξ_1 and Ξ_2 are known as the two pump waves, with Ξ_3 being the signal wave, and Ξ_4 being the generated conjugate wave.

If the pump waves and the signal wave are superimposed in a medium with a non-zero $\chi^{(3)}$, this produces a nonlinear polarization field:

$$P_{\text{NL}} = \varepsilon_0 \chi^{(3)}(\Xi_1 + \Xi_2 + \Xi_3)^3,$$

resulting in generation of waves with frequencies given by $\omega = \pm\omega_1 \pm \omega_2 \pm \omega_3$ in addition to third-harmonic generation waves with $\omega = 3\omega_1, 3\omega_2, 3\omega_3$.

As above, the phase-matching condition determines which of these waves is the dominant. By choosing conditions such that $\omega = \omega_1 + \omega_2 - \omega_3$ and $k = k_1 + k_2 - k_3$, this gives a polarization field:

$$P_\omega = \frac{1}{2}\chi^{(3)}\varepsilon_0 E_1 E_2 E_3^* e^{i(\omega t - \mathbf{k}\cdot\mathbf{x})} + \text{c.c.}$$

This is the generating field for the phase-conjugate beam, Ξ_4. Its direction is given by $k_4 = k_1 + k_2 - k_3$, and so if the two pump beams are counterpropagating ($k_1 = -k_2$), then the conjugate and signal beams propagate in opposite directions ($k_4 = -k_3$). This results in the retroreflecting property of the effect.

Further, it can be shown that for a medium with refractive index n and a beam interaction length l, the electric field amplitude of the conjugate beam is approximated by

$$E_4 = \frac{i\omega l}{2nc}\chi^{(3)}E_1 E_2 E_3^*,$$

where c is the speed of light. If the pump beams E_1 and E_2 are plane (counterpropagating) waves, then

$$E_4(\mathbf{x}) \propto E_3^*(\mathbf{x}),$$

that is, the generated beam amplitude is the complex conjugate of the signal beam amplitude. Since the imaginary part of the amplitude contains the phase of the beam, this results in the reversal of phase property of the effect.

Note that the constant of proportionality between the signal and conjugate beams can be greater than 1. This is effectively a mirror with a reflection coefficient greater than 100%, producing an amplified reflection. The power for this comes from the two pump beams, which are depleted by the process.

The frequency of the conjugate wave can be different from that of the signal wave. If the pump waves are of frequency $\omega_1 = \omega_2 = \omega$, and the signal wave is higher in frequency such that $\omega_3 = \omega + \Delta\omega$, then the conjugate wave is of frequency $\omega_4 = \omega - \Delta\omega$. This is known as *frequency flipping*.

Angular and Linear Momenta in Optical Phase Conjugation

Optical phase conjugation in the near field performs the reversal of classical rays, or retroreflection.

Classical Picture

In *classical Maxwell electrodynamics* a phase-conjugating mirror performs reversal of the Poynting vector:

$$\mathbf{S}_{out}(\mathbf{r},t) = -\mathbf{S}_{in}(\mathbf{r},t),$$

("in" means incident field, "out" means reflected field) where

$$\mathbf{S}(\mathbf{r},t) = \epsilon_0 c^2 \mathbf{E}(\mathbf{r},t) \times \mathbf{B}(\mathbf{r},t),$$

which is a linear momentum density of electromagnetic field. In the same way a phase-conjugated wave has an opposite angular momentum density vector $\mathbf{L}(\mathbf{r},t) = \mathbf{r} \times \mathbf{S}(\mathbf{r},t)$ with respect to incident field:

$$\mathbf{L}_{out}(\mathbf{r},t) = -\mathbf{L}_{in}(\mathbf{r},t).$$

The above identities are valid *locally*, i.e. in each space point $\mathbf{r}$ in a given moment t for an *ideal phase-conjugating mirror*.

Quantum Picture

In *quantum electrodynamics* the photon with energy $\hbar\omega$ also possesses linear momentum $\mathbf{P} = \hbar\mathbf{k}$

and angular momentum, whose projection on propagation axis is $L_z = \pm\hbar\ell,$, where ℓ is *topologi-*

cal charge of photon, or winding number, $\mathbf{z}$ is propagation axis. The angular momentum projection on propagation axis has *discrete values* $\pm\hbar\ell$.

In *quantum electrodynamics* the interpretation of phase conjugation is much simpler compared to *classical electrodynamics*. The photon reflected from phase conjugating-mirror (out) has opposite directions of linear and angular momenta with respect to incident photon (in):

$$\mathbf{P}_{out} = -\hbar\mathbf{k} = -\mathbf{P}_{in} = \hbar\mathbf{k},$$

$$L_{z\,out} = -\hbar\ell = -L_{z\,in} = \hbar\ell.$$

Common SHG Materials

Dark-red gallium selenide in its bulk form

Ordered by pump wavelength:

- 800 nm: BBO

- 806 nm: lithium iodate ($LiIO_3$)

- 860 nm: potassium niobate ($KNbO_3$)

- 980 nm: $KNbO_3$

- 1064 nm: monopotassium phosphate (KH_2PO_4, KDP), lithium triborate (LBO) and β-barium borate (BBO)

- 1300 nm: gallium selenide (GaSe)

- 1319 nm: $KNbO_3$, BBO, KDP, potassium titanyl phosphate (KTP), lithium niobate ($LiNbO_3$), $LiIO_3$, and ammonium dihydrogen phosphate (ADP)

- 1550 nm: potassium titanyl phosphate (KTP), lithium niobate ($LiNbO_3$)

Crystal Optics

Crystal optics is the branch of optics that describes the behaviour of light in *anisotropic media*, that is, media (such as crystals) in which light behaves differently depending on which direction the light is propagating. The index of refraction depends on both composition and crystal structure and can be calculated using the Gladstone–Dale relation. Crystals are often naturally anisotropic, and in some media (such as liquid crystals) it is possible to induce anisotropy by applying an external electric field.

Isotropic Media

Typical transparent media such as glasses are *isotropic*, which means that light behaves the same way no matter which direction it is travelling in the medium. In terms of Maxwell's equations in a dielectric, this gives a relationship between the electric displacement field D and the electric field E:

$$\mathbf{D} = \varepsilon_0 \mathbf{E} + \mathbf{P}$$

where ε_0 is the permittivity of free space and P is the electric polarization (the vector field corresponding to electric dipole moments present in the medium). Physically, the polarization field can be regarded as the response of the medium to the electric field of the light.

Electric susceptibility

In an isotropic and linear medium, this polarization field P is proportional to and parallel to the electric field E:

$$\mathbf{P} = \chi \varepsilon_0 \mathbf{E}$$

where χ is the *electric susceptibility* of the medium. The relation between D and E is thus:

$$\mathbf{D} = \varepsilon_0 \mathbf{E} + \chi \varepsilon_0 \mathbf{E} = \varepsilon_0 (1 + \chi) \mathbf{E} = \varepsilon \mathbf{E}$$

where

$$\varepsilon = \varepsilon_0 (1 + \chi)$$

is the dielectric constant of the medium. The value $1+\chi$ is called the *relative permittivity* of the medium, and is related to the refractive index n, for non-magnetic media, by

$$\varepsilon = \varepsilon_0 (1 + \chi)$$

Anisotropic Media

In an anisotropic medium, such as a crystal, the polarisation field P is not necessarily aligned with the electric field of the light E. In a physical picture, this can be thought of as the dipoles induced in the medium by the electric field having certain preferred directions, related to the physical structure of the crystal. This can be written as:

$$\mathbf{P} = \varepsilon_0 \div \mathbf{E}.$$

Here χ is not a number as before but a tensor of rank 2, the *electric susceptibility tensor*. In terms of components in 3 dimensions:

$$\begin{pmatrix} P_x \\ P_y \\ P_z \end{pmatrix} = \varepsilon_0 \begin{pmatrix} \chi_{xx} & \chi_{xy} & \chi_{xz} \\ \chi_{yx} & \chi_{yy} & \chi_{yz} \\ \chi_{zx} & \chi_{zy} & \chi_{zz} \end{pmatrix} \begin{pmatrix} E_x \\ E_y \\ E_z \end{pmatrix}$$

or using the summation convention:

$$P_i = \varepsilon_0 \sum_{j \in \{x,y,z\}} \chi_{ij} E_j \quad .$$

Since χ is a tensor, P is not necessarily colinear with E.

In nonmagnetic and transparent materials, $\chi_{ij} = \chi_{ji}$, i.e. the χ tensor is real and symmetric. In accordance with the spectral theorem, it is thus possible to diagonalise the tensor by choosing the appropriate set of coordinate axes, zeroing all components of the tensor except χ_{xx}, χ_{yy} and χ_{zz}. This gives the set of relations:

$$P_x = \varepsilon_0 \chi_{xx} E_x$$

$$P_y = \varepsilon_0 \chi_{yy} E_y$$

$$P_z = \varepsilon_0 \chi_{zz} E_z$$

The directions x, y and z are in this case known as the *principal axes* of the medium. Note that these axes will be orthogonal if all entries in the χ tensor are real, corresponding to a case in which the refractive index is real in all directions.

It follows that D and E are also related by a tensor:

$$\mathbf{D} = \varepsilon_0 \mathbf{E} + \mathbf{P} = \varepsilon_0 \mathbf{E} + \varepsilon_0 \ddot{\div} \mathbf{E} = \varepsilon_0 (I + \div) \mathbf{E} = \varepsilon_0 \ddot{a} \mathbf{E}.$$

Here ε is known as the *relative permittivity tensor* or *dielectric tensor*. Consequently, the refractive index of the medium must also be a tensor. Consider a light wave propagating along the z principal axis polarised such the electric field of the wave is parallel to the x-axis. The wave experiences a susceptibility χ_{xx} and a permittivity ε_{xx}. The refractive index is thus:

$$n_{xx} = (1 + \chi_{xx})^{1/2} = (\varepsilon_{xx})^{1/2}.$$

For a wave polarised in the y direction:

$$n_{yy} = (1 + \chi_{yy})^{1/2} = (\varepsilon_{yy})^{1/2}.$$

Thus these waves will see two different refractive indices and travel at different speeds. This phenomenon is known as *birefringence* and occurs in some common crystals such as calcite and quartz.

If $\chi_{xx} = \chi_{yy} \neq \chi_{zz}$, the crystal is known as uniaxial. (See Optic axis of a crystal.) If $\chi_{xx} \neq \chi_{yy}$ and $\chi_{xx} \neq \chi_{zz}$

the crystal is called biaxial. A uniaxial crystal exhibits two refractive indices, an "ordinary" index (n_o) for light polarised in the x or y directions, and an "extraordinary" index (n_e) for polarisation in the z direction. A uniaxial crystal is "positive" if $n_e > n_o$ and "negative" if $n_e < n_o$. Light polarised at some angle to the axes will experience a different phase velocity for different polarization components, and cannot be described by a single index of refraction. This is often depicted as an index ellipsoid.

Other Effects

Certain nonlinear optical phenomena such as the electro-optic effect cause a variation of a medium's permittivity tensor when an external electric field is applied, proportional (to lowest order) to the strength of the field. This causes a rotation of the principal axes of the medium and alters the behaviour of light travelling through it; the effect can be used to produce light modulators.

In response to a magnetic field, some materials can have a dielectric tensor that is complex-Hermitian; this is called a gyro-magnetic or magneto-optic effect. In this case, the principal axes are complex-valued vectors, corresponding to elliptically polarized light, and time-reversal symmetry can be broken. This can be used to design optical isolators, for example.

A dielectric tensor that is not Hermitian gives rise to complex eigenvalues, which corresponds to a material with gain or absorption at a particular frequency.

Near-field Optics

Near-field optics is that branch of optics that considers configurations that depend on the passage of light to, from, through, or near an element with subwavelength features, and the coupling of that light to a second element located a subwavelength distance from the first. The barrier of spatial resolution imposed by the very nature of light itself in conventional optical microscopy contributed significantly to the development of near-field optical devices, most notably the near-field scanning optical microscope, or NSOM.

Size Constraints

The limit of optical resolution in a conventional microscope, the so-called diffraction limit, is in the order of half the wavelength of the light used to image. Thus, when imaging at visible wavelengths, the smallest resolvable features are several hundred nanometers in size (although point-like sources, such as quantum dots, can be resolved quite readily). Using near-field optical techniques, researchers currently resolve features in the order of tens of nanometers in size. While other imaging techniques (e.g. atomic force microscopy and electron microscopy) can resolve features of much smaller size, the many advantages of optical microscopy make near-field optics a field of considerable interest.

History

The notion of developing a near-field optical device was first conceived by Edward Hutchinson Synge in 1928 but was not realized experimentally until the 1950s when several researchers demonstrated the feasibility of sub-wavelength resolution. Published images of sub-wavelength resolution appeared when Ash and Nichols examined gratings with line spacing less than one millimeter using microwaves of 3 cm wavelength. In 1982 Dieter Pohl at IBM in Zurich, Switzerland, first obtained sub-wavelength resolution at visible wavelengths using near-field optical techniques.

Lens (Optics)

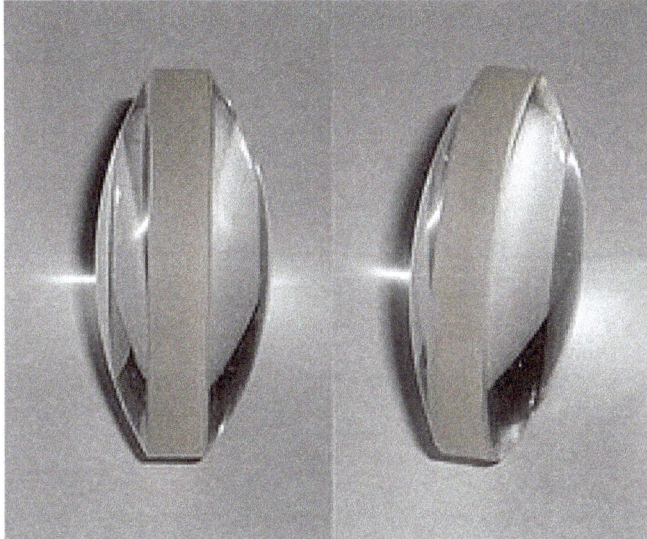

A biconvex lens

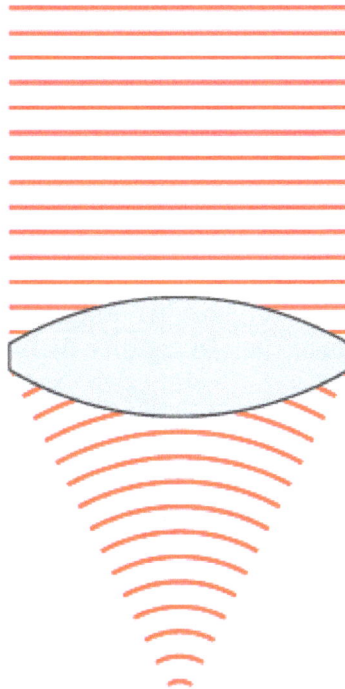

Lenses can be used to focus light

A lens is a transmissive optical device that focuses or disperses a light beam by means of refraction. A simple lens consists of a single piece of transparent material, while a compound lens consists of several simple lenses (*elements*), usually arranged along a common axis. Lenses are made from materials such as glass or plastic, and are ground and polished or moulded to a desired shape. A lens can focus light to form an image, unlike a prism, which refracts light without focusing. Devices that similarly focus or disperse radiation other than visible light are also called lenses, such as microwave lenses, electron lenses or acoustic lenses.

History

The Nimrud lens

The word *lens* comes from the Latin name of the lentil, because a double-convex lens is lentil-shaped. The genus of the lentil plant is *Lens*, and the most commonly eaten species is *Lens culinaris*. The lentil plant also gives its name to a geometric figure.

The variant spelling *lense* is sometimes seen. While it is listed as an alternative spelling in some dictionaries, most mainstream dictionaries do not list it as acceptable.

The oldest lens artifact is the Nimrud lens, dating back 2700 years (7th century B.C.) to ancient Assyria. David Brewster proposed that it may have been used as a magnifying glass, or as a burning-glass to start fires by concentrating sunlight. Another early reference to magnification dates back to ancient Egyptian hieroglyphs in the 8th century BC, which depict "simple glass meniscal lenses".

The earliest written records of lenses date to Ancient Greece, with Aristophanes' play *The Clouds* (424 BC) mentioning a burning-glass (a biconvex lens used to focus the sun's rays to produce fire). Some scholars argue that the archeological evidence indicates that there was widespread use of lenses in antiquity, spanning several millennia. Such lenses were used by artisans for fine work, and for authenticating seal impressions. The writings of Pliny the Elder (23–79) show that burning-glasses were known to the Roman Empire, and mentions what is arguably the earliest written reference to a corrective lens: Nero was said to watch the gladiatorial games using an emerald (presumably concave to correct for nearsightedness, though the reference is vague). Both Pliny and Seneca the Younger (3 BC–65) described the magnifying effect of a glass globe filled with water.

Excavations at the Viking harbour town of Fröjel, Gotland, Sweden discovered in 1999 the rock crystal Visby lenses, produced by turning on pole lathes at Fröjel in the 11th to 12th century, with an imaging quality comparable to that of 1950s aspheric lenses. The Viking lenses were capable of concentrating enough sunlight to ignite fires.

Between the 11th and 13th century "reading stones" were invented. Often used by monks to assist in illuminating manuscripts, these were primitive plano-convex lenses initially made by cutting a glass sphere in half. As the stones were experimented with, it was slowly understood that shallower lenses magnified more effectively.

Lenses came into widespread use in Europe with the invention of spectacles, probably in Italy in the 1280s. This was the start of the optical industry of grinding and polishing lenses for spectacles, first in Venice and Florence in the thirteenth century, and later in the spectacle-making centres in both the Netherlands and Germany. Spectacle makers created improved types of lenses for the correction of vision based more on empirical knowledge gained from observing the effects of the lenses (probably without the knowledge of the rudimentary optical theory of the day). The practical development and experimentation with lenses led to the invention of the compound optical microscope around 1595, and the refracting telescope in 1608, both of which appeared in the spectacle-making centres in the Netherlands.

With the invention of the telescope and microscope there was a great deal of experimentation with lens shapes in the 17th and early 18th centuries trying to correct chromatic errors seen in lenses. Opticians tried to construct lenses of varying forms of curvature, wrongly assuming errors arose from defects in the spherical figure of their surfaces. Optical theory on refraction and experimentation was showing no single-element lens could bring all colours to a focus. This led to the invention of the compound achromatic lens by Chester Moore Hall in England in 1733, an invention also claimed by fellow Englishman John Dollond in a 1758 patent.

Construction of Simple Lenses

Most lenses are *spherical lenses*: their two surfaces are parts of the surfaces of spheres. Each surface can be *convex* (bulging outwards from the lens), *concave* (depressed into the lens), or *planar* (flat). The line joining the centres of the spheres making up the lens surfaces is called the *axis* of the lens. Typically the lens axis passes through the physical centre of the lens, because of the way they are manufactured. Lenses may be cut or ground after manufacturing to give them a different shape or size. The lens axis may then not pass through the physical centre of the lens.

Toric or sphero-cylindrical lenses have surfaces with two different radii of curvature in two orthogonal planes. They have a different focal power in different meridians. This forms an astigmatic lens. An example is eyeglass lenses that are used to correct astigmatism in someone's eye.

More complex are aspheric lenses. These are lenses where one or both surfaces have a shape that is neither spherical nor cylindrical. The more complicated shapes allow such lenses to form images with less aberration than standard simple lenses, but they are more difficult and expensive to produce.

Types of Simple Lenses

Lenses are classified by the curvature of the two optical surfaces. A lens is *biconvex* (or *double convex*, or just *convex*) if both surfaces are convex. If both surfaces have the same radius of curvature, the lens is *equiconvex*. A lens with two concave surfaces is *biconcave* (or just *concave*). If one of the surfaces is flat, the lens is *plano-convex* or *plano-concave* depending on the curvature of the other surface. A lens with one convex and one concave side is *convex-concave* or *meniscus*. It is this type of lens that is most commonly used in corrective lenses.

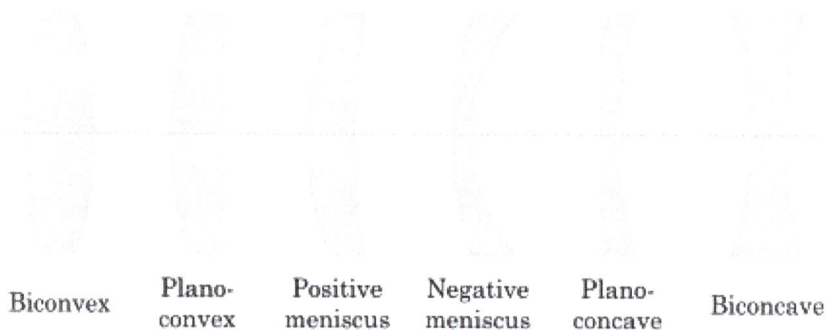

| Biconvex | Plano-convex | Positive meniscus | Negative meniscus | Plano-concave | Biconcave |

If the lens is biconvex or plano-convex, a collimated beam of light passing through the lens converges to a spot (a *focus*) behind the lens. In this case, the lens is called a *positive* or *converging* lens. The distance from the lens to the spot is the focal length of the lens, which is commonly abbreviated *f* in diagrams and equations.

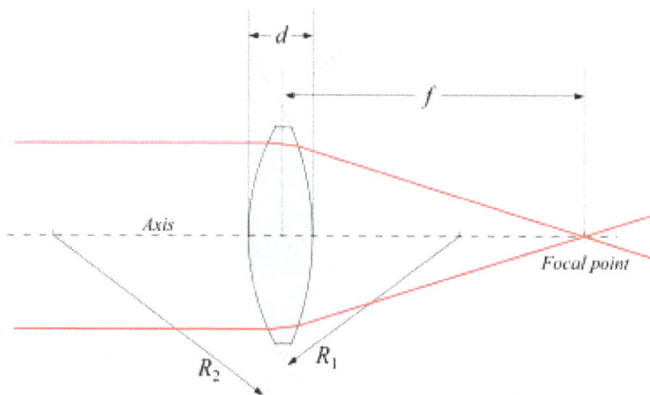

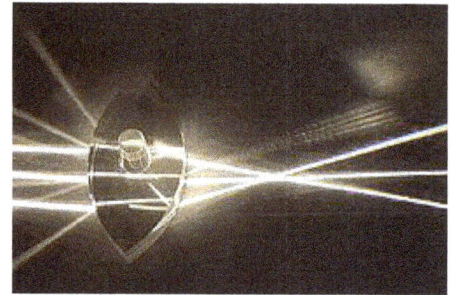

Positive (converging) lens

If the lens is biconcave or plano-concave, a collimated beam of light passing through the lens is diverged (spread); the lens is thus called a *negative* or *diverging* lens. The beam, after passing through the lens, appears to emanate from a particular point on the axis in front of the lens. The distance from this point to the lens is also known as the focal length, though it is negative with respect to the focal length of a converging lens.

Convex-concave (meniscus) lenses can be either positive or negative, depending on the relative curvatures of the two surfaces. A *negative meniscus* lens has a steeper concave surface and is thinner at the centre than at the periphery. Conversely, a *positive meniscus* lens has a steeper convex surface and is thicker at the centre than at the periphery. An ideal thin lens with two surfaces of equal curvature would have zero optical power, meaning that it would neither converge nor diverge light. All real lenses have nonzero thickness, however, which makes a real lens with identical curved surfaces slightly positive. To obtain exactly zero optical power, a meniscus lens must have slightly unequal curvatures to account for the effect of the lens' thickness.

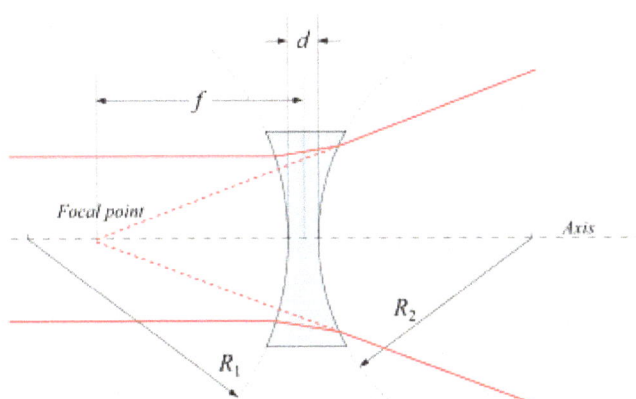

Negative (diverging) lens

Lensmaker's Equation

The focal length of a lens *in air* can be calculated from the lensmaker's equation:

$$\frac{1}{f} = (n-1)\left[\frac{1}{R_1} - \frac{1}{R_2} + \frac{(n-1)d}{nR_1R_2}\right],$$

where

f is the focal length of the lens,

n is the refractive index of the lens material,

R_1 is the radius of curvature (with sign) of the lens surface closest to the light source,

R_2 is the radius of curvature of the lens surface farthest from the light source, and

d is the thickness of the lens (the distance along the lens axis between the two surface vertices).

The focal length *f* is positive for converging lenses, and negative for diverging lenses. The reciprocal of the focal length, 1/*f*, is the optical power of the lens. If the focal length is in metres, this gives the optical power in dioptres (inverse metres).

Lenses have the same focal length when light travels from the back to the front as when light goes from the front to the back. Other properties of the lens, such as the aberrations are not the same in both directions.

Sign Convention for Radii of Curvature R_1 and R_2

The signs of the lens' radii of curvature indicate whether the corresponding surfaces are convex or concave. The sign convention used to represent this varies, but in this article a *positive R* indicates a surface's center of curvature is further along in the direction of the ray travel (right, in the accompanying diagrams), while *negative R* means that rays reaching the surface have already passed the center of curvature. Consequently, for external lens surfaces as diagrammed above, $R_1 > 0$ and $R_2 < 0$ indicate *convex* surfaces (used to converge light in a positive lens), while $R_1 < 0$ and $R_2 > 0$ indicate *concave* surfaces. The reciprocal of the radius of curvature is called the curvature. A flat surface has zero curvature, and its radius of curvature is infinity.

Thin Lens Approximation

If d is small compared to R_1 and R_2, then the *thin lens* approximation can be made. For a lens in air, f is then given by

$$\frac{1}{f} \approx (n-1)\left[\frac{1}{R_1} - \frac{1}{R_2}\right].$$

Imaging Properties

As mentioned above, a positive or converging lens in air focuses a collimated beam travelling along the lens axis to a spot (known as the focal point) at a distance f from the lens. Conversely, a point source of light placed at the focal point is converted into a collimated beam by the lens. These two cases are examples of image formation in lenses. In the former case, an object at an infinite distance (as represented by a collimated beam of waves) is focused to an image at the focal point of the lens. In the latter, an object at the focal length distance from the lens is imaged at infinity. The plane perpendicular to the lens axis situated at a distance f from the lens is called the *focal plane*.

If the distances from the object to the lens and from the lens to the image are S_1 and S_2 respectively, for a lens of negligible thickness, in air, the distances are related by the thin lens formula:

$$\frac{1}{S_1} + \frac{1}{S_2} = \frac{1}{f}.$$

This can also be put into the "Newtonian" form:

$$x_1 x_2 = f^2,$$

where $x_1 = S_1 - f$ and $x_2 = S_2 - f..$

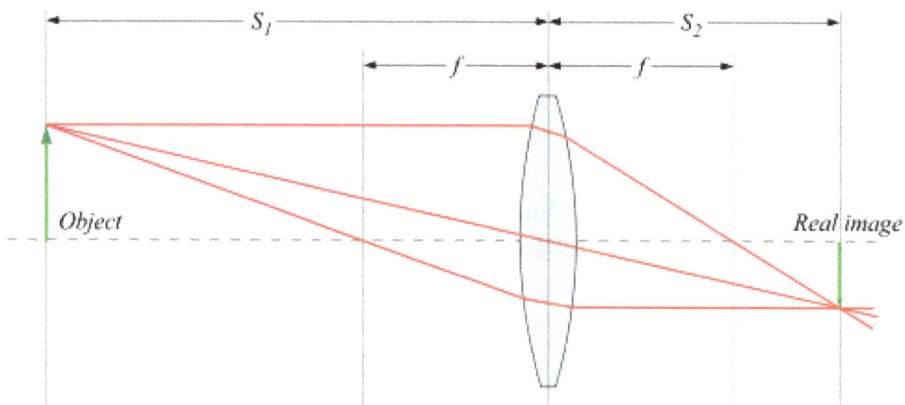

A camera lens forms a *real image* of a distant object.

Therefore, if an object is placed at a distance $S_1 > f$ from a positive lens of focal length f, we will find an image distance S_2 according to this formula. If a screen is placed at a distance S_2 on the opposite side of the lens, an image is formed on it. This sort of image, which can be projected onto a screen or image sensor, is known as a *real image*.

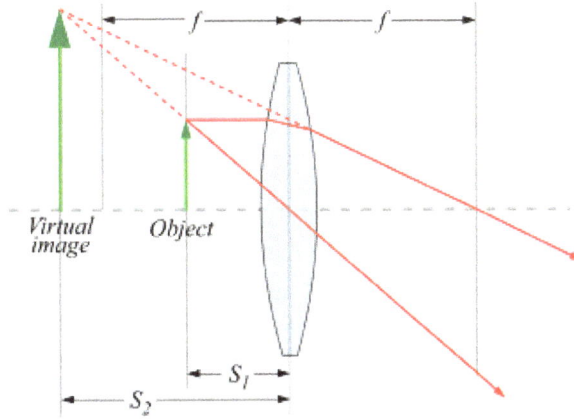

Virtual image formation using a positive lens as a magnifying glass.

This is the principle of the camera, and of the human eye. The focusing adjustment of a camera adjusts S_2, as using an image distance different from that required by this formula produces a de-focused (fuzzy) image for an object at a distance of S_1 from the camera. Put another way, modifying S_2 causes objects at a different S_1 to come into perfect focus.

In some cases S_2 is negative, indicating that the image is formed on the opposite side of the lens from where those rays are being considered. Since the diverging light rays emanating from the lens never come into focus, and those rays are not physically present at the point where they *appear* to form an image, this is called a virtual image. Unlike real images, a virtual image cannot be projected on a screen, but appears to an observer looking through the lens as if it were a real object at the location of that virtual image. Likewise, it appears to a subsequent lens as if it were an object at that location, so that second lens could again focus that light into a real image, S_1 then being measured from the virtual image location behind the first lens to the second lens. This is exactly what the eye does when looking through a magnifying glass. The magnifying glass creates a (magnified) virtual image behind the magnifying glass, but those rays are then re-imaged by the lens of the eye to create a *real image* on the retina.

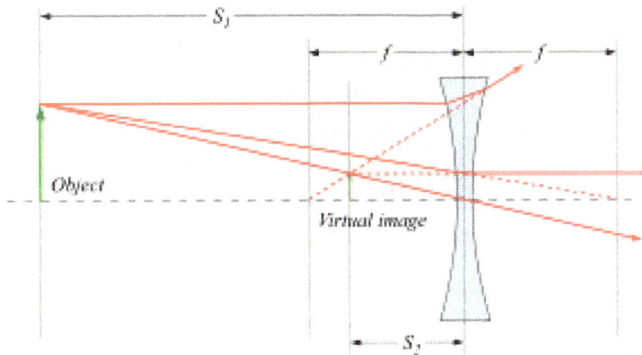

A *negative* lens produces a demagnified virtual image.

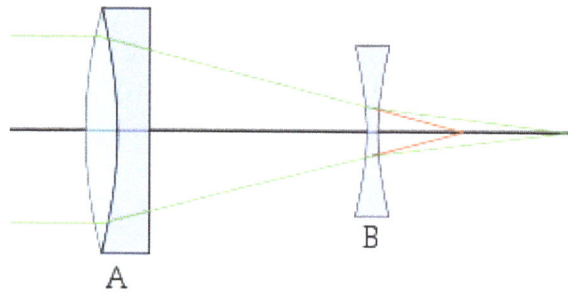

A Barlow lens (B) reimages a *virtual object* (focus of red ray path) into a magnified real image (green rays at focus)

Using a positive lens of focal length f, a virtual image results when $S_1 < f$, the lens thus being used a magnifying glass (rather than if $S_1 >> f$ as for a camera). Using a negative lens ($f < 0$) with a *real object* ($S_1 > 0$) can only produce a virtual image ($S_2 < 0$), according to the above formula. It is also possible for the object distance S_1 to be negative, in which case the lens sees a so-called *virtual object*. This happens when the lens is inserted into a converging beam (being focused by a previous lens) *before* the location of its real image. In that case even a negative lens can project a real image, as is done by a Barlow lens.

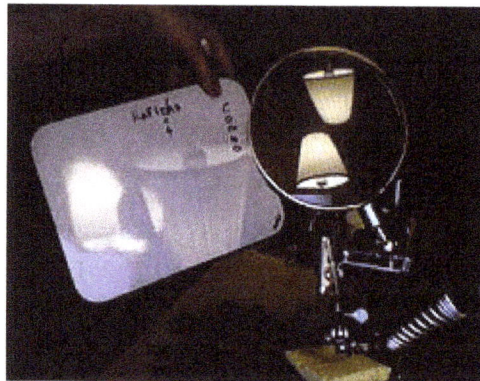

Real image of a lamp is projected onto a screen (inverted). Reflections of the lamp from both surfaces of the biconvex lens are visible.

A convex lens ($f << S_1$) forming a real, inverted image rather than the upright, virtual image as seen in a magnifying glass

For a thin lens, the distances S_1 and S_2 are measured from the object and image to the position of the lens, as described above. When the thickness of the lens is not much smaller than S_1 and S_2 or there are multiple lens elements (a compound lens), one must instead measure from the object and image to the principal planes of the lens. If distances S_1 or S_2 pass through a medium other than air or vacuum a more complicated analysis is required.

Magnification

The linear *magnification* of an imaging system using a single lens is given by

$$M = -\frac{S_2}{S_1} = \frac{f}{f - S_1},$$

where M is the magnification factor defined as the ratio of the size of an image compared to the size of the object. The sign convention here dictates that if M is negative, as it is for real images, the image is upside-down with respect to the object. For virtual images M is positive, so the image is upright.

Linear magnification M is not always the most useful measure of magnifying power. For instance, when characterizing a visual telescope or binoculars that produce only a virtual image, one would be more concerned with the angular magnification—which expresses how much larger a distant object appears through the telescope compared to the naked eye. In the case of a camera one would quote the plate scale, which compares the apparent (angular) size of a distant object to the size of the real image produced at the focus. The plate scale is the reciprocal of the focal length of the camera lens; lenses are categorized as long-focus lenses or wide-angle lenses according to their focal lengths.

Using an inappropriate measurement of magnification can be formally correct but yield a meaningless number. For instance, using a magnifying glass of 5 cm focal length, held 20 cm from the eye and 5 cm from the object, produces a virtual image at infinity of infinite linear size: $M = \infty$. But the *angular magnification* is 5, meaning that the object appears 5 times larger to the eye than without the lens. When taking a picture of the moon using a camera with a 50 mm lens, one is not concerned with the linear magnification $M \approx -50$ mm / 380000 km $= -1.3 \times 10^{-10}$. Rather, the plate scale of the camera is about 1°/mm, from which one can conclude that the 0.5 mm image on the film corresponds to an angular size of the moon seen from earth of about 0.5°.

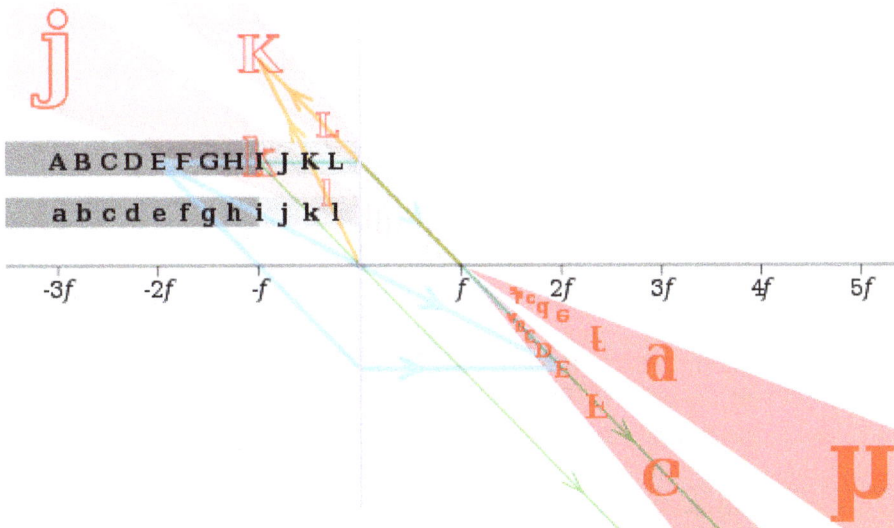

Images of black letters in a thin convex lens of focal length f are shown in red. Selected rays are shown for letters E, I and K in blue, green and orange, respectively. Note that E (at $2f$) has an equal-size, real and inverted image; I (at f) has its image at infinity; and K (at $f/2$) has a double-size, virtual and upright image.

In the extreme case where an object is an infinite distance away, $S_1 = \infty$, $S_2 = f$ and $M = -f/\infty = 0$, indicating that the object would be imaged to a single point in the focal plane. In fact, the diameter of the projected spot is not actually zero, since diffraction places a lower limit on the size of the point spread function. This is called the diffraction limit.

Aberrations

Lenses do not form perfect images, and a lens always introduces some degree of distortion or *aberration* that makes the image an imperfect replica of the object. Careful design of the lens system for a particular application minimizes the aberration. Several types of aberration affect image quality, including spherical aberration, coma, and chromatic aberration.

Spherical Aberration

Spherical aberration occurs because spherical surfaces are not the ideal shape for a lens, but are by far the simplest shape to which glass can be ground and polished, and so are often used. Spherical aberration causes beams parallel to, but distant from, the lens axis to be focused in a slightly different place than beams close to the axis. This manifests itself as a blurring of the image. Lenses in which closer-to-ideal, non-spherical surfaces are used are called *aspheric* lenses. These were formerly complex to make and often extremely expensive, but advances in technology have greatly reduced the manufacturing cost for such lenses. Spherical aberration can be minimised by carefully choosing the surface curvatures for a particular application. For instance, a plano-convex lens, which is used to focus a collimated beam, produces a sharper focal spot when used with the convex side towards the beam source.

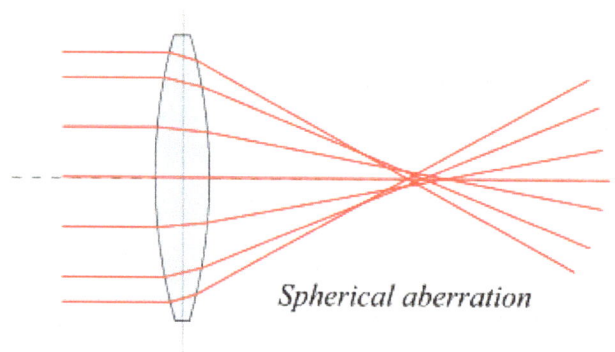

Spherical aberration

Coma

Coma, or *comatic aberration*, derives its name from the comet-like appearance of the aberrated image. Coma occurs when an object off the optical axis of the lens is imaged, where rays pass through the lens at an angle to the axis θ. Rays that pass through the centre of a lens of focal length f are focused at a point with distance $f \tan \theta$ from the axis. Rays passing through the outer margins of the lens are focused at different points, either further from the axis (positive coma) or closer to the axis (negative coma). In general, a bundle of parallel rays passing through the lens at a fixed distance from the centre of the lens are focused to a ring-shaped image in the focal plane, known as a *comatic circle*. The sum of all these circles results in a V-shaped or comet-like flare. As with spherical aberration, coma can be minimised (and in some cases eliminated) by choosing the curvature of the two lens surfaces to match the application. Lenses in which both spherical aberration and coma are minimised are called *bestform* lenses.

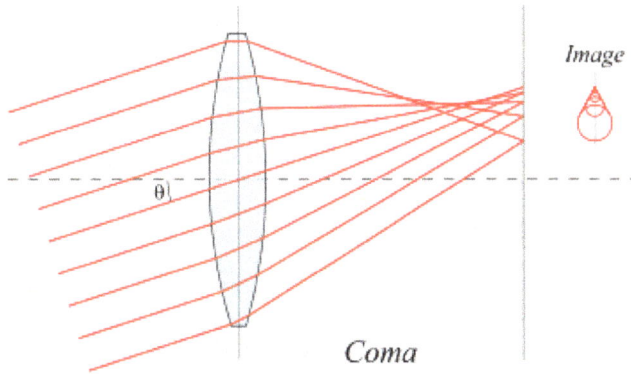

Coma

Chromatic aberration

Chromatic aberration is caused by the dispersion of the lens material—the variation of its refractive index, n, with the wavelength of light. Since, from the formulae above, f is dependent upon n, it follows that light of different wavelengths is focused to different positions. Chromatic aberration of a lens is seen as fringes of colour around the image. It can be minimised by using an achromatic doublet (or *achromat*) in which two materials with differing dispersion are bonded together to form a single lens. This reduces the amount of chromatic aberration over a certain range of wavelengths, though it does not produce perfect correction. The use of achromats was an important step in the development of the optical microscope. An apochromat is a lens or lens system with even better chromatic aberration correction, combined with improved spherical aberration correction. Apochromats are much more expensive than achromats.

Different lens materials may also be used to minimise chromatic aberration, such as specialised coatings or lenses made from the crystal fluorite. This naturally occurring substance has the highest known Abbe number, indicating that the material has low dispersion.

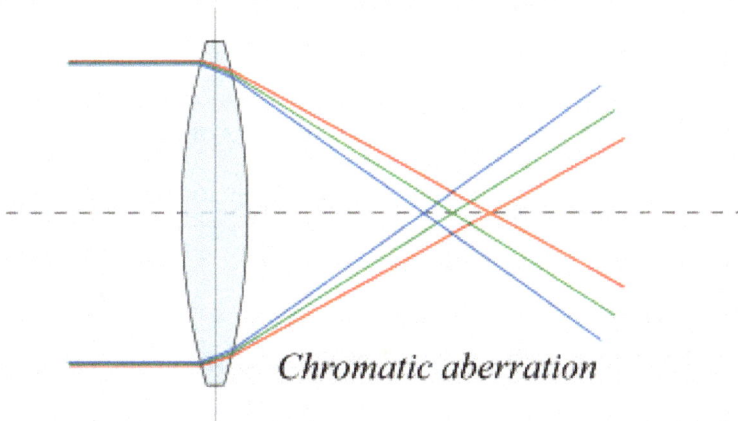

Chromatic aberration

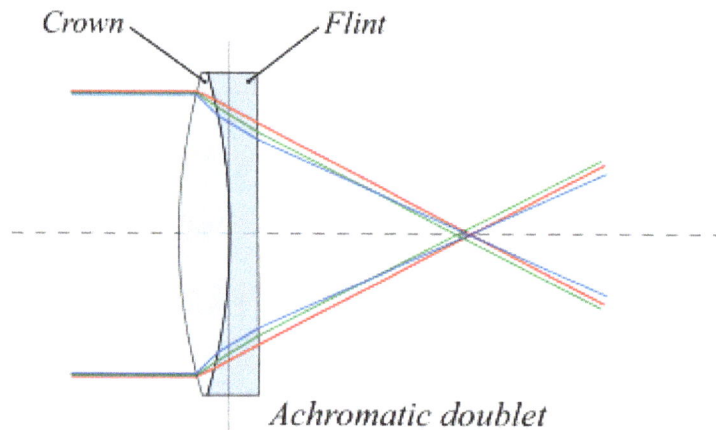

Achromatic doublet

Other types of aberration

Other kinds of aberration include *field curvature*, *barrel* and *pincushion distortion*, and *astigmatism*.

Aperture Diffraction

Even if a lens is designed to minimize or eliminate the aberrations described above, the image quality is still limited by the diffraction of light passing through the lens' finite aperture. A diffraction-limited lens is one in which aberrations have been reduced to the point where the image quality is primarily limited by diffraction under the design conditions.

Compound Lenses

Simple lenses are subject to the optical aberrations discussed above. In many cases these aberrations can be compensated for to a great extent by using a combination of simple lenses with complementary aberrations. A *compound lens* is a collection of simple lenses of different shapes and made of materials of different refractive indices, arranged one after the other with a common axis.

The simplest case is where lenses are placed in contact: if the lenses of focal lengths f_1 and f_2 are "thin", the combined focal length f of the lenses is given by

$$\frac{1}{f} = \frac{1}{f_1} + \frac{1}{f_2}.$$

Since $1/f$ is the power of a lens, it can be seen that the powers of thin lenses in contact are additive.

If two thin lenses are separated in air by some distance d, the focal length for the combined system is given by

$$\frac{1}{f} = \frac{1}{f_1} + \frac{1}{f_2} - \frac{d}{f_1 f_2}.$$

The distance from the front focal point of the combined lenses to the first lens is called the *front focal length* (FFL):

$$FFL = \frac{f_1(f_2 - d)}{(f_1 + f_2) - d}.$$

Similarly, the distance from the second lens to the rear focal point of the combined system is the *back focal length* (BFL):

$$\text{BFL} = \frac{f_2(d-f_1)}{d-(f_1+f_2)}.$$

As d tends to zero, the focal lengths tend to the value of f given for thin lenses in contact.

If the separation distance is equal to the sum of the focal lengths ($d = f_1+f_2$), the FFL and BFL are infinite. This corresponds to a pair of lenses that transform a parallel (collimated) beam into another collimated beam. This type of system is called an *afocal system*, since it produces no net convergence or divergence of the beam. Two lenses at this separation form the simplest type of optical telescope. Although the system does not alter the divergence of a collimated beam, it does alter the width of the beam. The magnification of such a telescope is given by

$$M = -\frac{f_2}{f_1},$$

which is the ratio of the output beam width to the input beam width. Note the sign convention: a telescope with two convex lenses ($f_1 > 0, f_2 > 0$) produces a negative magnification, indicating an inverted image. A convex plus a concave lens ($f_1 > 0 > f_2$) produces a positive magnification and the image is upright.

Other Types

Cylindrical lenses have curvature in only one direction. They are used to focus light into a line, or to convert the elliptical light from a laser diode into a round beam.

Close-up view of a flat Fresnel lens.

A Fresnel lens has its optical surface broken up into narrow rings, allowing the lens to be much thinner and lighter than conventional lenses. Durable Fresnel lenses can be molded from plastic and are inexpensive.

Lenticular lenses are arrays of microlenses that are used in lenticular printing to make images that have an illusion of depth or that change when viewed from different angles.

A gradient index lens has flat optical surfaces, but has a radial or axial variation in index of refraction that causes light passing through the lens to be focused.

An axicon has a conical optical surface. It images a point source into a line *along* the optic axis, or transforms a laser beam into a ring.

Diffractive optical elements can function as lenses.

Superlenses are made from negative index metamaterials and claim to produce images at spatial resolutions exceeding the diffraction limit. The first superlenses were made in 2004 using such a metamaterial for microwaves. Improved versions have been made by other researchers.As of 2014 the superlens has not yet been demonstrated at visible or near-infrared wavelengths.

A prototype flat ultrathin lens, with no curvature has been developed.

Uses

A single convex lens mounted in a frame with a handle or stand is a magnifying glass.

Lenses are used as prosthetics for the correction of visual impairments such as myopia, hyperopia, presbyopia, and astigmatism. Most lenses used for other purposes have strict axial symmetry; eyeglass lenses are only approximately symmetric. They are usually shaped to fit in a roughly oval, not circular, frame; the optical centres are placed over the eyeballs; their curvature may not be axially symmetric to correct for astigmatism. Sunglasses' lenses are designed to attenuate light; sunglass lenses that also correct visual impairments can be custom made.

Other uses are in imaging systems such as monoculars, binoculars, telescopes, microscopes, cameras and projectors. Some of these instruments produce a virtual image when applied to the human eye; others produce a real image that can be captured on photographic film or an optical sensor, or can be viewed on a screen. In these devices lenses are sometimes paired up with curved mirrors to make a catadioptric system where the lens's spherical aberration corrects the opposite aberration in the mirror (such as Schmidt and meniscus correctors).

Convex lenses produce an image of an object at infinity at their focus; if the sun is imaged, much of the visible and infrared light incident on the lens is concentrated into the small image. A large lens creates enough intensity to burn a flammable object at the focal point. Since ignition can be achieved even with a poorly made lens, lenses have been used as burning-glasses for at least 2400 years. A modern application is the use of relatively large lenses to concentrate solar energy on relatively small photovoltaic cells, harvesting more energy without the need to use larger and more expensive cells.

Radio astronomy and radar systems often use dielectric lenses, commonly called a lens antenna to refract electromagnetic radiation into a collector antenna.

Lenses can become scratched and abraded. Abrasion-resistant coatings are available to help control this.

References

- W.T. Welford and Roland Winston, The Optics of Nonimaging Concentrators: Light and Solar Energy, Academic Press, 1978 [ISBN 978-0-12-745350-7]

- Ralf Leutz and Akio Suzuki, Nonimaging Fresnel Lenses: Design and Performance of Solar Concentrators, Springer, 2001 [ISBN 978-3-642-07531-5]

- William Cassarly, Nonimaging Optics: Concentration and Illumination in Michael Bass, Handbook of optics, Third edition, Vol. II, Chapter 39, McGraw Hill (Sponsored by the Optical Society of America), 2010 [ISBN 978-0-07-149890-6]

- Tilton, Buck (2005). The Complete Book of Fire: Building Campfires for Warmth, Light, Cooking, and Survival. Menasha Ridge Press. p. 25. ISBN 0-89732-633-4.

- Glick, Thomas F.; Steven John Livesey; Faith Wallis (2005). Medieval science, technology, and medicine: an encyclopedia. Routledge. p. 167. ISBN 978-0-415-96930-7. Retrieved 24 April 2011.

- Paul S. Agutter; Denys N. Wheatley (12 December 2008). Thinking about Life: The History and Philosophy of Biology and Other Sciences. Springer. p. 17. ISBN 978-1-4020-8865-0. Retrieved 6 June 2012.

- Vincent Ilardi (2007). Renaissance Vision from Spectacles to Telescopes. American Philosophical Society. p. 210. ISBN 978-0-87169-259-7. Retrieved 6 June 2012.

- Fred Watson (1 October 2007). Stargazer: The Life and Times of the Telescope. Allen & Unwin. p. 55. ISBN 978-1-74175-383-7. Retrieved 6 June 2012.

Significant Topics of Optics

The significant topics of optics explained in this section are optical instruments and optical engineering. An optical instrument processes light waves to improve an image for viewing and optical engineering is the study of optics. The section strategically encompasses and incorporates the major components and significant topics of optics, providing a complete understanding.

Optical Instrument

An optical instrument either processes light waves to enhance an image for viewing, or analyzes light waves (or photons) to determine one of a number of characteristic properties.

Image Enhancement

An illustration of some of the optical devices available for laboratory work in England in 1858.

The first optical instruments were telescopes used for magnification of distant images, and microscopes used for magnifying very tiny images. Since the days of Galileo and Van Leeuwenhoek, these instruments have been greatly improved and extended into other portions of the electromagnetic spectrum. The binocular device is a generally compact instrument for both eyes designed for mobile use. A camera could be considered a type of optical instrument, with the pinhole camera and *camera obscura* being very simple examples of such devices.

Analysis

Another class of optical instrument is used to analyze the properties of light or optical materials. They include:

- Interferometer for measuring the interference properties of light waves

- Photometer for measuring light intensity

- Polarimeter for measuring dispersion or rotation of polarized light

- Reflectometer for measuring the reflectivity of a surface or object

- Refractometer for measuring refractive index of various materials, invented by Ernst Abbe

- Spectrometer or monochromator for generating or measuring a portion of the optical spectrum, for the purpose of chemical or material analysis

- Autocollimator which is used to measure angular deflections

- Vertometer which is used to determine refractive power of lenses such as glasses, contact lenses and magnifier lens

DNA sequencers can be considered optical instruments as they analyse the *color* and intensity of the light emitted by a fluorochrome attached to a specific nucleotide of a DNA strand.

Surface plasmon resonance-based instruments use refractometry to measure and analyze biomolecular interactions.

Other optical devices

- Polarization controller

- Camera

- Magic lantern

Optical Engineering

Optical engineering is the field of study that focuses on applications of optics. Optical engineers design components of optical instruments such as lenses, microscopes, telescopes, and other equipment that utilizes the properties of light. Other devices include optical sensors and measurement systems, lasers, fiber optic communication systems, optical disc systems (e.g. CD, DVD), etc.

Because optical engineers want to design and build devices that make light do something useful, they must understand and apply the science of optics in substantial detail, in order to know what is physically possible to achieve (physics and chemistry). However, they also must know what is practical in terms of available technology, materials, costs, design methods, etc. As with other fields of engineering, computers are important to many (perhaps most) optical engineers. They are used with instruments, for simulation, in design, and for many other applications. Engineers often use general computer tools such as spreadsheets and programming languages, and they make frequent use of specialized optical software designed specifically for their field.

Optical engineering metrology uses optical methods to measure micro-vibrations with instruments like the laser speckle interferometer or to measure the properties of the various masses with instruments measuring refraction.

Concepts of Optics

The concepts of optics explained in the following text are photometry, exposure, atmospheric optics, optical vortex, focus and dispersion. Photometry is the study of measuring light; there is a difference between photometry and radiometry. Photometry mainly studies the sensitivity of visible light on the human eye; it measures this by calculating the power of each wavelength by representing the sensitivity of eye at that particular wavelength. This section elucidates the crucial theories of optics.

Photometry (Optics)

Photometry is the science of the measurement of light, in terms of its *perceived* brightness to the human eye. It is distinct from radiometry, which is the science of measurement of radiant energy (including light) in terms of absolute power. In modern photometry, the radiant power at each wavelength is weighted by a luminosity function that models human brightness sensitivity. Typically, this weighting function is the photopic sensitivity function, although the scotopic function or other functions may also be applied in the same way.

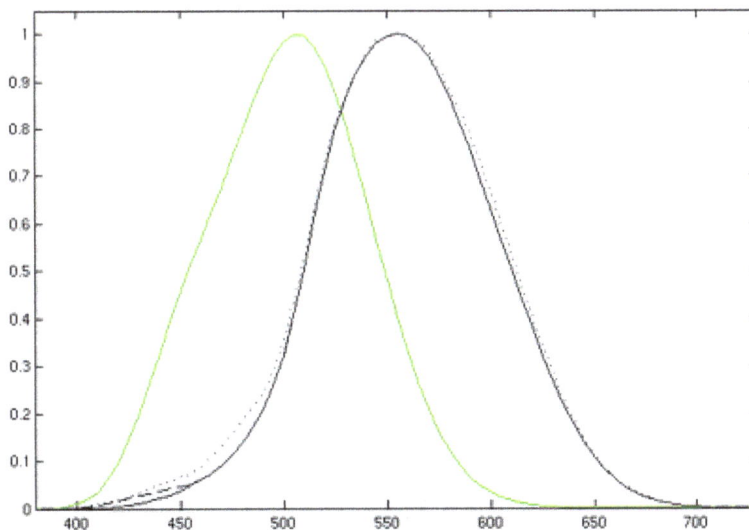

Photopic (daytime-adapted, black curve) and scotopic (darkness-adapted, green curve) luminosity functions. The photopic includes the CIE 1931 standard (solid), the Judd-Vos 1978 modified data (dashed), and the Sharpe, Stockman, Jagla & Jägle 2005 data (dotted). The horizontal axis is wavelength in nm.

Photometry and the Eye

The human eye is not equally sensitive to all wavelengths of visible light. Photometry attempts to account for this by weighing the measured power at each wavelength with a factor that represents how sensitive the eye is at that wavelength. The standardized model of the eye's response to light as a function of wavelength is given by the luminosity function. The eye has different responses as a function of wavelength when it is adapted to light conditions (photopic vision) and dark conditions (scotopic vision). Photometry is typically based on the eye's photopic response, and so photometric measurements may not accurately indicate the perceived brightness of sources in dim lighting conditions where colors are not discernible, such as under just moonlight or starlight. Photopic vision is characteristic of the eye's response at luminance levels over three candela per square metre. Scotopic vision occurs below 2×10^{-5} cd/m². Mesopic vision occurs between these limits and is not well characterised for spectral response.

Photometric Quantities

Measurement of the effects of electromagnetic radiation became a field of study as early as the end of 18th century. Measurement techniques varied depending on the effects under study and gave rise to different nomenclature. The total heating effect of infrared radiation as measured by thermometers led to development of radiometric units in terms of total energy and power. Use of the human eye as a detector led to photometric units, weighted by the eye's response characteristic. Study of the chemical effects of ultraviolet radiation led to characterization by the total dose or actinometric units expressed in photons per second.

Many different units of measure are used for photometric measurements. People sometimes ask why there need to be so many different units, or ask for conversions between units that can't be converted (lumens and candelas, for example). We are familiar with the idea that the adjective "heavy" can refer to weight or density, which are fundamentally different things. Similarly, the adjective "bright" can refer to a light source which delivers a high luminous flux (measured in lumens), or to a light source which concentrates the luminous flux it has into a very narrow beam (candelas), or to a light source that is seen against a dark background. Because of the ways in which light propagates through three-dimensional space — spreading out, becoming concentrated, reflecting off shiny or matte surfaces — and because light consists of many different wavelengths, the number of fundamentally different kinds of light measurement that can be made is large, and so are the numbers of quantities and units that represent them.

For example, offices are typically "brightly" illuminated by an array of many recessed fluorescent lights for a combined high luminous flux. A laser pointer has very low luminous flux (it could not illuminate a room) but is blindingly bright in one direction (high luminous intensity in that direction).

Table 1. SI Photometry Quantities

Quantity		Unit		Dimension	Notes
Name	Symbol[nb 1]	Name	Symbol	Symbol	
Luminous energy	Q_v [nb 2]	lumen second	lm·s	T·J [nb 3]	Units are sometimes called *talbots*.
Luminous flux / luminous power	Φ_v [nb 2]	lumen (= cd·sr)	lm	J [nb 3]	Luminous energy per unit time.

Luminous intensity	I_v	candela (= lm/sr)	cd	J [nb 3]	Luminous power per unit solid angle.
Luminance	L_v	candela per square metre	cd/m²	$L^{-2} \cdot J$	Luminous power per unit solid angle per unit *projected* source area. Units are sometimes called *nits*.
Illuminance	E_v	lux (= lm/m²)	lx	$L^{-2} \cdot J$	Luminous power *incident* on a surface.
Luminous exitance / luminous emittance	M_v	lux	lx	$L^{-2} \cdot J$	Luminous power *emitted* from a surface.
Luminous exposure	H_v	lux second	lx·s	$L^{-2} \cdot T \cdot J$	
Luminous energy density	ω_v	lumen second per cubic metre	lm·s·m⁻³	$L^{-3} \cdot T \cdot J$	
Luminous efficacy	η [nb 2]	lumen per watt	lm/W	$M^{-1} \cdot L^{-2} \cdot T^3 \cdot J$	Ratio of luminous flux to radiant flux or power consumption, depending on context.
Luminous efficiency / luminous coefficient	V			1	

Photometric Versus Radiometric Quantities

There are two parallel systems of quantities known as photometric and radiometric quantities. Every quantity in one system has an analogous quantity in the other system. Some examples of parallel quantities include:

- Luminance (photometric) and radiance (radiometric)

- Luminous flux (photometric) and radiant flux (radiometric)

- Luminous intensity (photometric) and radiant intensity (radiometric)

In photometric quantities every wavelength is weighted according to how sensitive the human eye is to it, while radiometric quantities use unweighted absolute power. For example, the eye responds much more strongly to green light than to red, so a green source will have greater luminous flux than a red source with the same radiant flux would. Radiant energy outside the visible spectrum does not contribute to photometric quantities at all, so for example a 1000 watt space heater may put out a great deal of radiant flux (1000 watts, in fact), but as a light source it puts out very few lumens (because most of the energy is in the infrared, leaving only a dim red glow in the visible).

Table 2. SI Radiometry Units

Quantity		Unit		Dimension	Notes
Name	**Symbol[nb 4]**	**Name**	**Symbol**	**Symbol**	**Notes**
Radiant energy	Q_e [nb 5]	joule	J	$M \cdot L^2 \cdot T^{-2}$	Energy of electromagnetic radiation.
Radiant energy density	w_e	joule per cubic metre	J/m^3	$M \cdot L^{-1} \cdot T^{-2}$	Radiant energy per unit volume.
Radiant flux	Φ_e [nb 5]	watt	W *or* J/s	$M \cdot L^2 \cdot T^{-3}$	Radiant energy emitted, reflected, transmitted or received, per unit time. This is sometimes also called "radiant power".
Spectral flux	$\Phi_{e,\nu}$ [nb 6] *or* $\Phi_{e,\lambda}$ [nb 7]	watt per hertz *or* watt per metre	W/Hz *or* W/m	$M \cdot L^2 \cdot T^{-2}$ *or* $M \cdot L \cdot T^{-3}$	Radiant flux per unit frequency or wavelength. The latter is commonly measured in $W \cdot sr^{-1} \cdot m^{-2} \cdot nm^{-1}$.
Radiant intensity	$I_{e,\Omega}$ [nb 8]	watt per steradian	W/sr	$M \cdot L^2 \cdot T^{-3}$	Radiant flux emitted, reflected, transmitted or received, per unit solid angle. This is a *directional* quantity.
Spectral intensity	$I_{e,\Omega,\nu}$ [nb 6] *or* $I_{e,\Omega,\lambda}$ [nb 7]	watt per steradian per hertz *or* watt per steradian per metre	$W \cdot sr^{-1} \cdot Hz^{-1}$ *or* $W \cdot sr^{-1} \cdot m^{-1}$	$M \cdot L^2 \cdot T^{-2}$ *or* $M \cdot L \cdot T^{-3}$	Radiant intensity per unit frequency or wavelength. The latter is commonly measured in $W \cdot sr^{-1} \cdot m^{-2} \cdot nm^{-1}$. This is a *directional* quantity.
Radiance	$L_{e,\Omega}$ [nb 8]	watt per steradian per square metre	$W \cdot sr^{-1} \cdot m^{-2}$	$M \cdot T^{-3}$	Radiant flux emitted, reflected, transmitted or received by a *surface*, per unit solid angle per unit projected area. This is a *directional* quantity. This is sometimes also confusingly called "intensity".
Spectral radiance	$L_{e,\Omega,\nu}$ [nb 6] *or* $L_{e,\Omega,\lambda}$ [nb 7]	watt per steradian per square metre per hertz *or* watt per steradian per square metre, per metre	$W \cdot sr^{-1} \cdot m^{-2} \cdot Hz^{-1}$ *or* $W \cdot sr^{-1} \cdot m^{-3}$	$M \cdot T^{-2}$ *or* $M \cdot L^{-1} \cdot T^{-3}$	Radiance of a *surface* per unit frequency or wavelength. The latter is commonly measured in $W \cdot sr^{-1} \cdot m^{-2} \cdot nm^{-1}$. This is a *directional* quantity. This is sometimes also confusingly called "spectral intensity".
Irradiance	E_e [nb 5]	watt per square metre	W/m^2	$M \cdot T^{-3}$	Radiant flux *received* by a *surface* per unit area. This is sometimes also confusingly called "intensity".

Spectral irradiance	$E_{e,\nu}$ [nb 6] or $E_{e,\lambda}$ [nb 7]	watt per square metre per hertz *or* watt per square metre, per metre	$W{\cdot}m^{-2}{\cdot}Hz^{-1}$ *or* W/m^3	$M{\cdot}T^{-2}$ *or* $M{\cdot}L^{-1}{\cdot}T^{-3}$	Irradiance of a *surface* per unit frequency or wavelength. The terms spectral flux density or more confusingly "spectral intensity" are also used. Non-SI units of spectral irradiance include Jansky $= 10^{-26}\ W{\cdot}m^{-2}{\cdot}Hz^{-1}$ and solar flux unit (1SFU $= 10^{-22}$ $W{\cdot}m^{-2}{\cdot}Hz^{-1}$).
Radiosity	J_e [nb 5]	watt per square metre	W/m^2	$M{\cdot}T^{-3}$	Radiant flux *leaving* (emitted, reflected and transmitted by) a *surface* per unit area. This is sometimes also confusingly called "intensity".
Spectral radiosity	$J_{e,\nu}$ [nb 6] or $J_{e,\lambda}$ [nb 7]	watt per square metre per hertz *or* watt per square metre, per metre	$W{\cdot}m^{-2}{\cdot}Hz^{-1}$ *or* W/m^3	$M{\cdot}T^{-2}$ *or* $M{\cdot}L^{-1}{\cdot}T^{-3}$	Radiosity of a *surface* per unit frequency or wavelength. The latter is commonly measured in $W{\cdot}m^{-2}{\cdot}nm^{-1}$. This is sometimes also confusingly called "spectral intensity".
Radiant exitance	M_e [nb 5]	watt per square metre	W/m^2	$M{\cdot}T^{-3}$	Radiant flux *emitted* by a *surface* per unit area. This is the emitted component of radiosity. "Radiant emittance" is an old term for this quantity. This is sometimes also confusingly called "intensity".
Spectral exitance	$M_{e,\nu}$ [nb 6] or $M_{e,\lambda}$ [nb 7]	watt per square metre per hertz *or* watt per square metre, per metre	$W{\cdot}m^{-2}{\cdot}Hz^{-1}$ *or* W/m^3	$M{\cdot}T^{-2}$ *or* $M{\cdot}L^{-1}{\cdot}T^{-3}$	Radiant exitance of a *surface* per unit frequency or wavelength. The latter is commonly measured in $W{\cdot}m^{-2}{\cdot}nm^{-1}$. "Spectral emittance" is an old term for this quantity. This is sometimes also confusingly called "spectral intensity".
Radiant exposure	H_e	joule per square metre	J/m^2	$M{\cdot}T^{-2}$	Radiant energy received by a *surface* per unit area, or equivalently irradiance of a *surface* integrated over time of irradiation. This is sometimes also called "radiant fluence".
Spectral exposure	$H_{e,\nu}$ [nb 6] or $H_{e,\lambda}$ [nb 7]	joule per square metre per hertz *or* joule per square metre, per metre	$J{\cdot}m^{-2}{\cdot}Hz^{-1}$ *or* J/m^3	$M{\cdot}T^{-1}$ *or* $M{\cdot}L^{-1}{\cdot}T^{-2}$	Radiant exposure of a *surface* per unit frequency or wavelength. The latter is commonly measured in $J{\cdot}m^{-2}{\cdot}nm^{-1}$. This is sometimes also called "spectral fluence".

Hemispherical emissivity	ε			1	Radiant exitance of a *surface*, divided by that of a *black body* at the same temperature as that surface.
Spectral hemispherical emissivity	ε_v or ε_λ			1	Spectral exitance of a *surface*, divided by that of a *black body* at the same temperature as that surface.
Directional emissivity	ε_Ω			1	Radiance *emitted* by a *surface*, divided by that emitted by a *black body* at the same temperature as that surface.
Spectral directional emissivity	$\varepsilon_{\Omega,v}$ or $\varepsilon_{\Omega,\lambda}$			1	Spectral radiance *emitted* by a *surface*, divided by that of a *black body* at the same temperature as that surface.
Hemispherical absorptance	A			1	Radiant flux *absorbed* by a *surface*, divided by that received by that surface. This should not be confused with "absorbance".
Spectral hemispherical absorptance	A_v or A_λ			1	Spectral flux *absorbed* by a *surface*, divided by that received by that surface. This should not be confused with "spectral absorbance".
Directional absorptance	A_Ω			1	Radiance *absorbed* by a *surface*, divided by the radiance incident onto that surface. This should not be confused with "absorbance".
Spectral directional absorptance	$A_{\Omega,v}$ or $A_{\Omega,\lambda}$			1	Spectral radiance *absorbed* by a *surface*, divided by the spectral radiance incident onto that surface. This should not be confused with "spectral absorbance".
Hemispherical reflectance	R			1	Radiant flux *reflected* by a *surface*, divided by that received by that surface.
Spectral hemispherical reflectance	R_v or R_λ			1	Spectral flux *reflected* by a *surface*, divided by that received by that surface.
Directional reflectance	R_Ω			1	Radiance *reflected* by a *surface*, divided by that received by that surface.
Spectral directional reflectance	$R_{\Omega,v}$ or $R_{\Omega,\lambda}$			1	Spectral radiance *reflected* by a *surface*, divided by that received by that surface.

Hemispherical transmittance	T			1	Radiant flux *transmitted* by a *surface*, divided by that received by that surface.
Spectral hemispherical transmittance	T_v or T_λ			1	Spectral flux *transmitted* by a *surface*, divided by that received by that surface.
Directional transmittance	T_Ω			1	Radiance *transmitted* by a *surface*, divided by that received by that surface.
Spectral directional transmittance	$T_{\Omega,\text{v}}$ or $T_{\Omega,\lambda}$			1	Spectral radiance *transmitted* by a *surface*, divided by that received by that surface.
Hemispherical attenuation coefficient	μ	reciprocal metre	m^{-1}	L^{-1}	Radiant flux *absorbed* and *scattered* by a *volume* per unit length, divided by that received by that volume.
Spectral hemispherical attenuation coefficient	μ_v or μ_λ	reciprocal metre	m^{-1}	L^{-1}	Spectral radiant flux *absorbed* and *scattered* by a *volume* per unit length, divided by that received by that volume.
Directional attenuation coefficient	μ_Ω	reciprocal metre	m^{-1}	L^{-1}	Radiance *absorbed* and *scattered* by a *volume* per unit length, divided by that received by that volume.
Spectral directional attenuation coefficient	$\mu_{\Omega,\text{v}}$ or $\mu_{\Omega,\lambda}$	reciprocal metre	m^{-1}	L^{-1}	Spectral radiance *absorbed* and *scattered* by a *volume* per unit length, divided by that received by that volume.

Watts Versus Lumens

Watts are units of radiant flux while lumens are units of luminous flux. A comparison of the watt and the lumen illustrates the distinction between radiometric and photometric units.

The watt is a unit of power. We are accustomed to thinking of light bulbs in terms of power in watts. This power is not a measure of the amount of light output, but rather indicates how much energy the bulb will use. Because incandescent bulbs sold for "general service" all have fairly similar characteristics (same spectral power distribution), power consumption provides a rough guide to the light output of incandescent bulbs.

Watts can also be a direct measure of output. In a radiometric sense, an incandescent light bulb is about 80% efficient: 20% of the energy is lost (e.g. by conduction through the lamp base). The remainder is emitted as radiation, mostly in the infrared. Thus, a 60 watt light bulb emits a total radiant flux of about 45 watts. Incandescent bulbs are, in fact, sometimes used as heat sources (as in a chick incubator), but usually they are used for the purpose of providing light. As such, they are very inefficient, because most of the radiant energy they emit is invisible infrared. A compact fluorescent lamp can provide light comparable to a 60 watt incandescent while consuming as little as 15 watts of electricity.

The lumen is the photometric unit of light output. Although most consumers still think of light in terms of power consumed by the bulb, in the U.S. it has been a trade requirement for several decades that light bulb packaging give the output in lumens. The package of a 60 watt incandescent bulb indicates that it provides about 900 lumens, as does the package of the 15 watt compact fluorescent.

The lumen is defined as amount of light given into one steradian by a point source of one candela strength; while the candela, a base SI unit, is defined as the luminous intensity of a source of monochromatic radiation, of frequency 540 terahertz, and a radiant intensity of 1/683 watts per steradian. (540 THz corresponds to about 555 nanometres, the wavelength, in the green, to which the human eye is most sensitive. The number 1/683 was chosen to make the candela about equal to the standard candle, the unit which it superseded).

Combining these definitions, we see that 1/683 watt of 555 nanometre green light provides one lumen.

The relation between watts and lumens is not just a simple scaling factor. We know this already, because the 60 watt incandescent bulb and the 15 watt compact fluorescent can both provide 900 lumens.

The definition tells us that 1 watt of pure green 555 nm light is "worth" 683 lumens. It does not say anything about other wavelengths. Because lumens are photometric units, their relationship to watts depends on the wavelength according to how visible the wavelength is. Infrared and ultraviolet radiation, for example, are invisible and do not count. One watt of infrared radiation (which is where most of the radiation from an incandescent bulb falls) is worth zero lumens. Within the visible spectrum, wavelengths of light are weighted according to a function called the "photopic spectral luminous efficiency." According to this function, 700 nm red light is only about 0.4% as efficient as 555 nm green light. Thus, one watt of 700 nm red light is "worth" only 2.7 lumens.

Because of the summation over the visual portion of the EM spectrum that is part of this weighting, the unit of "lumen" is color-blind: there is no way to tell what color a lumen will appear. This is equivalent to evaluating groceries by number of bags: there is no information about the specific content, just a number that refers to the total weighted quantity.

Photometric Measurement Techniques

Photometric measurement is based on photodetectors, devices (of several types) that produce an electric signal when exposed to light. Simple applications of this technology include switching luminaires on and off based on ambient light conditions, and light meters, used to measure the total amount of light incident on a point.

More complex forms of photometric measurement are used frequently within the lighting industry. Spherical photometers can be used to measure the directional luminous flux produced by lamps, and consist of a large-diameter globe with a lamp mounted at its center. A photocell rotates about the lamp in three axes, measuring the output of the lamp from all sides.

Lamps and lighting fixtures are tested using goniophotometers and rotating mirror photometers, which keep the photocell stationary at a sufficient distance that the luminaire can be considered a point source. Rotating mirror photometers use a motorized system of mirrors to reflect light emanating from the luminaire in all directions to the distant photocell; goniophotometers use a rotating 2-axis table to change the orientation of the luminaire with respect to the photocell. In either case, luminous intensity is tabulated from this data and used in lighting design.

Non-SI Photometry Units

Luminance

- Footlambert
- Millilambert
- Stilb

Illuminance

- Foot-candle
- Phot

Exposure (Photography)

A long exposure showing stars rotating around the southern and northern celestial poles. Credit: European Southern Observatory

In photography, exposure is the amount of light per unit area (the image plane illuminance times the exposure time) reaching a photographic film or electronic image sensor, as determined by shutter speed, lens aperture and scene luminance. Exposure is measured in lux seconds, and can be computed from exposure value (EV) and scene luminance in a specified region.

In photographic jargon, *an exposure* generally refers to a single shutter cycle. For example: a long exposure refers to a single, protracted shutter cycle to capture enough low-intensity light, whereas a multiple exposure involves a series of relatively brief shutter cycles; effectively layering a series of photographs in one image. For the same film speed, the accumulated *photometric exposure* (H_v) should be similar in both cases.

A photograph of the sea after sunset with an exposure time of 15 seconds. The swell from the waves appears as fog.

Definitions

Underexposure and Overexposure

A photograph may be described as overexposed when it has a loss of highlight detail, that is, when important bright parts of an image are "washed out" or effectively all white, known as "blown-out highlights" or "clipped whites". A photograph may be described as underexposed when it has a loss of shadow detail, that is, when important dark areas are "muddy" or indistinguishable from black, known as "blocked-up shadows" (or sometimes "crushed shadows", "crushed blacks", or "clipped blacks", especially in video). As the image to the right shows, these terms are technical ones. There are three types of settings they are manual, automatic and exposure compensation.

Radiant Exposure

Radiant exposure of a *surface*, denoted H_e ("e" for "energetic", to avoid confusion with photometric quantities) and measured in J/m², is given by

$$H_e = E_e t,$$

where

- E_e is the irradiance of the surface, measured in W/m²;

- t is the exposure duration, measured in s.

Luminous Exposure

Luminous exposure of a *surface*, denoted H_v ("v" for "visual", to avoid confusion with radiometric quantities) and measured in lx·s, is given by

$$H_v = E_v t,$$

where

- E_v is the illuminance of the surface, measured in lx;

- t is the exposure duration, measured in s.

If the measurement is adjusted to account only for light that reacts with the photo-sensitive surface, that is, weighted by the appropriate spectral sensitivity, the exposure is still measured in radiometric units (joules per square meter), rather than photometric units (weighted by the nominal sensitivity of the human eye). Only in this appropriately weighted case does the H measure the effective amount of light falling on the film, such that the characteristic curve will be correct independent of the spectrum of the light.

Many photographic materials are also sensitive to "invisible" light, which can be a nuisance, or a benefit. The use of radiometric units is appropriate to characterize such sensitivity to invisible light.

In sensitometric data, such as characteristic curves, the *log exposure* is conventionally expressed as $\log_{10}(H)$. Photographers more familiar with base-2 logarithmic scales (such as exposure values) can convert using $\log_2(H) \approx 3.32 \log_{10}(H)$.

Table 2. SI Radiometry Units

Quantity		Unit		Dimension	Notes
Name	**Symbol**[nb 1]	**Name**	**Symbol**	**Symbol**	**Notes**
Radiant energy	Q_e [nb 2]	joule	J	$M \cdot L^2 \cdot T^{-2}$	Energy of electromagnetic radiation.
Radiant energy density	w_e	joule per cubic metre	J/m³	$M \cdot L^{-1} \cdot T^{-2}$	Radiant energy per unit volume.
Radiant flux	Φ_e [nb 2]	watt	W *or* J/s	$M \cdot L^2 \cdot T^{-3}$	Radiant energy emitted, reflected, transmitted or received, per unit time. This is sometimes also called "radiant power".
Spectral flux	$\Phi_{e,v}$ [nb 3] or $\Phi_{e,\lambda}$ [nb 4]	watt per hertz *or* watt per metre	W/Hz *or* W/m	$M \cdot L^2 \cdot T^{-2}$ *or* $M \cdot L \cdot T^{-3}$	Radiant flux per unit frequency or wavelength. The latter is commonly measured in $W \cdot sr^{-1} \cdot m^{-2} \cdot nm^{-1}$.
Radiant intensity	$I_{e,\Omega}$ [nb 5]	watt per steradian	W/sr	$M \cdot L^2 \cdot T^{-3}$	Radiant flux emitted, reflected, transmitted or received, per unit solid angle. This is a *directional* quantity.
Spectral intensity	$I_{e,\Omega,v}$ [nb 3] or $I_{e,\Omega,\lambda}$ [nb 4]	watt per steradian per hertz *or* watt per steradian per metre	$W \cdot sr^{-1} \cdot Hz^{-1}$ *or* $W \cdot sr^{-1} \cdot m^{-1}$	$M \cdot L^2 \cdot T^{-2}$ *or* $M \cdot L \cdot T^{-3}$	Radiant intensity per unit frequency or wavelength. The latter is commonly measured in $W \cdot sr^{-1} \cdot m^{-2} \cdot nm^{-1}$. This is a *directional* quantity.

Radiance	$L_{e,\Omega}$ [nb 5]	watt per steradian per square metre	$\mathrm{W{\cdot}sr^{-1}{\cdot}m^{-2}}$	$\mathrm{M{\cdot}T^{-3}}$	Radiant flux emitted, reflected, transmitted or received by a *surface*, per unit solid angle per unit projected area. This is a *directional* quantity. This is sometimes also confusingly called "intensity".
Spectral radiance	$L_{e,\Omega,\nu}$ [nb 3] or $L_{e,\Omega,\lambda}$ [nb 4]	watt per steradian per square metre per hertz *or* watt per steradian per square metre, per metre	$\mathrm{W{\cdot}sr^{-1}{\cdot}m^{-2}{\cdot}Hz^{-1}}$ *or* $\mathrm{W{\cdot}sr^{-1}{\cdot}m^{-3}}$	$\mathrm{M{\cdot}T^{-2}}$ *or* $\mathrm{M{\cdot}L^{-1}{\cdot}T^{-3}}$	Radiance of a *surface* per unit frequency or wavelength. The latter is commonly measured in $\mathrm{W{\cdot}sr^{-1}{\cdot}m^{-2}{\cdot}nm^{-1}}$. This is a *directional* quantity. This is sometimes also confusingly called "spectral intensity".
Irradiance	E_e [nb 2]	watt per square metre	$\mathrm{W/m^2}$	$\mathrm{M{\cdot}T^{-3}}$	Radiant flux *received* by a *surface* per unit area. This is sometimes also confusingly called "intensity".
Spectral irradiance	$E_{e,\nu}$ [nb 3] or $E_{e,\lambda}$ [nb 4]	watt per square metre per hertz *or* watt per square metre, per metre	$\mathrm{W{\cdot}m^{-2}{\cdot}Hz^{-1}}$ *or* $\mathrm{W/m^3}$	$\mathrm{M{\cdot}T^{-2}}$ *or* $\mathrm{M{\cdot}L^{-1}{\cdot}T^{-3}}$	Irradiance of a *surface* per unit frequency or wavelength. The terms spectral flux density or more confusingly "spectral intensity" are also used. Non-SI units of spectral irradiance include Jansky = 10^{-26} $\mathrm{W{\cdot}m^{-2}{\cdot}Hz^{-1}}$ and solar flux unit (1SFU = 10^{-22} $\mathrm{W{\cdot}m^{-2}{\cdot}Hz^{-1}}$).
Radiosity	J_e [nb 2]	watt per square metre	$\mathrm{W/m^2}$	$\mathrm{M{\cdot}T^{-3}}$	Radiant flux *leaving* (emitted, reflected and transmitted by) a *surface* per unit area. This is sometimes also confusingly called "intensity".
Spectral radiosity	$J_{e,\nu}$ [nb 3] or $J_{e,\lambda}$ [nb 4]	watt per square metre per hertz *or* watt per square metre, per metre	$\mathrm{W{\cdot}m^{-2}{\cdot}Hz^{-1}}$ *or* $\mathrm{W/m^3}$	$\mathrm{M{\cdot}T^{-2}}$ *or* $\mathrm{M{\cdot}L^{-1}{\cdot}T^{-3}}$	Radiosity of a *surface* per unit frequency or wavelength. The latter is commonly measured in $\mathrm{W{\cdot}m^{-2}{\cdot}nm^{-1}}$. This is sometimes also confusingly called "spectral intensity".
Radiant exitance	M_e [nb 2]	watt per square metre	$\mathrm{W/m^2}$	$\mathrm{M{\cdot}T^{-3}}$	Radiant flux *emitted* by a *surface* per unit area. This is the emitted component of radiosity. "Radiant emittance" is an old term for this quantity. This is sometimes also confusingly called "intensity".

Spectral exitance	$M_{e,\nu}$ [nb 3] or $M_{e,\lambda}$ [nb 4]	watt per square metre per hertz *or* watt per square metre, per metre	$W \cdot m^{-2} \cdot Hz^{-1}$ *or* W/m^3	$M \cdot T^{-2}$ *or* $M \cdot L^{-1} \cdot T^{-3}$	Radiant exitance of a *surface* per unit frequency or wavelength. The latter is commonly measured in $W \cdot m^{-2} \cdot nm^{-1}$. "Spectral emittance" is an old term for this quantity. This is sometimes also confusingly called "spectral intensity".
Radiant exposure	H_e	joule per square metre	J/m^2	$M \cdot T^{-2}$	Radiant energy received by a *surface* per unit area, or equivalently irradiance of a *surface* integrated over time of irradiation. This is sometimes also called "radiant fluence".
Spectral exposure	$H_{e,\nu}$ [nb 3] or $H_{e,\lambda}$ [nb 4]	joule per square metre per hertz *or* joule per square metre, per metre	$J \cdot m^{-2} \cdot Hz^{-1}$ *or* J/m^3	$M \cdot T^{-1}$ *or* $M \cdot L^{-1} \cdot T^{-2}$	Radiant exposure of a *surface* per unit frequency or wavelength. The latter is commonly measured in $J \cdot m^{-2} \cdot nm^{-1}$. This is sometimes also called "spectral fluence".
Hemispherical emissivity	ε			1	Radiant exitance of a *surface*, divided by that of a *black body* at the same temperature as that surface.
Spectral hemispherical emissivity	ε_ν or ε_λ			1	Spectral exitance of a *surface*, divided by that of a *black body* at the same temperature as that surface.
Directional emissivity	ε_Ω			1	Radiance *emitted* by a *surface*, divided by that emitted by a *black body* at the same temperature as that surface.
Spectral directional emissivity	$\varepsilon_{\Omega,\nu}$ or $\varepsilon_{\Omega,\lambda}$			1	Spectral radiance *emitted* by a *surface*, divided by that of a *black body* at the same temperature as that surface.
Hemispherical absorptance	A			1	Radiant flux *absorbed* by a *surface*, divided by that received by that surface. This should not be confused with "absorbance".
Spectral hemispherical absorptance	A_ν or A_λ			1	Spectral flux *absorbed* by a *surface*, divided by that received by that surface. This should not be confused with "spectral absorbance".
Directional absorptance	A_Ω			1	Radiance *absorbed* by a *surface*, divided by the radiance incident onto that surface. This should not be confused with "absorbance".

Spectral directional absorptance	$A_{\Omega,v}$ or $A_{\Omega,\lambda}$			1	Spectral radiance *absorbed* by a *surface*, divided by the spectral radiance incident onto that surface. This should not be confused with "spectral absorbance".
Hemispherical reflectance	R			1	Radiant flux *reflected* by a *surface*, divided by that received by that surface.
Spectral hemispherical reflectance	R_v or R_λ			1	Spectral flux *reflected* by a *surface*, divided by that received by that surface.
Directional reflectance	R_Ω			1	Radiance *reflected* by a *surface*, divided by that received by that surface.
Spectral directional reflectance	$R_{\Omega,v}$ or $R_{\Omega,\lambda}$			1	Spectral radiance *reflected* by a *surface*, divided by that received by that surface.
Hemispherical transmittance	T			1	Radiant flux *transmitted* by a *surface*, divided by that received by that surface.
Spectral hemispherical transmittance	T_v or T_λ			1	Spectral flux *transmitted* by a *surface*, divided by that received by that surface.
Directional transmittance	T_Ω			1	Radiance *transmitted* by a *surface*, divided by that received by that surface.
Spectral directional transmittance	$T_{\Omega,v}$ or $T_{\Omega,\lambda}$			1	Spectral radiance *transmitted* by a *surface*, divided by that received by that surface.
Hemispherical attenuation coefficient	μ	reciprocal metre	m^{-1}	L^{-1}	Radiant flux *absorbed* and *scattered* by a *volume* per unit length, divided by that received by that volume.
Spectral hemispherical attenuation coefficient	μ_v or μ_λ	reciprocal metre	m^{-1}	L^{-1}	Spectral radiant flux *absorbed* and *scattered* by a *volume* per unit length, divided by that received by that volume.
Directional attenuation coefficient	μ_Ω	reciprocal metre	m^{-1}	L^{-1}	Radiance *absorbed* and *scattered* by a *volume* per unit length, divided by that received by that volume.
Spectral directional attenuation coefficient	$\mu_{\Omega,v}$ or $\mu_{\Omega,\lambda}$	reciprocal metre	m^{-1}	L^{-1}	Spectral radiance *absorbed* and *scattered* by a *volume* per unit length, divided by that received by that volume.

Table 1. SI Photometry Quantities

Quantity		Unit		Dimension	Notes
Name	Symbol[nb 6]	Name	Symbol	Symbol	
Luminous energy	Q_v [nb 7]	lumen second	lm·s	T·J [nb 8]	Units are sometimes called *talbots*.
Luminous flux / luminous power	Φ_v [nb 7]	lumen (= cd·sr)	lm	J [nb 8]	Luminous energy per unit time.
Luminous intensity	I_v	candela (= lm/sr)	cd	J [nb 8]	Luminous power per unit solid angle.
Luminance	L_v	candela per square metre	cd/m²	L^{-2}·J	Luminous power per unit solid angle per unit *projected* source area. Units are sometimes called *nits*.
Illuminance	E_v	lux (= lm/m²)	lx	L^{-2}·J	Luminous power *incident* on a surface.
Luminous exitance / luminous emittance	M_v	lux	lx	L^{-2}·J	Luminous power *emitted* from a surface.
Luminous exposure	H_v	lux second	lx·s	L^{-2}·T·J	
Luminous energy density	ω_v	lumen second per cubic metre	lm·s·m⁻³	L^{-3}·T·J	
Luminous efficacy	η [nb 7]	lumen per watt	lm/W	M^{-1}·L^{-2}·T³·J	Ratio of luminous flux to radiant flux or power consumption, depending on context.
Luminous efficiency / luminous coefficient	V			1	

Optimum Exposure

"Correct" exposure may be defined as an exposure that achieves the effect the photographer intended.

A more technical approach recognises that a photographic film (or sensor) has a physically limited useful exposure range, sometimes called its dynamic range. If, for any part of the photograph, the actual exposure is outside this range, the film cannot record it accurately. In a very simple model, for example, out-of-range values would be recorded as "black" (underexposed) or "white" (overexposed) rather than the precisely graduated shades of colour and tone required to describe "detail". Therefore, the purpose of exposure adjustment (and/or lighting adjustment) is to control the physical amount of light from the subject that is allowed to fall on the film, so that 'significant' areas of shadow and highlight detail do not exceed the film's useful exposure range. This ensures that no 'significant' information is lost during capture.

It is worth noting that the photographer may carefully overexpose or underexpose the photograph to *eliminate* "insignificant" or "unwanted" detail; to make, for example, a white altar cloth appear immaculately clean, or to emulate the heavy, pitiless shadows of film noir. However, it is technically much easier to discard recorded information during post processing than to try to 're-create' unrecorded information.

In a scene with strong or harsh lighting, the *ratio* between highlight and shadow luminance values may well be larger than the *ratio* between the film's maximum and minimum useful exposure values. In this case, adjusting the camera's exposure settings (which only applies changes to the whole image, not selectively to parts of the image) only allows the photographer to choose between underexposed shadows or overexposed highlights; it cannot bring both into the useful exposure range at the same time. Methods for dealing with this situation include: using some kind of fill lighting to gently increase the illumination in shadow areas; using a graduated ND filter or gobo to reduce the amount of light coming from the highlight areas; or varying the exposure between multiple, otherwise identical, photographs (exposure bracketing) and then combining them afterwards in some kind of HDRI process.

Overexposure and Underexposure

White chair: Deliberate use of overexposure for aesthetic purposes.

A photograph may be described as *overexposed* when it has a loss of highlight detail, that is, when important bright parts of an image are "washed out" or effectively all white, known as "blown-out highlights" or "clipped whites". A photograph may be described as *underexposed* when it has a loss of shadow detail, that is, when important dark areas are "muddy" or indistinguishable from black, known as "blocked-up shadows" (or sometimes "crushed shadows", "crushed blacks", or "clipped blacks", especially in video). As the image to the right shows, these terms are technical ones rather than artistic judgments; an overexposed or underexposed image may be "correct" in that it provides the effect that the photographer intended. Intentionally over- or underexposing (relative to a standard or the camera's automatic exposure) is casually referred to as "shooting to the right" or "shooting to the left" respectively, as these shift the histogram of the image to the right or left.

Exposure Settings

Two similar images, one taken in auto mode (overexposed), the other with manual settings.

Manual Exposure

In manual mode, the photographer adjusts the lens aperture and/or shutter speed to achieve the desired exposure. Many photographers choose to control aperture and shutter independently because opening up the aperture increases exposure, but also decreases the depth of field, and a slower shutter increases exposure but also increases the opportunity for motion blur.

"Manual" exposure calculations may be based on some method of light metering with a working knowledge of exposure values, the APEX system and/or the Zone System.

Automatic Exposure

Houses photographed with an exposure time of 0,005 s

A camera in automatic exposure (abbreviation: AE) mode automatically calculates and adjusts exposure settings to match (as closely as possible) the subject's mid-tone to the mid-tone of the photograph. For most cameras this means using an on-board TTL exposure meter.

Aperture priority mode (commonly abbreviated to Av) gives the photographer manual control of the aperture, whilst the camera automatically adjusts the shutter speed to achieve the exposure specified by the TTL meter. Shutter priority mode (commonly abbreviated to TV) gives manual shutter control, with automatic aperture compensation. In each case, the actual exposure level is still determined by the camera's exposure meter.

Exposure Compensation

A street view of Taka-Töölö, Helsinki, Finland, during a very sunny winter day. The image has been deliberately overexposed by +1 EV to compensate for the bright sunlight and the exposure time calculated by the camera's program automatic metering is still 1/320 s.

The purpose of an exposure meter is to estimate the subject's mid-tone luminance and indicate the camera exposure settings required to record this as a mid-tone. In order to do this it has to make a number of assumptions which, under certain circumstances, will be wrong. If the exposure setting indicated by an exposure meter is taken as the "reference" exposure, the photographer may wish to deliberately *overexpose* or *underexpose* in order to compensate for known or anticipated metering inaccuracies.

Cameras with any kind of internal exposure meter usually feature an exposure compensation setting which is intended to allow the photographer to simply offset the exposure level from the internal meter's estimate of appropriate exposure. Frequently calibrated in stops, also known as EV units, a "+1" exposure compensation setting indicates one stop more (twice as much) exposure and "−1" means one stop less (half as much) exposure.

Exposure compensation is particularly useful in combination with auto-exposure mode, as it allows the photographer to *bias* the exposure level without resorting to full manual exposure and losing the flexibility of auto exposure. On low-end video camcorders, exposure compensation may be the only manual exposure control available.

Exposure Control

A 1/30 s exposure showing motion blur on fountain at Royal Botanic Gardens, Kew

An appropriate exposure for a photograph is determined by the sensitivity of the medium used. For photographic film, sensitivity is referred to as film speed and is measured on a scale published by the International Organization for Standardization (ISO). Faster film, that is, film with a higher ISO rating, requires less exposure to make a good image. Digital cameras usually have variable ISO settings that provide additional flexibility. Exposure is a combination of the length of time and the illuminance at the photosensitive material. Exposure time is controlled in a camera by shutter speed, and the illuminance depends on the lens aperture and the scene luminance. Slower shutter speeds (exposing the medium for a longer period of time), greater lens apertures (admitting more light), and higher-luminance scenes produce greater exposures.

A 1/320 s exposure showing individual drops on fountain at Royal Botanic Gardens, Kew

An approximately correct exposure will be obtained on a sunny day using ISO 100 film, an aperture of $f/16$ and a shutter speed of 1/100 of a second. This is called the sunny 16 rule: at an aperture of $f/16$ on a sunny day, a suitable shutter speed will be one over the film speed (or closest equivalent).

A scene can be exposed in many ways, depending on the desired effect a photographer wishes to convey.

Reciprocity

An important principle of exposure is reciprocity. If one exposes the film or sensor for a longer period, a reciprocally smaller aperture is required to reduce the amount of light hitting the film to obtain the same exposure. For example, the photographer may prefer to make his sunny-16 shot at an aperture of $f/5.6$ (to obtain a shallow depth of field). As $f/5.6$ is 3 stops "faster" than $f/16$, with each stop meaning double the amount of light, a new shutter speed of $(1/125)/(2 \cdot 2 \cdot 2) = 1/1000$ s is needed. Once the photographer has determined the exposure, aperture stops can be traded for halvings or doublings of speed, within limits.

A demonstration of the effect of exposure in night photography. Longer shutter speeds result in increased exposure.

The true characteristic of most photographic emulsions is not actually linear, but it is close enough over the exposure range of about 1 second to 1/1000 of a second. Outside of this range, it becomes necessary to increase the exposure from the calculated value to account for this characteristic of the emulsion. This characteristic is known as *reciprocity failure*. The film manufacturer's data sheets should be consulted to arrive at the correction required, as different emulsions have different characteristics.

Digital camera image sensors can also be subject to a form of reciprocity failure.

Determining Exposure

A fair ride taken with a 2/5 second exposure.

The Zone System is another method of determining exposure and development combinations to achieve a greater tonality range over conventional methods by varying the contrast of the film to fit the print contrast capability. Digital cameras can achieve similar results (high dynamic range) by combining several different exposures (varying shutter or diaphram) made in quick succession.

Today, most cameras automatically determine the correct exposure at the time of taking a photograph by using a built-in light meter, or multiple point meters interpreted by a built-in computer.

Negative/Print film tends to bias for exposing for the shadow areas (film dislikes being starved of light), with digital favouring exposure for highlights.

Latitude

Latitude is the degree by which one can over, or under expose an image, and still recover an acceptable level of quality from an exposure. Typically negative film has a better ability to record a range of brightness than slide/transparency film or digital. Digital should be considered to be the reverse of print film, with a good latitude in the shadow range, and a narrow one in the highlight area; in contrast to film's large highlight latitude, and narrow shadow latitude. Slide/Transparency film has a narrow latitude in both highlight and shadow areas, requiring greater exposure accuracy.

Negative film's latitude increases somewhat with high ISO material, in contrast digital tends to narrow on latitude with high ISO settings.

Highlights

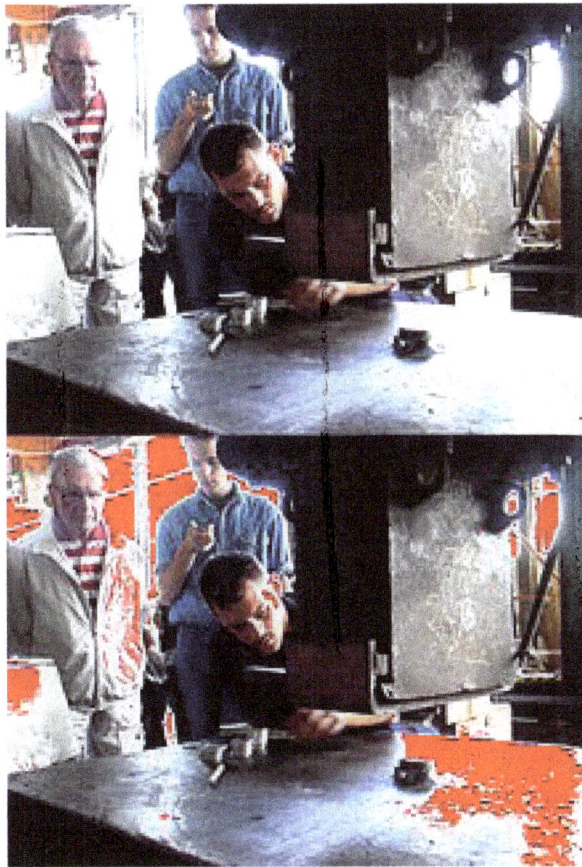

Example image exhibiting blown-out highlights. Top: original image, bottom: blown-out areas marked red

Areas of a photo where information is lost due to extreme brightness are described as having "blown-out highlights" or "flared highlights".

In digital images this information loss is often irreversible, though small problems can be made less noticeable using photo manipulation software. Recording to RAW format can correct this problem to some degree, as can using a digital camera with a better sensor.

Film can often have areas of extreme overexposure but still record detail in those areas. This information is usually somewhat recoverable when printing or transferring to digital.

A loss of highlights in a photograph is usually undesirable, but in some cases can be considered to "enhance" appeal. Examples include black-and-white photography and portraits with an out-of-focus background.

Blacks

Areas of a photo where information is lost due to extreme darkness are described as "crushed blacks". Digital capture tends to be more tolerant of underexposure, allowing better recovery of shadow detail, than same-ISO negative print film.

Crushed blacks cause loss of detail, but can be used for artistic effect.

Atmospheric Optics

A colorful sky is often due to scattering of light off particulates, as in this photograph of a sunset during the October 2007 California wildfires.

Atmospheric optics deals with how the unique optical properties of the Earth's atmosphere cause a wide range of spectacular optical phenomena. The blue color of the sky is a direct result of Rayleigh scattering which redirects higher frequency (blue) sunlight back into the field of view of the observer. Because blue light is scattered more easily than red light, the sun takes on a reddish hue when it is observed through a thick atmosphere, as during a sunrise or sunset. Additional particulate matter in the sky can scatter different colors at different angles creating colorful glowing skies at dusk and dawn. Scattering off of ice crystals and other particles in the atmosphere are responsible for halos, afterglows, coronas, rays of sunlight, and sun dogs. The variation in these kinds of phenomena is due to different particle sizes and geometries.

Mirages are optical phenomena in which light rays are bent due to thermal variations in the refraction index of air, producing displaced or heavily distorted images of distant objects. Other optical phenomena associated with this include the Novaya Zemlya effect where the sun appears to rise earlier or set later than predicted with a distorted shape. A spectacular form of refraction occurs with a temperature inversion called the Fata Morgana where objects on the horizon or even beyond the horizon, such as islands, cliffs, ships or icebergs, appear elongated and elevated, like "fairy tale castles".

Rainbows are the result of a combination of internal reflection and dispersive refraction of light in raindrops. Because rainbows are seen on the opposite side of the sky as the sun, rainbows are more prominent the closer the sun is to the horizon due to their greater distance apart.

Sun and Moon Size

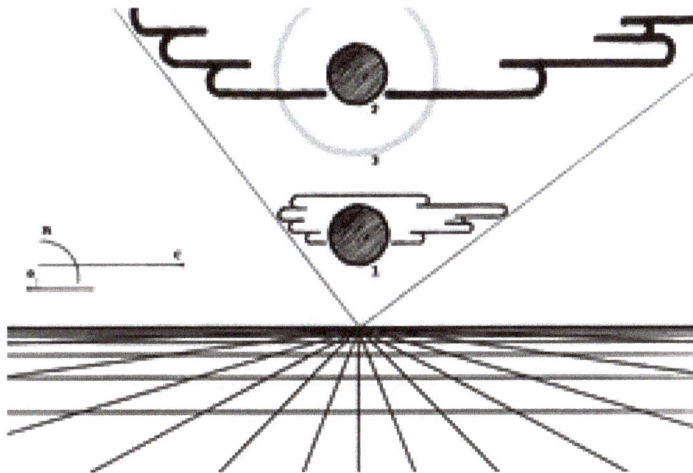

Comparison between the relative sizes of the Moon and a cloud at different points in the sky

In the Book of Optics (1011–1022 A.D.), Ibn al-Haytham argued that vision occurs in the brain, and that personal experience has an effect on what people see and how they see, and that vision and perception are subjective. Arguing against Ptolemy's refraction theory for why people perceive the sun and moon larger at the horizon than when they are higher in the sky, he redefined the problem in terms of perceived, rather than real, enlargement. He said that judging the distance of an object depends on there being an uninterrupted sequence of intervening bodies between the object and the observer. With the Moon, however, there are no intervening objects. Therefore, since the size of an object depends on its observed distance, which is in this case inaccurate, the Moon appears larger on the horizon. Through works by Roger Bacon, John Pecham and Witelo based on Ibn al-Haytham's explanation, the Moon illusion gradually came to be accepted as a psychological phenomenon, with Ptolemy's theory being rejected in the 17th century. For over 100 years, research on the Moon illusion has been conducted by vision scientists who invariably have been psychologists specializing in human perception. After reviewing the many different explanations in their 2002 book *The Mystery of the Moon Illusion*, Ross and Plug conclude "No single theory has emerged victorious".

Sky Coloration

Light from the sky is a result of the scattering of sunlight, which results in a blue color perceived by the human eye. On a sunny day Rayleigh scattering gives the sky a blue gradient — dark in the zenith, light near the horizon. Light that comes in from overhead encounters 1/38th of the air mass that light coming along a horizon path encounters. So, fewer particles scatter the zenith sunbeam, and therefore the light remains a darker blue. The blueness is at the horizon because the blue light coming from great distances is also preferentially scattered. This results in a red shift of the far lightsources that is compensated by the blue hue of the scattered light in the line of sight. In other words, the red light scatters also; if it does so at a point a great distance from the observer it has a much higher chance of reaching the observer than blue light. At distances nearing infinity the scattered light is therefore white. Far away clouds or snowy mountaintops will seem yellow for that reason; that effect is not obvious on clear days, but very pronounced when clouds are covering the line of sight reducing the blue hue from scattered sunlight.

When seen from altitude, as here from an airplane, the sky's color varies from pale to dark at elevations approaching the zenith

The scattering due to molecule sized particles (as in air) is greater in the forward and backward directions than it is in the lateral direction. Individual water droplets exposed to white light will create a set of colored rings. If a cloud is thick enough, scattering from multiple water droplets will wash out the set of colored rings and create a washed out white color. Dust from the Sahara moves around the southern periphery of the subtropical ridge moves into the southeastern United States during the summer, which changes the sky from a blue to a white appearance and leads to an increase in red sunsets. Its presence negatively impacts air quality during the summer since it adds to the count of airborne particulates.

Purple sky on the La Silla Observatory.

The sky can turn a multitude of colors such as red, orange, pink and yellow (especially near sunset or sunrise) and black at night. Scattering effects also partially polarize light from the sky, most pronounced at an angle 90° from the sun.

Sky luminance distribution models have been recommended by the International Commission on Illumination (CIE) for the design of daylighting schemes. Recent developments relate to "all sky models" for modelling sky luminance under weather conditions ranging from clear sky to overcast.

Cloud Coloration

An occurrence of altocumulus and cirrocumulus cloud iridescence

The color of a cloud, as seen from the Earth, tells much about what is going on inside the cloud. Dense deep tropospheric clouds exhibit a high reflectance (70% to 95%) throughout the visible spectrum. Tiny particles of water are densely packed and sunlight cannot penetrate far into the cloud before it is reflected out, giving a cloud its characteristic white color, especially when viewed from the top. Cloud droplets tend to scatter light efficiently, so that the intensity of the solar radiation decreases with depth into the gases. As a result, the cloud base can vary from a very light to very dark grey depending on the cloud's thickness and how much light is being reflected or transmitted back to the observer. Thin clouds may look white or appear to have acquired the color of their environment or background. High tropospheric and non-tropospheric clouds appear mostly white if composed entirely of ice crystals and/or supercooled water droplets.

As a tropospheric cloud matures, the dense water droplets may combine to produce larger droplets, which may combine to form droplets large enough to fall as rain. By this process of accumulation, the space between droplets becomes increasingly larger, permitting light to penetrate farther into the cloud. If the cloud is sufficiently large and the droplets within are spaced far enough apart, it may be that a percentage of the light which enters the cloud is not reflected back out before it is absorbed. A simple example of this is being able to see farther in heavy rain than in heavy fog. This process of reflection/absorption is what causes the range of cloud color from white to black.

Other colors occur naturally in clouds. Bluish-grey is the result of light scattering within the cloud. In the visible spectrum, blue and green are at the short end of light's visible wavelengths, while red and yellow are at the long end. The short rays are more easily scattered by water droplets, and the long rays are more likely to be absorbed. The bluish color is evidence that such scattering is being produced by rain-sized droplets in the cloud. A greenish tinge to a cloud is produced when sunlight is scattered by ice. A cumulonimbus cloud emitting green is a sign that it is a severe thunderstorm, capable of heavy rain, hail, strong winds and possible tornadoes. Yellowish clouds may occur in the late spring through early fall months during forest fire season. The yellow color is due to the presence of pollutants in the smoke. Yellowish clouds caused by the presence of nitrogen dioxide are sometimes seen in urban areas with high air pollution levels.

Sunset reflecting shades of pink onto grey stratocumulus clouds.

Red, orange and pink clouds occur almost entirely at sunrise and sunset and are the result of the scattering of sunlight by the atmosphere. When the angle between the sun and the horizon is less than 10 percent, as it is just after sunrise or just prior to sunset, sunlight becomes too red due to refraction for any colors other than those with a reddish hue to be seen. The clouds do not become that color; they are reflecting long and unscattered rays of sunlight, which are predominant at those hours. The effect is much like if one were to shine a red spotlight on a white sheet. In combination with large, mature thunderheads this can produce blood-red clouds. Clouds look darker in the near-infrared because water absorbs solar radiation at those wavelengths.

Halos

A halo (λως; also known as a nimbus, icebow or gloriole) is an optical phenomenon produced by the interaction of light from the sun or moon with ice crystals in the atmosphere, resulting in colored or white arcs, rings or spots in the sky. Many halos are positioned near the sun or moon, but others are elsewhere and even in the opposite part of the sky. They can also form around artificial lights in very cold weather when ice crystals called diamond dust are floating in the nearby air.

There are many types of ice halos. They are produced by the ice crystals in cirrus or cirrostratus clouds high in the upper troposphere, at an altitude of 5 kilometres (3.1 mi) to 10 kilometres (6.2 mi), or, during very cold weather, by ice crystals called diamond dust drifting in the air at low levels. The particular shape and orientation of the crystals are responsible for the types of halo observed. Light is reflected and refracted by the ice crystals and may split into colors because of dispersion. The crystals behave like prisms and mirrors, refracting and reflecting sunlight between their faces, sending shafts of light in particular directions. For circular halos, the preferred angular distance are 22 and 46 degrees from the ice crystals which create them. Atmospheric phenomena such as halos have been used as part of weather lore as an empirical means of weather forecasting, with their presence indicating an approach of a warm front and its associated rain.

A man in front of a complex halo display.

Sun Dogs

Very bright sundogs in Fargo, North Dakota. Note the halo arcs passing through each sun dog.

Sun dogs are a common type of halo, with the appearance of two subtly-colored bright spots to the left and right of the sun, at a distance of about 22° and at the same elevation above the horizon.

They are commonly caused by plate-shaped hexagonal ice crystals. These crystals tend to become horizontally aligned as they sink through the air, causing them to refract the sunlight to the left and right, resulting in the two sun dogs.

As the sun rises higher, the rays passing through the crystals are increasingly skewed from the horizontal plane. Their angle of deviation increases and the sundogs move further from the sun. However, they always stay at the same elevation as the sun. Sun dogs are red-colored at the side nearest the sun. Farther out the colors grade to blue or violet. However, the colors overlap considerably and so are muted, rarely pure or saturated. The colors of the sun dog finally merge into the white of the parhelic circle (if the latter is visible).

It is theoretically possible to predict the forms of sun dogs as would be seen on other planets and moons. Mars might have sundogs formed by both water-ice and CO_2-ice. On the giant gas planets — Jupiter, Saturn, Uranus and Neptune — other crystals form the clouds of ammonia, methane, and other substances that can produce halos with four or more sundogs.

Glory

Solar glory at the steam from a hot spring

A common optical phenomenon involving water droplets is the glory. A glory is an optical phenomenon, appearing much like an iconic Saint's halo about the head of the observer, produced by light backscattered (a combination of diffraction, reflection and refraction) towards its source by a cloud of uniformly sized water droplets. A glory has multiple colored rings, with red colors on the outermost ring and blue/violet colors on the innermost ring.

The angular distance is much smaller than a rainbow, ranging between 5° and 20°, depending on the size of the droplets. The glory can only be seen when the observer is directly between the sun and cloud of refracting water droplets. Hence, it is commonly observed while airborne, with the glory surrounding the airplane's shadow on clouds (this is often called *The Glory of the Pilot*). Glories can also be seen from mountains and tall buildings, when there are clouds or fog below the level of the observer, or on days with ground fog. The glory is related to the optical phenomenon anthelion.

Rainbow

Double rainbow and supernumerary rainbows on the inside of the primary arc. The shadow of the photographer's head marks the centre of the rainbow circle (antisolar point).

A rainbow is an optical and meteorological phenomenon that causes a spectrum of light to appear in the sky when the Sun shines on to droplets of moisture in the Earth's atmosphere. It takes the form of a multicolored arc. Rainbows caused by sunlight always appear in the section of sky directly opposite the sun, but originate no further than 42 degrees above the horizon for observers on the ground. To see them at higher angles, an observer would need to be in an airplane or near a mountaintop since the rainbow would otherwise be below the horizon. The bigger the droplets which formed the rainbow, the brighter it will be. Rainbows are most common near afternoon thunderstorms during the summer.

A single reflection off the backs of an array of raindrops produces a rainbow with an angular size on the sky that ranges from 40° to 42° with red on the outside. Double rainbows are produced by two internal reflections with angular size of 50.5° to 54° with violet on the outside. Within the "primary rainbow" (the lowest, and also normally the brightest rainbow) the arc of a rainbow shows red on the outer (or upper) part of the arc, and violet on the inner section. This rainbow is caused by light being reflected once in droplets of water. In a double rainbow, a second arc may be seen above and outside the primary arc, and has the order of its colors reversed (red faces inward toward the other rainbow, in both rainbows). This second rainbow is caused by light reflecting twice inside water droplets. The region between a double rainbow is dark. The reason for this dark band is that, while light *below* the primary rainbow comes from droplet reflection, and light *above* the upper (secondary) rainbow also comes from droplet reflection, there is no mechanism for the region *between* a double rainbow to show any light reflected from water drops, at all.

A rainbow spans a continuous spectrum of colors; the distinct bands (including the number of bands) are an artifact of human color vision, and no banding of any type is seen in a black-and-white photograph of a rainbow (only a smooth gradation of intensity to a maxima, then fading to a minima at the other side of the arc). For colors seen by a normal human eye, the most commonly cited and remembered sequence, in English, is Newton's sevenfold red, orange, yellow, green, blue, indigo and violet (popularly memorized by mnemonics like Roy G. Biv). However, color-blind persons will see fewer colors.

Rainbows can be caused by many forms of airborne water. These include not only rain, but also mist, spray, and airborne dew.

Mirage

Various kinds of mirages in one location taken over the course of six minutes. The uppermost inset frame shows an inferior mirage of the Farallon Islands. The second inset frame shows a green flash on the left-hand side. The two lower frames and the main frame all show superior mirages of the Farallon Islands. In these three frames, the superior mirage evolves from a 3-image mirage to a 5-image mirage, and back to a 2-image mirage. Such a display is consistent with a Fata Morgana.

A mirage is a naturally occurring optical phenomenon in which light rays are bent to produce a displaced image of distant objects or the sky. The word comes to English via the French *mirage*, from the Latin *mirare*, meaning "to look at, to wonder at". This is the same root as for "mirror" and "to admire". Also, it has its roots in the Arabic *mirage*.

In contrast to a hallucination, a mirage is a real optical phenomenon which can be captured on camera, since light rays actually are refracted to form the false image at the observer's location. What the image appears to represent, however, is determined by the interpretive faculties of the human mind. For example, inferior images on land are very easily mistaken for the reflections from a small body of water.

Mirages can be categorized as "inferior" (meaning lower), "superior" (meaning higher) and "Fata Morgana", one kind of superior mirage consisting of a series of unusually elaborate, vertically stacked images, which form one rapidly changing mirage.

Green flashes and green rays are optical phenomena that occur shortly after sunset or before sunrise, when a green spot is visible, usually for no more than a second or two, above the sun, or a green ray shoots up from the sunset point. Green flashes are actually a group of phenomena stemming from different causes, and some are more common than others. Green flashes can be observed from any altitude (even from an aircraft). They are usually seen at an unobstructed horizon, such as over the ocean, but are possible over cloud tops and mountain tops as well.

A green flash from the moon and bright planets at the horizon, including Venus and Jupiter, can also be observed.

Fata Morgana

A Fata Morgana of a boat

This optical phenomenon occurs because rays of light are strongly bent when they pass through air layers of different temperatures in a steep thermal inversion where an atmospheric duct has formed. A thermal inversion is an atmospheric condition where warmer air exists in a well-defined layer above a layer of significantly cooler air. This temperature inversion is the opposite of what is normally the case; air is usually warmer close to the surface, and cooler higher up. In calm weather, a layer of significantly warmer air can rest over colder dense air, forming an atmospheric duct which acts like a refracting lens, producing a series of both inverted and erect images.

A Fata Morgana is an unusual and very complex form of mirage, a form of superior mirage, which, like many other kinds of superior mirages, is seen in a narrow band right above the horizon. It is an Italian phrase derived from the vulgar Latin for "fairy" and the Arthurian sorcerer Morgan le Fay, from a belief that the mirage, often seen in the Strait of Messina, were fairy castles in the air, or false land designed to lure sailors to their death created by her witchcraft. Although the term Fata Morgana is sometimes incorrectly applied to other, more common kinds of mirages, the true Fata Morgana is not the same as an ordinary superior mirage, and is certainly not the same as an inferior mirage.

Fata Morgana mirages tremendously distort the object or objects which they are based on, such that the object often appears to be very unusual, and may even be transformed in such a way that it is completely unrecognizable. A Fata Morgana can be seen on land or at sea, in polar regions or in deserts. This kind of mirage can involve almost any kind of distant object, including such things as boats, islands, and coastline.

A Fata Morgana is not only complex, but also rapidly changing. The mirage comprises several inverted (upside down) and erect (right side up) images that are stacked on top of one another. Fata Morgana mirages also show alternating compressed and stretched zones.

Novaya Zemlya Effect

The Novaya Zemlya effect is a polar mirage caused by high refraction of sunlight between atmospheric thermoclines. The Novaya Zemlya effect will give the impression that the sun is rising earlier or setting later than it actually should (astronomically speaking). Depending on the meteorological situation the effect will present the sun as a line or a square (which is sometimes referred to as the "rectangular sun"), made up of flattened hourglass shapes. The mirage requires rays of sunlight to have an inversion layer for hundreds of kilometres, and depends on the inversion layer's temperature gradient. The sunlight must bend to the Earth's curvature at least 400 kilometres (250 mi) to allow an elevation rise of 5 degrees for sight of the sun disk.

The first person to record the phenomenon was Gerrit de Veer, a member of Willem Barentsz' ill-fated third expedition into the polar region. Novaya Zemlya, the archipelago where de Veer first observed the phenomenon, lends its name to the effect.

Crepuscular Rays

Crepuscular rays, taken near Waterberg Plateau, Namibia

Crepuscular rays are near-parallel rays of sunlight moving through the Earth's atmosphere, but appear to diverge because of linear perspective. They often occur when objects such as mountain peaks or clouds partially shadow the sun's rays like a cloud cover. Various airborne compounds scatter the sunlight and make these rays visible, due to diffraction, reflection, and scattering.

Crepuscular rays can also occasionally be viewed underwater, particularly in arctic areas, appearing from ice shelves or cracks in the ice. Also they are also viewed in days when the sun hits the clouds in a perfect angle shining upon the area.

There are three primary forms of crepuscular rays:

- Rays of light penetrating holes in low clouds (also called "Jacob's Ladder").

- Beams of light diverging from behind a cloud.

- Pale, pinkish or reddish rays that radiate from below the horizon. These are often mistaken for light pillars.

They are commonly seen near sunrise and sunset, when tall clouds such as cumulonimbus and mountains can be most effective at creating these rays

Anticrepuscular Rays

Anticrepuscular rays while parallel in reality are sometimes visible in the sky in the direction opposite the sun. They appear to converge again at the distant horizon.

Atmospheric Refraction

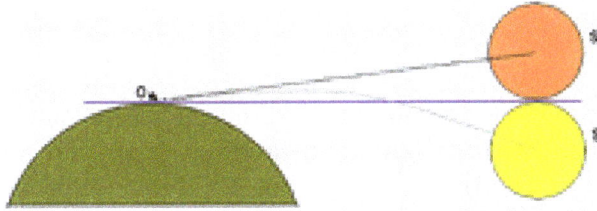

Diagram showing displacement of the Sun's image at sunrise and sunset

Atmospheric refraction influences the apparent position of astronomical and terrestrial objects, usually causing them to appear higher than they actually are. For this reason navigators, astronomers, and surveyors observe positions when these effects are minimal. Sailors will only shoot a star when 20° or more above the horizon, astronomers try to schedule observations when an object is highest in the sky, and surveyors try to observe in the afternoon when refraction is minimum.

Wave–particle Duality

Wave–particle duality is the concept that every elementary particle or quantic entity may be partly described in terms not only of particles, but also of waves. It expresses the inability of the classical concepts "particle" or "wave" to fully describe the behavior of quantum-scale objects. As Albert Einstein wrote: *"It seems as though we must use sometimes the one theory and sometimes the other, while at times we may use either. We are faced with a new kind of difficulty. We have two contradictory pictures of reality; separately neither of them fully explains the phenomena of light, but together they do"*.

Through the work of Max Planck, Einstein, Louis de Broglie, Arthur Compton, Niels Bohr and many others, current scientific theory holds that all particles also have a wave nature (and vice versa). This phenomenon has been verified not only for elementary particles, but also for compound particles like atoms and even molecules. For macroscopic particles, because of their extremely short wavelengths, wave properties usually cannot be detected.

Although the use of the wave-particle duality has worked well in physics, the *meaning* or *interpretation* has not been satisfactorily resolved.

Niels Bohr regarded the "duality paradox" as a fundamental or metaphysical fact of nature. A given kind of quantum object will exhibit sometimes wave, sometimes particle, character, in respectively different physical settings. He saw such duality as one aspect of the concept of complementarity. Bohr regarded renunciation of the cause-effect relation, or complementarity, of the space-time picture, as essential to the quantum mechanical account.

Werner Heisenberg considered the question further. He saw the duality as present for all quantic entities, but not quite in the usual quantum mechanical account considered by Bohr. He saw it in what is called second quantization, which generates an entirely new concept of fields which exist in ordinary space-time, causality still being visualizable. Classical field values (e.g. the electric and magnetic field strengths of Maxwell) are replaced by an entirely new kind of field value, as considered in quantum field theory. Turning the reasoning around, ordinary quantum mechanics can be deduced as a specialized consequence of quantum field theory.

Brief History of Wave and Particle Viewpoints

Democritus—the original *atomist*—argued that all things in the universe, including light, are composed of indivisible sub-components (light being some form of solar atom). At the beginning of the 11th Century, the Arabic scientist Alhazen wrote the first comprehensive treatise on optics; describing refraction, reflection, and the operation of a pinhole lens via rays of light traveling from the point of emission to the eye. He asserted that these rays were composed of particles of light. In 1630, René Descartes popularized and accredited the opposing wave description in his treatise on light, showing that the behavior of light could be re-created by modeling wave-like disturbances in a universal medium ("plenum"). Beginning in 1670 and progressing over three decades, Isaac Newton developed and championed his corpuscular hypothesis, arguing that the perfectly straight lines of reflection demonstrated light's particle nature; only particles could travel in such straight lines. He explained refraction by positing that particles of light accelerated laterally upon entering a denser medium. Around the same time, Newton's contemporaries Robert Hooke and Christiaan Huygens—and later Augustin-Jean Fresnel—mathematically refined the wave viewpoint, showing that if light traveled at different speeds in different media (such as water and air), refraction could be easily explained as the medium-dependent propagation of light waves. The resulting Huygens–Fresnel principle was extremely successful at reproducing light's behavior and was subsequently supported by Thomas Young's 1803 discovery of double-slit interference. The wave view did not immediately displace the ray and particle view, but began to dominate scientific thinking about light in the mid 19th century, since it could explain polarization phenomena that the alternatives could not.

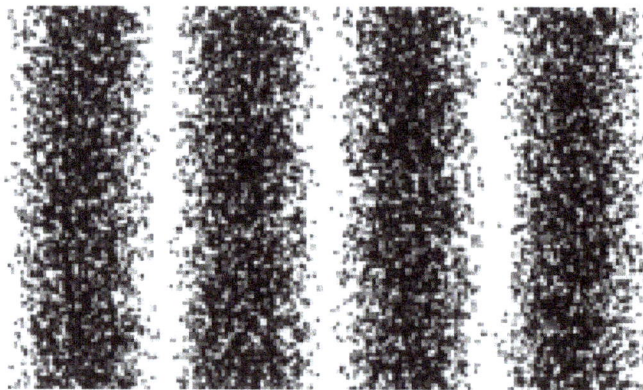

Particle impacts make visible the interference pattern of waves.

James Clerk Maxwell discovered that he could combine four simple equations, which had been previously discovered, along with a slight modification to describe self-propagating waves of oscillating electric and magnetic fields. When the propagation speed of these electromagnetic waves was calculated, the speed of light fell out. It quickly became apparent that visible light, ultraviolet light, and infrared light (phenomena thought previously to be unrelated) were all electromagnetic waves of differing frequency. The wave theory had prevailed—or at least it seemed to.

While the 19th century had seen the success of the wave theory at describing light, it had also witnessed the rise of the atomic theory at describing matter. Antoine Lavoisier deduced the law of conservation of mass and categorized many new chemical elements and compounds; and Joseph Louis

Proust advanced chemistry towards the atom by showing that elements combined in definite proportions. This led John Dalton to propose that elements were invisible sub components; Amedeo Avogadro discovered diatomic gases and completed the basic atomic theory, allowing the correct molecular formulae of most known compounds—as well as the correct weights of atoms—to be deduced and categorized in a consistent manner. Dimitri Mendeleev saw an order in recurring chemical properties, and created a table presenting the elements in unprecedented order and symmetry.

A quantum particle is represented by a wave packet.

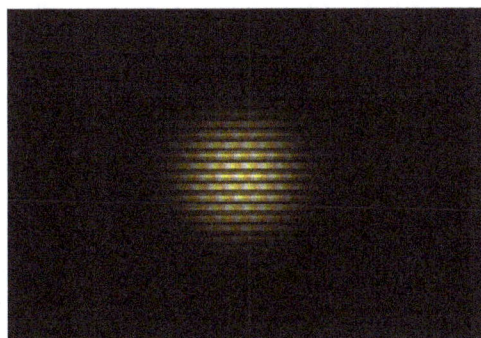

Interference of a quantum particle with itself.

Turn of the 20th century and the paradigm shift

Particles of Electricity

At the close of the 19th century, the reductionism of atomic theory began to advance into the atom itself; determining, through physics, the nature of the atom and the operation of chemical reactions. Electricity, first thought to be a fluid, was now understood to consist of particles called electrons. This was first demonstrated by J. J. Thomson in 1897 when, using a cathode ray tube, he found that an electrical charge would travel across a vacuum (which would possess infinite resistance in classical theory). Since the vacuum offered no medium for an electric fluid to travel, this discovery could only be explained via a particle carrying a negative charge and moving through the vacuum. This *electron* flew in the face of classical electrodynamics, which had successfully treated electricity as a fluid for many years (leading to the invention of batteries, electric motors, dynamos, and arc lamps). More importantly, the intimate relation between electric charge and electromagnetism had been well documented following the discoveries of Michael Faraday and James Clerk Maxwell. Since electromagnetism was *known* to be a wave generated by a changing electric or magnetic *field* (a continuous, wave-like entity itself) an atomic/particle description of electricity and charge was a non sequitur. Furthermore, classical electrodynamics was not the only classical theory rendered incomplete.

Radiation Quantization

In 1901, Max Planck published an analysis that succeeded in reproducing the observed spectrum of light emitted by a glowing object. To accomplish this, Planck had to make an ad hoc mathematical assumption of quantized energy of the oscillators (atoms of the black body) that emit radiation. It was Einstein who later proposed that it is the electromagnetic radiation itself that is quantized, and not the energy of radiating atoms.

Black-body radiation, the emission of electromagnetic energy due to an object's heat, could not be explained from classical arguments alone. The equipartition theorem of classical mechanics, the basis of all classical thermodynamic theories, stated that an object's energy is partitioned equally among the object's vibrational modes. But applying the same reasoning to the electromagnetic emission of such a thermal object was not so successful. It had been long known that thermal objects emit light. Since light was known to be waves of electromagnetism, physicists hoped to describe this emission via classical laws. This became known as the black body problem. Since the equipartition theorem worked so well in describing the vibrational modes of the thermal object itself, it was natural to assume that it would perform equally well in describing the radiative emission of such objects. But a problem quickly arose: if each mode received an equal partition of energy, the short wavelength modes would consume all the energy. This became clear when plotting the Rayleigh–Jeans law which, while correctly predicting the intensity of long wavelength emissions, predicted infinite total energy as the intensity diverges to infinity for short wavelengths. This became known as the ultraviolet catastrophe.

In 1900, Max Planck hypothesized that the frequency of light emitted by the black body depended on the frequency of the *oscillator* that emitted it, and the energy of these oscillators increased linearly with frequency (according to his constant h, where $E = h\nu$). This was not an unsound proposal considering that macroscopic oscillators operate similarly: when studying five simple harmonic oscillators of equal amplitude but different frequency, the oscillator with the highest frequency possesses the highest energy (though this relationship is not linear like Planck's). By demanding that high-frequency light must be emitted by an oscillator of equal frequency, and further requiring that this oscillator occupy higher energy than one of a lesser frequency, Planck avoided any catastrophe; giving an equal partition to high-frequency oscillators produced successively fewer oscillators and less emitted light. And as in the Maxwell–Boltzmann distribution, the low-frequency, low-energy oscillators were suppressed by the onslaught of thermal jiggling from higher energy oscillators, which necessarily increased their energy and frequency.

The most revolutionary aspect of Planck's treatment of the black body is that it inherently relies on an integer number of oscillators in thermal equilibrium with the electromagnetic field. These oscillators *give* their entire energy to the electromagnetic field, creating a quantum of light, as often as they are *excited* by the electromagnetic field, absorbing a quantum of light and beginning to oscillate at the corresponding frequency. Planck had intentionally created an atomic theory of the black body, but had unintentionally generated an atomic theory of light, where the black body never generates quanta of light at a given frequency with an energy less than $h\nu$. However, once realizing that he had quantized the electromagnetic field, he denounced particles of light as a limitation of his approximation, not a property of reality.

Photoelectric Effect Illuminated

While Planck had solved the ultraviolet catastrophe by using atoms and a quantized electromagnetic field, most contemporary physicists agreed that Planck's "light quanta" represented only flaws in his model. A more-complete derivation of black body radiation would yield a fully continuous and 'wave-like' electromagnetic field with no quantization. However, in 1905 Albert Einstein took Planck's black body model to produce his solution to another outstanding problem of the day: the photoelectric effect, wherein electrons are emitted from atoms when they absorb energy from light. Since their discovery eight years previously, electrons had been studied in physics laboratories worldwide.

In 1902 Philipp Lenard discovered that the energy of these ejected electrons did *not* depend on the intensity of the incoming light, but instead on its *frequency*. So if one shines a little low-frequency light upon a metal, a few low energy electrons are ejected. If one now shines a very intense beam of low-frequency light upon the same metal, a whole slew of electrons are ejected; however they possess the same low energy, there are merely *more of them*. The more light there is, the more electrons are ejected. Whereas in order to get high energy electrons, one must illuminate the metal with high-frequency light. Like blackbody radiation, this was at odds with a theory invoking continuous transfer of energy between radiation and matter. However, it can still be explained using a fully classical description of light, as long as matter is quantum mechanical in nature.

If one used Planck's energy quanta, and demanded that electromagnetic radiation at a given frequency could only transfer energy to matter in integer multiples of an energy quantum $h\nu$, then the photoelectric effect could be explained very simply. Low-frequency light only ejects low-energy electrons because each electron is excited by the absorption of a single photon. Increasing the intensity of the low-frequency light (increasing the number of photons) only increases the number of excited electrons, not their energy, because the energy of each photon remains low. Only by increasing the frequency of the light, and thus increasing the energy of the photons, can one eject electrons with higher energy. Thus, using Planck's constant h to determine the energy of the photons based upon their frequency, the energy of ejected electrons should also increase linearly with frequency; the gradient of the line being Planck's constant. These results were not confirmed until 1915, when Robert Andrews Millikan, who had previously determined the charge of the electron, produced experimental results in perfect accord with Einstein's predictions. While the energy of ejected electrons reflected Planck's constant, the existence of photons was not explicitly proven until the discovery of the photon antibunching effect, of which a modern experiment can be performed in undergraduate-level labs. This phenomenon could only be explained via photons, and not through any semi-classical theory (which could alternatively explain the photoelectric effect). When Einstein received his Nobel Prize in 1921, it was not for his more difficult and mathematically laborious special and general relativity, but for the simple, yet totally revolutionary, suggestion of quantized light. Einstein's "light quanta" would not be called photons until 1925, but even in 1905 they represented the quintessential example of wave-particle duality. Electromagnetic radiation propagates following linear wave equations, but can only be emitted or absorbed as discrete elements, thus acting as a wave and a particle simultaneously.

Einstein's Explanation of the Photoelectric Effect

In 1905, Albert Einstein provided an explanation of the photoelectric effect, a hitherto troubling experiment that the wave theory of light seemed incapable of explaining. He did so by postulating the existence of photons, quanta of light energy with particulate qualities.

In the photoelectric effect, it was observed that shining a light on certain metals would lead to an electric current in a circuit. Presumably, the light was knocking electrons out of the metal, causing current to flow. However, using the case of potassium as an example, it was also observed that while a dim blue light was enough to cause a current, even the strongest, brightest red light available with the technology of the time caused no current at all. According to the classical theory of light and matter, the strength or amplitude of a light wave was in proportion to its brightness: a bright light should have been easily strong enough to create a large current. Yet, oddly, this was not so.

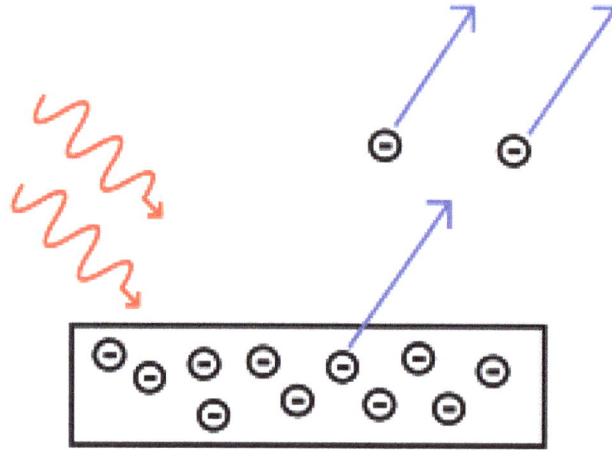

The photoelectric effect. Incoming photons on the left strike a metal plate (bottom), and eject electrons, depicted as flying off to the right.

Einstein explained this conundrum by postulating that the electrons can receive energy from electromagnetic field only in discrete portions (quanta that were called photons): an amount of energy E that was related to the frequency f of the light by

$$E = hf$$

where h is Planck's constant (6.626×10^{-34} J seconds). Only photons of a high enough frequency (above a certain *threshold* value) could knock an electron free. For example, photons of blue light had sufficient energy to free an electron from the metal, but photons of red light did not. One photon of light above the threshold frequency could release only one electron; the higher the frequency of a photon, the higher the kinetic energy of the emitted electron, but no amount of light (using technology available at the time) below the threshold frequency could release an electron. To "violate" this law would require extremely high-intensity lasers which had not yet been invented. Intensity-dependent phenomena have now been studied in detail with such lasers.

Einstein was awarded the Nobel Prize in Physics in 1921 for his discovery of the law of the photoelectric effect.

De Broglie's Wavelength

In 1924, Louis-Victor de Broglie formulated the de Broglie hypothesis, claiming that *all* matter, not just light, has a wave-like nature; he related wavelength (denoted as λ), and momentum (denoted as p):

$$\lambda = \frac{h}{p}$$

This is a generalization of Einstein's equation above, since the momentum of a photon is given by $p = \frac{E}{c}$ and the wavelength (in a vacuum) by $\lambda = \frac{c}{f}$, where c is the speed of light in vacuum.

$$\Psi = Ae^{i(px - \omega t)}$$

$$\Psi = \sum_{n} A_n e^{i(p_n x - \omega_n t)}$$

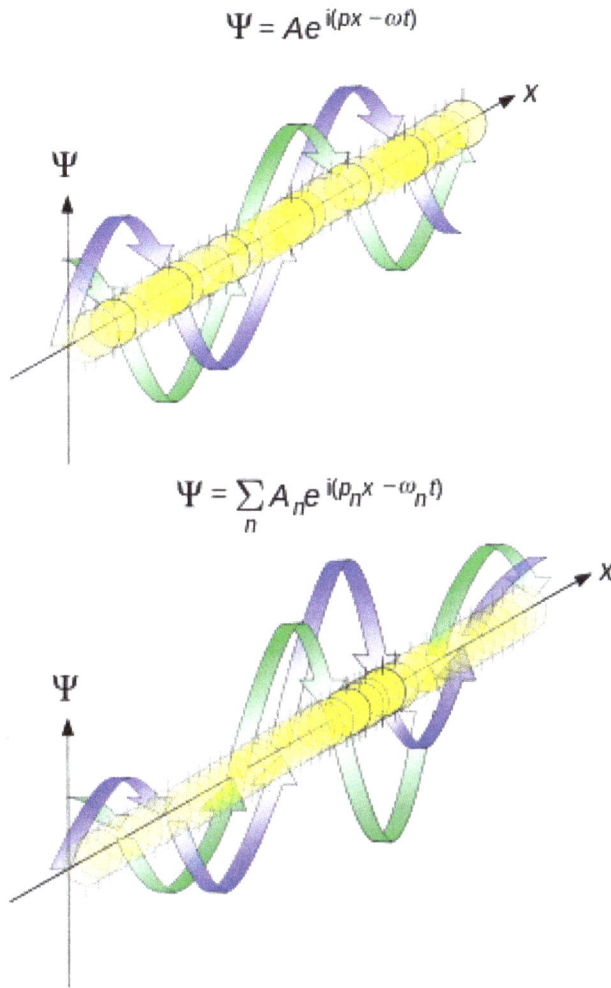

Propagation of de Broglie waves in 1d—real part of the complex amplitude is blue, imaginary part is green. The probability (shown as the colour opacity) of finding the particle at a given point x is spread out like a waveform; there is no definite position of the particle. As the amplitude increases above zero the curvature decreases, so the amplitude decreases again, and vice versa—the result is an alternating amplitude: a wave. Top: Plane wave. Bottom: Wave packet.

De Broglie's formula was confirmed three years later for electrons (which differ from photons in having a rest mass) with the observation of electron diffraction in two independent experiments. At the University of Aberdeen, George Paget Thomson passed a beam of electrons through a thin metal film and observed the predicted interference patterns. At Bell Labs, Clinton Joseph Davisson and Lester Halbert Germer guided their beam through a crystalline grid.

De Broglie was awarded the Nobel Prize for Physics in 1929 for his hypothesis. Thomson and Davisson shared the Nobel Prize for Physics in 1937 for their experimental work.

Heisenberg's uncertainty principle

In his work on formulating quantum mechanics, Werner Heisenberg postulated his uncertainty principle, which states:

$$\Delta x \Delta p \geq \frac{\hbar}{2}$$

where

Δ here indicates standard deviation, a measure of spread or uncertainty;

x and p are a particle's position and linear momentum respectively.

$\hbar$ is the reduced Planck's constant (Planck's constant divided by 2π).

Heisenberg originally explained this as a consequence of the process of measuring: Measuring position accurately would disturb momentum and vice versa, offering an example (the "gamma-ray microscope") that depended crucially on the de Broglie hypothesis. It is now thought, however, that this only partly explains the phenomenon, but that the uncertainty also exists in the particle itself, even before the measurement is made.

In fact, the modern explanation of the uncertainty principle, extending the Copenhagen interpretation first put forward by Bohr and Heisenberg, depends even more centrally on the wave nature of a particle: Just as it is nonsensical to discuss the precise location of a wave on a string, particles do not have perfectly precise positions; likewise, just as it is nonsensical to discuss the wavelength of a "pulse" wave traveling down a string, particles do not have perfectly precise momenta (which corresponds to the inverse of wavelength). Moreover, when position is relatively well defined, the wave is pulse-like and has a very ill-defined wavelength (and thus momentum). And conversely, when momentum (and thus wavelength) is relatively well defined, the wave looks long and sinusoidal, and therefore it has a very ill-defined position.

De Broglie–Bohm Theory

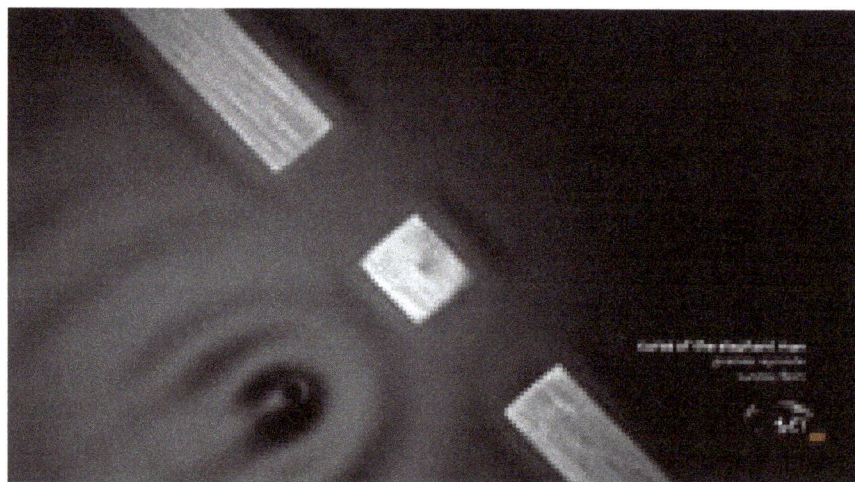

Couder experiments, "materializing" the *pilot wave* model.

De Broglie himself had proposed a pilot wave construct to explain the observed wave-particle duality. In this view, each particle has a well-defined position and momentum, but is guided by a wave function derived from Schrödinger's equation. The pilot wave theory was initially rejected because it generated non-local effects when applied to systems involving more than one particle. Non-locality, however, soon became established as an integral feature of quantum theory, and David Bohm extended de Broglie's model to explicitly include it.

In the resulting representation, also called the de Broglie–Bohm theory or Bohmian mechanics, the wave-particle duality vanishes, and explains the wave behaviour as a scattering with wave appearance, because the particle's motion is subject to a guiding equation or quantum potential.

"This idea seems to me so natural and simple, to resolve the wave-particle dilemma in such a clear and ordinary way, that it is a great mystery to me that it was so generally ignored", J.S. Bell.

The best illustration of the *pilot-wave model* was given by Couder's 2010 "walking droplets" experiments, demonstrating the pilot-wave behaviour in a macroscopic mechanical analog.

Wave Behavior of Large Objects

Since the demonstrations of wave-like properties in photons and electrons, similar experiments have been conducted with neutrons and protons. Among the most famous experiments are those of Estermann and Otto Stern in 1929. Authors of similar recent experiments with atoms and molecules, described below, claim that these larger particles also act like waves. A wave is basically a group of particles which moves in a particular form of motion, i.e. to and fro. If we break that flow by an object it will convert into radiants.

A dramatic series of experiments emphasizing the action of gravity in relation to wave–particle duality was conducted in the 1970s using the neutron interferometer. Neutrons, one of the components of the atomic nucleus, provide much of the mass of a nucleus and thus of ordinary matter. In the neutron interferometer, they act as quantum-mechanical waves directly subject to the force of gravity. While the results were not surprising since gravity was known to act on everything, including light, the self-interference of the quantum mechanical wave of a massive fermion in a gravitational field had never been experimentally confirmed before.

In 1999, the diffraction of C_{60} fullerenes by researchers from the University of Vienna was reported. Fullerenes are comparatively large and massive objects, having an atomic mass of about 720 u. The de Broglie wavelength of the incident beam was about 2.5 pm, whereas the diameter of the molecule is about 1 nm, about 400 times larger. In 2012, these far-field diffraction experiments could be extended to phthalocyanine molecules and their heavier derivatives, which are composed of 58 and 114 atoms respectively. In these experiments the build-up of such interference patterns could be recorded in real time and with single molecule sensitivity.

In 2003, the Vienna group also demonstrated the wave nature of tetraphenylporphyrin—a flat biodye with an extension of about 2 nm and a mass of 614 u. For this demonstration they employed a near-field Talbot Lau interferometer. In the same interferometer they also found interference fringes for $C_{60}F_{48}$, a fluorinated buckyball with a mass of about 1600 u, composed of 108 atoms. Large molecules are already so complex that they give experimental access to some aspects of the quantum-classical interface, i.e., to certain decoherence mechanisms. In 2011, the interference of molecules as heavy as 6910 u could be demonstrated in a Kapitza–Dirac–Talbot–Lau interferometer. In 2013, the interference of molecules beyond 10,000 u has been demonstrated.

Whether objects heavier than the Planck mass (about the weight of a large bacterium) have a de Broglie wavelength is theoretically unclear and experimentally unreachable; above the Planck mass a particle's Compton wavelength would be smaller than the Planck length and its own Schwarzschild radius, a scale at which current theories of physics may break down or need to be replaced by more general ones.

Recently Couder, Fort, *et al.* showed that we can use macroscopic oil droplets on a vibrating surface as a model of wave–particle duality—localized droplet creates periodical waves around and interaction with them leads to quantum-like phenomena: interference in double-slit experiment, unpredictable tunneling (depending in complicated way on practically hidden state of field), orbit quantization (that particle has to 'find a resonance' with field perturbations it creates—after one orbit, its internal phase has to return to the initial state) and Zeeman effect.

Treatment in Modern Quantum Mechanics

Wave–particle duality is deeply embedded into the foundations of quantum mechanics. In the formalism of the theory, all the information about a particle is encoded in its *wave function*, a complex-valued function roughly analogous to the amplitude of a wave at each point in space. This function evolves according to a differential equation (generically called the Schrödinger equation). For particles with mass this equation has solutions that follow the form of the wave equation. Propagation of such waves leads to wave-like phenomena such as interference and diffraction. Particles without mass, like photons, have no solutions of the Schrödinger equation so have another wave.

The particle-like behavior is most evident due to phenomena associated with measurement in quantum mechanics. Upon measuring the location of the particle, the particle will be forced into a more localized state as given by the uncertainty principle. When viewed through this formalism, the measurement of the wave function will randomly "collapse", or rather "decohere", to a sharply peaked function at some location. For particles with mass the likelihood of detecting the particle at any particular location is equal to the squared amplitude of the wave function there. The measurement will return a well-defined position, (subject to uncertainty), a property traditionally associated with particles. It is important to note that a measurement is only a particular type of interaction where some data is recorded and the measured quantity is forced into a particular eigenstate. The act of measurement is therefore not fundamentally different from any other interaction.

Following the development of quantum field theory the ambiguity disappeared. The field permits solutions that follow the wave equation, which are referred to as the wave functions. The term particle is used to label the irreducible representations of the Lorentz group that are permitted by the field. An interaction as in a Feynman diagram is accepted as a calculationally convenient approximation where the outgoing legs are known to be simplifications of the propagation and the internal lines are for some order in an expansion of the field interaction. Since the field is non-local and quantized, the phenomena which previously were thought of as paradoxes are explained. Within the limits of the wave-particle duality the quantum field theory gives the same results.

Visualization

There are two ways to visualize the wave-particle behaviour: by the "standard model", described below; and by the Broglie–Bohm model, where no duality is perceived.

Below is an illustration of wave–particle duality as it relates to De Broglie's hypothesis and Heisenberg's uncertainty principle (above), in terms of the position and momentum space wavefunctions for one spinless particle with mass in one dimension. These wavefunctions are Fourier transforms of each other.

The more localized the position-space wavefunction, the more likely the particle is to be found with the position coordinates in that region, and correspondingly the momentum-space wavefunction is less localized so the possible momentum components the particle could have are more widespread.

Conversely the more localized the momentum-space wavefunction, the more likely the particle is to be found with those values of momentum components in that region, and correspondingly the less localized the position-space wavefunction, so the position coordinates the particle could occupy are more widespread.

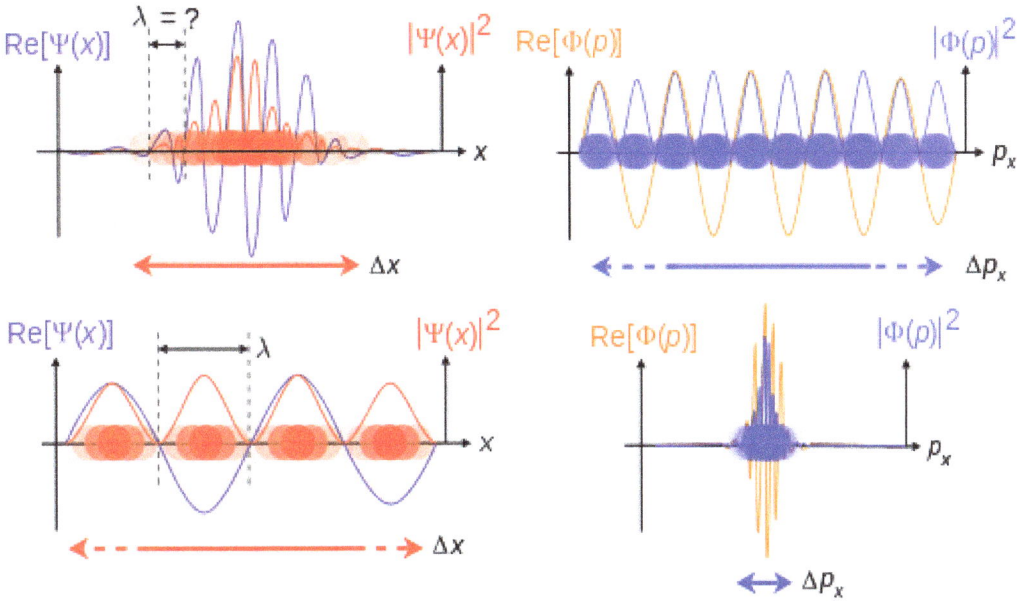

Position x and momentum p wavefunctions corresponding to quantum particles. The colour opacity (%) of the particles corresponds to the probability density of finding the particle with position x or momentum component p.

Top: If wavelength λ is unknown, so are momentum p, wave-vector k and energy E (de Broglie relations). As the particle is more localized in position space, Δx is smaller than for Δp_x. Bottom: If λ is known, so are p, k, and E. As the particle is more localized in momentum space, Δp is smaller than for Δx.

Alternative Views

Wave–particle duality is an ongoing conundrum in modern physics. Most physicists accept wave-particle duality as the best explanation for a broad range of observed phenomena; however, it is not without controversy. Alternative views are also presented here. These views are not generally accepted by mainstream physics, but serve as a basis for valuable discussion within the community.

Both-particle-and-wave View

The pilot wave model, originally developed by Louis de Broglie and further developed by David Bohm into the hidden variable theory proposes that there is no duality, but rather a system exhibits both particle properties and wave properties simultaneously, and particles are guided, in a deterministic fashion, by the pilot wave (or its "quantum potential") which will direct them to areas of constructive interference in preference to areas of destructive interference. This idea is held by a significant minority within the physics community.

At least one physicist considers the "wave-duality" as not being an incomprehensible mystery. L.E. Ballentine, *Quantum Mechanics, A Modern Development*, p. 4, explains:

When first discovered, particle diffraction was a source of great puzzlement. Are "particles" really "waves?" In the early experiments, the diffraction patterns were detected holistically by means of a photographic plate, which could not detect individual particles. As a result, the notion grew that particle and wave properties were mutually incompatible, or complementary, in the sense that different measurement apparatuses would be required to observe them. That idea, however, was only an unfortunate generalization from a technological limitation. Today it is possible to detect the arrival of individual electrons, and to see the diffraction pattern emerge as a statistical pattern made up of many small spots (Tonomura et al., 1989). Evidently, quantum particles are indeed particles, but whose behaviour is very different from classical physics would have us to expect.

It has been claimed that the Afshar experiment (2007) shows that it is possible to simultaneously observe both wave and particle properties of photons. This claim is, however, rejected by other scientists

Wave-only View

At least one scientist proposes that the duality can be replaced by a "wave-only" view. In his book *Collective Electrodynamics: Quantum Foundations of Electromagnetism* (2000), Carver Mead purports to analyze the behavior of electrons and photons purely in terms of electron wave functions, and attributes the apparent particle-like behavior to quantization effects and eigenstates. According to reviewer David Haddon:

Mead has cut the Gordian knot of quantum complementarity. He claims that atoms, with their neutrons, protons, and electrons, are not particles at all but pure waves of matter. Mead cites as the gross evidence of the exclusively wave nature of both light and matter the discovery between 1933 and 1996 of ten examples of pure wave phenomena, including the ubiquitous laser of CD players, the self-propagating electrical currents of superconductors, and the Bose–Einstein condensate of atoms.

Albert Einstein, who, in his search for a Unified Field Theory, did not accept wave-particle duality, wrote:

This double nature of radiation (and of material corpuscles)...has been interpreted by quantum-mechanics in an ingenious and amazingly successful fashion. This interpretation...appears to me as only a temporary way out...

The many-worlds interpretation (MWI) is sometimes presented as a waves-only theory, including by its originator, Hugh Everett who referred to MWI as "the wave interpretation".

The *Three Wave Hypothesis* of R. Horodecki relates the particle to wave. The hypothesis implies that a massive particle is an intrinsically spatially as well as temporally extended wave phenomenon by a nonlinear law.

Particle-only View

Still in the days of the old quantum theory, a pre-quantum-mechanical version of wave–particle duality was pioneered by William Duane, and developed by others including Alfred Landé. Duane explained diffraction of x-rays by a crystal in terms solely of their particle aspect. The deflection of the trajectory of each diffracted photon was explained as due to quantized momentum transfer from the spatially regular structure of the diffracting crystal.

Neither-wave-nor-particle View

It has been argued that there are never exact particles or waves, but only some compromise or intermediate between them. For this reason, in 1928 Arthur Eddington coined the name "*wavicle*" to describe the objects although it is not regularly used today. One consideration is that zero-dimensional mathematical points cannot be observed. Another is that the formal representation of such points, the Dirac delta function is unphysical, because it cannot be normalized. Parallel arguments apply to pure wave states. Roger Penrose states:

"Such 'position states' are idealized wavefunctions in the opposite sense from the momentum states. Whereas the momentum states are infinitely spread out, the position states are infinitely concentrated. Neither is normalizable [...]."

Relational Approach to Wave–particle Duality

Relational quantum mechanics is developed which regards the detection event as establishing a relationship between the quantized field and the detector. The inherent ambiguity associated with applying Heisenberg's uncertainty principle and thus wave–particle duality is subsequently avoided.

Applications

Although it is difficult to draw a line separating wave–particle duality from the rest of quantum mechanics, it is nevertheless possible to list some applications of this basic idea.

- Wave–particle duality is exploited in electron microscopy, where the small wavelengths associated with the electron can be used to view objects much smaller than what is visible using visible light.

- Similarly, neutron diffraction uses neutrons with a wavelength of about 0.1 nm, the typical spacing of atoms in a solid, to determine the structure of solids.

- Photos are now able to show this dual nature, perhaps this will lead to new ways of examining and recording this behaviour.

Optical Vortex

An optical vortex (also known as a screw dislocation or phase singularity) is a zero of an optical field, a point of zero intensity. Research into the properties of vortices has thrived since a comprehensive paper by John Nye and Michael Berry, in 1974, described the basic properties of "dislocations in wave trains". The research that followed became the core of what is now known as "singular optics".

Explanation

In an optical vortex, light is twisted like a corkscrew around its axis of travel. Because of the twisting, the light waves at the axis itself cancel each other out. When projected onto a flat surface, an optical vortex looks like a ring of light, with a dark hole in the center. This corkscrew of light, with darkness at the center, is called an optical vortex.

The vortex is given a number, called the topological charge, according to how many twists the light does in one wavelength. The number is always an integer, and can be positive or negative, depending on the direction of the twist. The higher the number of the twist, the faster the light is spinning around the axis. This spinning carries orbital angular momentum with the wave train, and will induce torque on an electric dipole.

This orbital angular momentum of light can be observed in the orbiting motion of trapped particles. Interfering an optical vortex with a plane wave of light reveals the spiral phase as concentric spirals. The number of arms in the spiral equals the topological charge.

Optical vortices are studied by creating them in the lab in various ways. They can be generated directly in a laser, or a laser beam can be twisted into vortex using any of several methods, such as computer-generated holograms, spiral-phase delay structures, or birefringent vortices in materials.

Properties

An optical singularity is a zero of an optical field. The phase in the field circulates around these points of zero intensity (giving rise to the name *vortex*). Vortices are points in 2D fields and lines in 3D fields (as they have codimension two). Integrating the phase of the field around a path enclosing a vortex yields an integer multiple of 2π. This integer is known as the topological charge, or strength, of the vortex.

A hypergeometric-Gaussian mode (HyGG) has an optical vortex in its center. The beam, which has the form

$$\psi \propto e^{im\phi} e^{-r^2},$$

is a solution to the paraxial wave equation consisting of the Bessel function. Photons in a hypergeometric-Gaussian beam have an orbital angular momentum of $m\hbar$. The integer m also gives the strength of the vortex at the beam's centre. Spin angular momentum of circularly polarized light can be converted into orbital angular momentum.

Creation

Several methods exist to create Hypergeometric-Gaussian modes, including with a spiral phase plate, computer-generated holograms, mode conversion, a q-plate, or a spatial light modulator.

- Static spiral phase plates (SPPs) are spiral-shaped pieces of crystal or plastic that are engineered specifically to the desired topological charge and incident wavelength. They are efficient, yet expensive. Adjustable SPPs can be made by moving a wedge between two sides of a cracked piece of plastic.

- Computer-generated holograms (CGHs) are the calculated interferogram between a plane wave and a Laguerre-Gaussian beam which is transferred to film. The CGH resembles a common Ronchi linear diffraction grating, save a "fork" dislocation. An incident laser beam creates a diffraction pattern with vortices whose topological charge increases with diffraction order. The zero order is Gaussian, and the vortices have opposite helicity on either side of this undiffracted beam. The number of prongs in the CGH fork is directly related to the topological charge of the first diffraction order vortex. The CGH can be blazed to direct more intensity into the first order. Bleaching transforms it from an intensity grating to a phase grating, which increases efficiency.

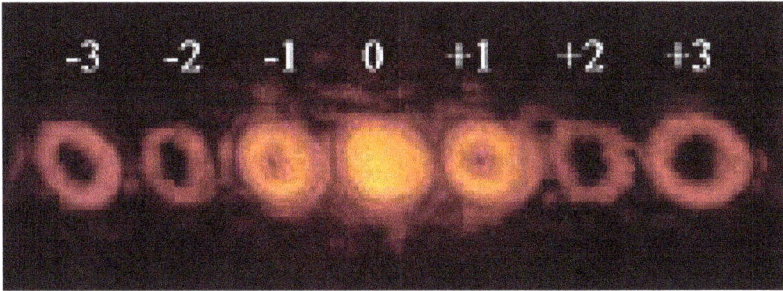

Vortices created by CGH

- Mode conversion requires Hermite-Gaussian (HG) modes, which can easily be made inside the laser cavity or externally by less accurate means. A pair of astigmatic lenses introduces a Gouy phase shift which creates an LG beam with azimuthal and radial indices dependent upon the input HG.

- A spatial light modulator is a computer-controlled electronic liquid-crystal device which can create dynamic vortices, arrays of vortices, and other types of beams by creating a hologram of varying refractive indices. This hologram may be a fork pattern, a spiral phase plate, or some similar pattern with non-zero topological charge.

- Deformable mirror made of segments can be used to dynamically (with a rate of up to a few kHz) create vortices, even if illuminated by high power lasers.

- A q-plate is a birefringent liquid crystal plate with an azimuthal distribution of the local optical axis, which has a topological charge q at its center defect. The q-plate with topological charge q can generate a $\pm 2q$ charge vortex based on the input beam polarization.

- An s-plate is a similar technology to a q-plate, using a high-intensity UV laser to permanently etch a birefringent pattern into silica glass with an azimuthal variation in the fast axis with topological charge of s. Unlike a q-plate, which may be wavelength tuned by adjusting the bias voltage on the liquid crystal, an s-plate only works for one wavelength of light.

- At radio frequencies it is trivial to produce a (non optical) vortex. Simply arrange a one wavelength or greater diameter ring of antennas such that the phase shift of the broadcast antennas varies an integral multiple of 2π around the ring.

Applications

There are a broad variety of applications of optical vortices in diverse areas of communications and imaging.

- Extrasolar planets have only recently been directly detected, as their parent star is so bright.

Progress has been made in creating an optical vortex coronagraph to directly observe planets with too low a contrast ratio to their parent to be observed with other techniques.

- Optical vortices are used in optical tweezers to manipulate micrometer-sized particles such as cells. Such particles can be rotated in orbits around the axis of the beam using OAM. Micro-motors have also been created using optical vortex tweezers.

- Optical vortices can significantly improve communication bandwidth. For instance, twisted radio beams could increase radio spectral efficiency by using the large number of vortical states. The amount of phase front 'twisting' indicates the orbital angular momentum state number, and beams with different orbital angular momentum are orthogonal. Such orbital angular momentum based multiplexing can potentially increase the system capacity and spectral efficiency of millimetre-wave wireless communication.

- Similarly, early experimental results for orbital angular momentum multiplexing in the optical domain have shown results over short distances, but longer distance demonstrations are still forthcoming. The main challenge that these demonstrations have faced is that conventional optical fibers change the spin angular momentum of vortices as they propagate, and may change the orbital angular momentum when bent or stressed. So far stable propagation of up to 50 meters has been demonstrated in specialty optical fibers.

- Current computers use electrons which have two states, zero and one. Quantum computing could use light to encode and store information. Optical vortices theoretically have an infinite number of states in free space, as there is no limit to the topological charge. This could allow for faster data manipulation. The cryptography community is also interested in optical vortices for the promise of higher bandwidth communication discussed above.

- In optical microscopy, optical vortices may be used to achieve spatial resolution beyond normal diffraction limits using a technique called Stimulated Emission Depletion (STED) Microscopy. This technique takes advantage of the low intensity at the singularity in the center of the beam to deplete the fluorophores around a desired area with a high-intensity optical vortex beam without depleting fluorophores in the desired target area.

Focus (Optics)

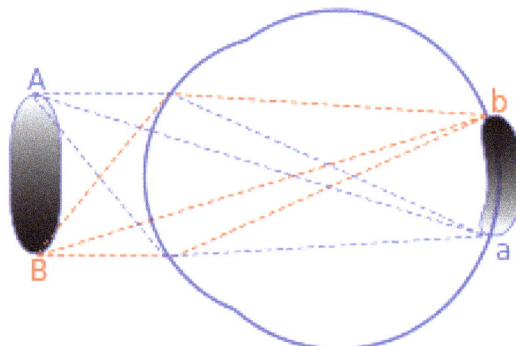

Eye focusing ideally collects all light rays from a point on an object into a corresponding point on the retina.

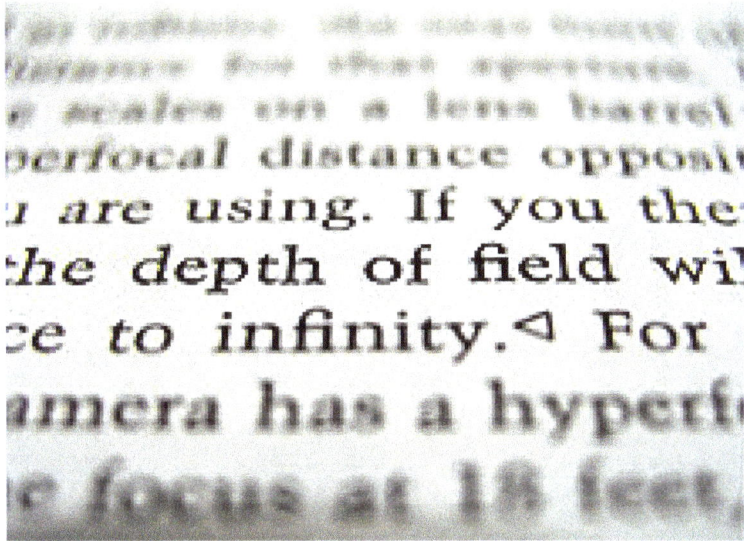

An image that is partially in focus, but mostly out of focus in varying degrees.

In geometrical optics, a focus, also called an image point, is the point where light rays originating from a point on the object converge. Although the focus is conceptually a point, physically the focus has a spatial extent, called the *blur circle*. This non-ideal focusing may be caused by aberrations of the imaging optics. In the absence of significant aberrations, the smallest possible blur circle is the Airy disc, which is caused by diffraction from the optical system's aperture. Aberrations tend to get worse as the aperture diameter increases, while the Airy circle is smallest for large apertures.

An image, or image point or region, is *in focus* if light from object points is converged almost as much as possible in the image, and *out of focus* if light is not well converged. The border between these is sometimes defined using a circle of confusion criterion.

A principal focus or focal point is a special focus:

- For a lens, or a spherical or parabolic mirror, it is a point onto which collimated light parallel to the axis is focused. Since light can pass through a lens in either direction, a lens has two focal points—one on each side. The distance in air from the lens or mirror's principal plane to the focus is called the *focal length*.

- Elliptical mirrors have two focal points: light that passes through one of these before striking the mirror is reflected such that it passes through the other.

- The focus of a hyperbolic mirror is either of two points which have the property that light from one is reflected as if it came from the other.

Diverging (negative) lenses and convex mirrors do not focus a collimated beam to a point. Instead, the focus is the point from which the light appears to be emanating, after it travels through the lens or reflects from the mirror. A convex parabolic mirror will reflect a beam of collimated light to make it appear as if it were radiating from the focal point, or conversely, reflect rays directed toward the focus as a collimated beam. A convex elliptical mirror will reflect light directed towards one focus as if it were radiating from the other focus, both of which are behind the mirror. A convex hyperbolic mirror will reflect rays emanating from the focal point in front of the mirror as if they were emanating from the focal point behind the mirror. Conversely, it can focus rays directed at the focal point that is behind the mirror towards the focal point that is in front of the mirror as in a Cassegrain telescope.

Focal blur is simulated in this computer generated image of glasses, which was rendered in POV-Ray.

Dispersion (Optics)

In a dispersive prism, material dispersion (a wavelength-dependent refractive index) causes different colors to refract at different angles, splitting white light into a rainbow.

A compact fluorescent lamp seen through an Amici prism.

In optics, dispersion is the phenomenon in which the phase velocity of a wave depends on its frequency. Media having this common property may be termed *dispersive media*. Sometimes the term *chromatic* dispersion is used for specificity. Although the term is used in the field of optics to describe light and other electromagnetic waves, dispersion in the same sense can apply to any sort of wave motion such as acoustic dispersion in the case of sound and seismic waves, in gravity waves (ocean waves), and for telecommunication signals propagating along transmission lines (such as coaxial cable) or optical fiber.

In optics, one important and familiar consequence of dispersion is the change in the angle of refraction of different colors of light, as seen in the spectrum produced by a dispersive prism and in chromatic aberration of lenses. Design of compound achromatic lenses, in which chromatic aberration is largely cancelled, uses a quantification of a glass's dispersion given by its Abbe number V, where *lower* Abbe numbers correspond to *greater* dispersion over the visible spectrum. In some applications such as telecommunications, the absolute phase of a wave is often not important but only the propagation of wave packets or "pulses"; in that case one is interested only in variations of group velocity with frequency, so-called group-velocity dispersion (GVD).

Examples of Dispersion

The most familiar example of dispersion is probably a rainbow, in which dispersion causes the spatial separation of a white light into components of different wavelengths (different colors). However, dispersion also has an effect in many other circumstances: for example, group velocity dispersion (GVD) causes pulses to spread in optical fibers, degrading signals over long distances; also, a cancellation between group-velocity dispersion and nonlinear effects leads to soliton waves.

Material and waveguide dispersion

Most often, chromatic dispersion refers to bulk material dispersion, that is, the change in refractive index with optical frequency. However, in a waveguide there is also the phenomenon of *waveguide dispersion*, in which case a wave's phase velocity in a structure depends on its frequency simply due to the structure's geometry. More generally, "waveguide" dispersion can occur for waves propagating through any inhomogeneous structure (e.g., a photonic crystal), whether or not the waves are confined to some region. In a waveguide, *both* types of dispersion will generally be present, although they are not strictly additive. For example, in fiber optics the material and waveguide dispersion can effectively cancel each other out to produce a Zero-dispersion wavelength, important for fast Fiber-optic communication.

Material Dispersion in Optics

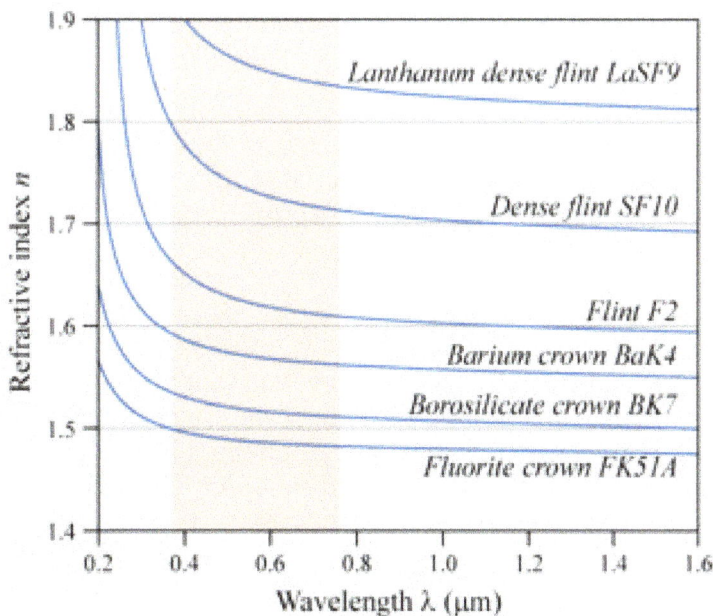

The variation of refractive index vs. vacuum wavelength for various glasses. The wavelengths of visible l
ight are shaded in red.

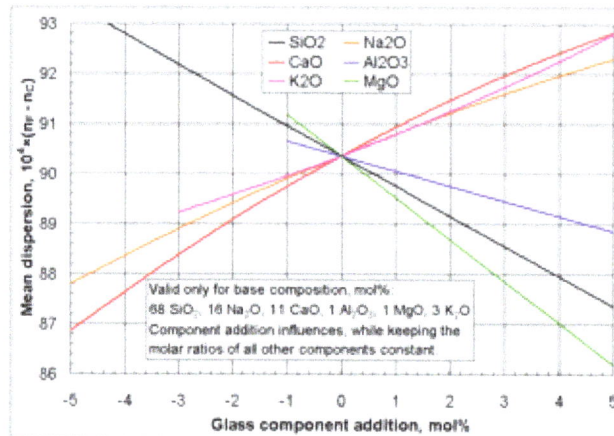

Influences of selected glass component additions on the mean dispersion of a specific base glass (n_F valid for $\lambda = 486$ nm (blue), n_C valid for $\lambda = 656$ nm (red))

Material dispersion can be a desirable or undesirable effect in optical applications. The dispersion of light by glass prisms is used to construct spectrometers and spectroradiometers. Holographic gratings are also used, as they allow more accurate discrimination of wavelengths. However, in lenses, dispersion causes chromatic aberration, an undesired effect that may degrade images in microscopes, telescopes and photographic objectives.

The *phase velocity*, v, of a wave in a given uniform medium is given by

$$v = \frac{c}{n}$$

where c is the speed of light in a vacuum and n is the refractive index of the medium.

In general, the refractive index is some function of the frequency f of the light, thus $n = n(f)$, or alternatively, with respect to the wave's wavelength $n = n(\lambda)$. The wavelength dependence of a material's refractive index is usually quantified by its Abbe number or its coefficients in an empirical formula such as the Cauchy or Sellmeier equations.

Because of the Kramers–Kronig relations, the wavelength dependence of the real part of the refractive index is related to the material absorption, described by the imaginary part of the refractive index (also called the extinction coefficient). In particular, for non-magnetic materials ($\mu = \mu_0$), the susceptibility χ that appears in the Kramers–Kronig relations is the electric susceptibility $\chi_e = n^2 - 1$.

The most commonly seen consequence of dispersion in optics is the separation of white light into a color spectrum by a prism. From Snell's law it can be seen that the angle of refraction of light in a prism depends on the refractive index of the prism material. Since that refractive index varies with wavelength, it follows that the angle that the light is refracted by will also vary with wavelength, causing an angular separation of the colors known as *angular dispersion*.

For visible light, refraction indices n of most transparent materials (e.g., air, glasses) decrease with increasing wavelength λ:

$$1 < n(\lambda_{red}) < n(\lambda_{yellow}) < n(\lambda_{blue}),$$

or alternatively:

$$\frac{dn}{d\lambda} < 0.$$

In this case, the medium is said to have *normal dispersion*. Whereas, if the index increases with increasing wavelength (which is typically the case for X-rays), the medium is said to have *anomalous dispersion*.

At the interface of such a material with air or vacuum (index of ~1), Snell's law predicts that light incident at an angle θ to the normal will be refracted at an angle arcsin(sin θ/n). Thus, blue light, with a higher refractive index, will be bent more strongly than red light, resulting in the well-known rainbow pattern.

Group and Phase Velocity

Another consequence of dispersion manifests itself as a temporal effect. The formula $v = c/n$ calculates the *phase velocity* of a wave; this is the velocity at which the *phase* of any one frequency component of the wave will propagate. This is not the same as the *group velocity* of the wave, that is the rate at which changes in amplitude (known as the *envelope* of the wave) will propagate. For a homogeneous medium, the group velocity v_g is related to the phase velocity v by (here λ is the wavelength in vacuum, not in the medium):

$$v_g = \frac{c}{n - \lambda \dfrac{dn}{d\lambda}} = \frac{v}{1 - \dfrac{\lambda}{n}\dfrac{dn}{d\lambda}}.$$

The group velocity v_g is often thought of as the velocity at which energy or information is conveyed along the wave. In most cases this is true, and the group velocity can be thought of as the *signal velocity* of the waveform. In some unusual circumstances, called cases of anomalous dispersion, the rate of change of the index of refraction with respect to the wavelength changes sign (becoming negative), in which case it is possible for the group velocity to exceed the speed of light ($v_g > c$). Anomalous dispersion occurs, for instance, where the wavelength of the light is close to an absorption resonance of the medium. When the dispersion is anomalous, however, group velocity is no longer an indicator of signal velocity. Instead, a signal travels at the speed of the wavefront, which is c irrespective of the index of refraction. Recently, it has become possible to create gases in which the group velocity is not only larger than the speed of light, but even negative. In these cases, a pulse can appear to exit a medium before it enters. Even in these cases, however, a signal travels at, or less than, the speed of light, as demonstrated by Stenner, et al.

The group velocity itself is usually a function of the wave's frequency. This results in group velocity dispersion (GVD), which causes a short pulse of light to spread in time as a result of different frequency components of the pulse travelling at different velocities. GVD is often quantified as the *group delay dispersion parameter* (again, this formula is for a uniform medium only):

$$D = -\frac{\lambda}{c}\frac{d^2 n}{d\lambda^2}.$$

If D is less than zero, the medium is said to have *positive dispersion*. If D is greater than zero, the medium has *negative dispersion*. If a light pulse is propagated through a normally dispersive medium, the result is the higher frequency components travel slower than the lower frequency components. The pulse therefore becomes *positively chirped*, or *up-chirped*, increasing in frequency with time. Conversely, if a pulse travels through an anomalously dispersive medium, high frequency components travel faster than the lower ones, and the pulse becomes *negatively chirped*, or *down-chirped*, decreasing in frequency with time.

The result of GVD, whether negative or positive, is ultimately temporal spreading of the pulse. This makes dispersion management extremely important in optical communications systems based on optical fiber, since if dispersion is too high, a group of pulses representing a bit-stream will spread in time and merge, rendering the bit-stream unintelligible. This limits the length of fiber that a signal can be sent down without regeneration. One possible answer to this problem is to send signals down the optical fibre at a wavelength where the GVD is zero (e.g., around 1.3–1.5 µm in silica fibres), so pulses at this wavelength suffer minimal spreading from dispersion. In practice, however, this approach causes more problems than it solves because zero GVD unacceptably amplifies other nonlinear effects (such as four wave mixing). Another possible option is to use soliton pulses in the regime of anomalous dispersion, a form of optical pulse which uses a nonlinear optical effect to self-maintain its shape. Solitons have the practical problem, however, that they require a certain power level to be maintained in the pulse for the nonlinear effect to be of the correct strength. Instead, the solution that is currently used in practice is to perform dispersion compensation, typically by matching the fiber with another fiber of opposite-sign dispersion so that the dispersion effects cancel; such compensation is ultimately limited by nonlinear effects such as self-phase modulation, which interact with dispersion to make it very difficult to undo.

Dispersion control is also important in lasers that produce short pulses. The overall dispersion of the optical resonator is a major factor in determining the duration of the pulses emitted by the laser. A pair of prisms can be arranged to produce net negative dispersion, which can be used to balance the usually positive dispersion of the laser medium. Diffraction gratings can also be used to produce dispersive effects; these are often used in high-power laser amplifier systems. Recently, an alternative to prisms and gratings has been developed: chirped mirrors. These dielectric mirrors are coated so that different wavelengths have different penetration lengths, and therefore different group delays. The coating layers can be tailored to achieve a net negative dispersion.

Dispersion in Waveguides

Waveguides are highly dispersive due to their geometry (rather than just to their material composition). Optical fibers are a sort of waveguide for optical frequencies (light) widely used in modern telecommunications systems. The rate at which data can be transported on a single fiber is limited by pulse broadening due to chromatic dispersion among other phenomena.

In general, for a waveguide mode with an angular frequency $\omega(\beta)$ at a propagation constant β (so that the electromagnetic fields in the propagation direction z oscillate proportional to $e^{i(\beta z - \omega t)}$), the group-velocity dispersion parameter D is defined as:

$$D = -\frac{2\pi c}{\lambda^2}\frac{d^2\beta}{d\omega^2} = \frac{2\pi c}{v_g^2 \lambda^2}\frac{dv_g}{d\omega}$$

where $\lambda = 2\pi c/\omega$ is the vacuum wavelength and $v_g = d\omega/d\beta$ is the group velocity. This formula generalizes the one in the previous section for homogeneous media, and includes both waveguide dispersion and material dispersion. The reason for defining the dispersion in this way is that $|D|$ is the (asymptotic) temporal pulse spreading Δt per unit bandwidth $\Delta\lambda$ per unit distance travelled, commonly reported in ps / nm km for optical fibers.

In the case of multi-mode optical fibers, so-called modal dispersion will also lead to pulse broadening. Even in single-mode fibers, pulse broadening can occur as a result of polarization mode dispersion (since there are still two polarization modes). These are *not* examples of chromatic dispersion as they are not dependent on the wavelength or bandwidth of the pulses propagated.

Higher-order Dispersion Over Broad Bandwidths

When a broad range of frequencies (a broad bandwidth) is present in a single wavepacket, such as in an ultrashort pulse or a chirped pulse or other forms of spread spectrum transmission, it may not be accurate to approximate the dispersion by a constant over the entire bandwidth, and more complex calculations are required to compute effects such as pulse spreading.

In particular, the dispersion parameter D defined above is obtained from only one derivative of the group velocity. Higher derivatives are known as *higher-order dispersion*. These terms are simply a Taylor series expansion of the dispersion relation $\beta(\omega)$ of the medium or waveguide around some particular frequency. Their effects can be computed via numerical evaluation of Fourier transforms of the waveform, via integration of higher-order slowly varying envelope approximations, by a split-step method (which can use the exact dispersion relation rather than a Taylor series), or by direct simulation of the full Maxwell's equations rather than an approximate envelope equation.

Dispersion in Gemology

In the technical terminology of gemology, *dispersion* is the difference in the refractive index of a material at the B and G (686.7 nm and 430.8 nm) or C and F (656.3 nm and 486.1 nm) Fraunhofer wavelengths, and is meant to express the degree to which a prism cut from the gemstone demonstrates "fire." Fire is a colloquial term used by gemologists to describe a gemstone's dispersive nature or lack thereof. Dispersion is a material property. The amount of fire demonstrated by a given gemstone is a function the gemstone's facet angles, the polish quality, the lighting environment, the material's refractive index, the saturation of color, and the orientation of the viewer relative to the gemstone.

Dispersion in Imaging

In photographic and microscopic lenses, dispersion causes chromatic aberration, which causes the different colors in the image not to overlap properly. Various techniques have been developed to counteract this, such as the use of achromats, multielement lenses with glasses of different dispersion. They are constructed in such a way that the chromatic aberrations of the different parts cancel out.

Dispersion in Pulsar Timing

Pulsars are spinning neutron stars that emit pulses at very regular intervals ranging from milliseconds to seconds. Astronomers believe that the pulses are emitted simultaneously over a wide range of frequencies. However, as observed on Earth, the components of each pulse emitted at higher radio frequencies arrive before those emitted at lower frequencies. This dispersion occurs because of the ionized component of the interstellar medium, mainly the free electrons, which make the group velocity frequency dependent. The extra delay added at a frequency v is

$$t = k_{DM} \times \left(\frac{DM}{v^2} \right)$$

where the dispersion constant k_{DM} is given by

$$k_{DM} = \frac{e^2}{2\pi m_e c} \simeq 4.149 \, \text{GHz}^2 \, \text{pc}^{-1} \, \text{cm}^3 \, \text{ms,}^{[1]}$$

and the dispersion measure (DM) is the column density of free electrons (total electron content) — i.e. the number density of electrons n_e (electrons/cm³) integrated along the path traveled by the photon from the pulsar to the Earth — and is given by

$$DM = \int_0^d n_e \, dl$$

with units of parsecs per cubic centimetre (1 pc/cm³ = 30.857 × 10²¹ m⁻²).

Typically for astronomical observations, this delay cannot be measured directly, since the emission time is unknown. What *can* be measured is the difference in arrival times at two different frequencies. The delay Δt between a high frequency ν_{hi} and a low frequency ν_{lo} component of a pulse will be

$$\Delta t = k_{DM} \times DM \times \left(\frac{1}{\nu_{lo}^2} - \frac{1}{\nu_{hi}^2} \right)$$

Rewriting the above equation in terms of Δt allows one to determine the DM by measuring pulse arrival times at multiple frequencies. This in turn can be used to study the interstellar medium, as well as allow for observations of pulsars at different frequencies to be combined.

Opacity (Optics)

Opacity is the measure of impenetrability to electromagnetic or other kinds of radiation, especially visible light. In radiative transfer, it describes the absorption and scattering of radiation in a medium, such as a plasma, dielectric, shielding material, glass, etc. An opaque object is neither transparent (allowing all light to pass through) nor translucent (allowing some light to pass through). When light strikes an interface between two substances, in general some may be reflected, some absorbed, some scattered, and the rest transmitted. Reflection can be diffuse, for example light reflecting off a white wall, or specular, for example light reflecting off a mirror. An opaque substance transmits no light, and therefore reflects, scatters, or absorbs all of it. Both mirrors and carbon black are opaque. Opacity depends on the frequency of the light being considered. For instance, some kinds of glass, while transparent in the visual range, are largely opaque to ultraviolet light. More extreme frequency-dependence is visible in the absorption lines of cold gases. Opacity can be quantified in many ways; for example, the article mathematical descriptions of opacity.

Different processes can lead to opacity including absorption, reflection, and scattering.

Radiopacity

Radiopacity is preferentially used to describe opacity of X-rays. In modern medicine, radiodense substances are those that will not allow X-rays or similar radiation to pass. Radiographic imaging has been revolutionized by radiodense contrast media, which can be passed through the bloodstream, the gastrointestinal tract, or into the cerebral spinal fluid and utilized to highlight CT scan or X-ray images. Radiopacity is one of the key considerations in the design of various devices such as guidewires or stents that are used during radiological intervention. The radiopacity of a given endovascular device is important since it allows the device to be tracked during the interventional procedure.

Quantitative Definition

The words "opacity" and "opaque" are often used as colloquial terms for objects or media with the properties described above. However, there is also a specific, quantitative definition of "opacity", used in astronomy, plasma physics, and other fields, given here.

In this use, "opacity" is another term for the mass attenuation coefficient (or, depending on context, mass absorption coefficient, the difference is described here) κ_ν at a particular frequency ν of electromagnetic radiation.

More specifically, if a beam of light with frequency ν travels through a medium with opacity κ_ν and mass density ρ, both constant, then the intensity will be reduced with distance x according to the formula

$$I(x) = I_0 e^{-\kappa_\nu \rho x}$$

where

- x is the distance the light has traveled through the medium
- $I(x)$ is the intensity of light remaining at distance x
- I_0 is the initial intensity of light, at $x = 0$

For a given medium at a given frequency, the opacity has a numerical value that may range between 0 and infinity, with units of length²/mass.

Planck and Rosseland opacities

It is customary to define the average opacity, calculated using a certain weighting scheme. Planck opacity uses normalized Planck black body radiation energy density distribution, $B_\nu(T)$, as the weighting function, and averages κ_ν directly:

$$\kappa_{Pl} = \frac{\int_0^\infty \kappa_\nu B_\nu(T) d\nu}{\int_0^\infty B_\nu(T) d\nu} = \left(\frac{\pi}{\sigma T^4}\right) \int_0^\infty \kappa_\nu B_\nu(T) d\nu,$$

where σ is the Stefan-Boltzmann constant.

Rosseland opacity (after Svein Rosseland), on the other hand, uses a temperature derivative of

Planck distribution, $u(v,T) = \partial B_v(T)/\partial T$, as the weighting function, and averages κ_v^{-1},

$$\frac{1}{\kappa} = \frac{\int_0^\infty \kappa_v^{-1} u(v,T)dv}{\int_0^\infty u(v,T)dv}.$$

The photon mean free path is $\lambda_v = (\kappa_v \rho)^{-1}$. The Rosseland opacity is derived in the diffusion approximation to the radiative transport equation. It is valid whenever the radiation field is isotropic over distances comparable to or less than a radiation mean free path, such as in local thermal equilibrium. In practice, the mean opacity for Thomson electron scattering is:

$$\kappa_{es} = 0.20(1+X)\,\mathrm{cm^2\,g^{-1}}$$

where X is the hydrogen mass fraction. For nonrelativistic thermal bremsstrahlung, or free-free transitions, assuming solar metallicity, it is:

$$\kappa_{ff}(\rho,T) = 0.64 \times 10^{23} (\rho[\mathrm{g\,cm^{-3}}])(T[\mathrm{K}])^{-7/2}\,\mathrm{cm^2\,g^{-1}}.$$

The Rosseland mean attenuation coefficient is:

$$\frac{1}{\kappa} = \frac{\int_0^\infty (\kappa_{v,es} + \kappa_{v,ff})^{-1} u(v,T)dv}{\int_0^\infty u(v,T)dv}.\,[\,.$$

Fourier optics

Fourier optics is the study of classical optics using Fourier transforms, (FT) in which the wave is regarded as a superposition of plane waves that are not related to *any* identifiable sources; instead they are the natural modes of the propagation medium itself. Fourier optics can be seen as the dual of the Huygens–Fresnel principle, in which the wave is regarded as a superposition of expanding spherical waves which radiate outward from actual (physically identifiable) current sources via a Green's function relationship.

A curved phasefront may be synthesized from an infinite number of these "natural modes" i.e., from plane wave phasefronts oriented in different directions in space. Far from its sources, an expanding spherical wave is locally tangent to a planar phase front (a single plane wave out of the infinite spectrum), which is transverse to the radial direction of propagation. In this case, a Fraunhofer diffraction pattern is created, which emanates from a single spherical wave phase center. In the near field, no single well-defined spherical wave phase center exists, so the wavefront isn't locally tangent to a spherical ball. In this case, a Fresnel diffraction pattern would be created, which emanates from an *extended* source, consisting of a distribution of (physically identifiable) spherical wave sources in space. In the near field, a full spectrum of plane waves is necessary to represent the Fresnel near-field wave, *even locally*. A "wide" wave moving forward (like an expanding ocean wave coming toward the shore) can be regarded as an infinite number of "plane wave modes", all of which could (when they collide with something in the way) scatter independently of one other. These mathematical simplifications and calculations are the realm of Fourier analysis and synthesis – together, they can describe what happens when light passes through various slits, lenses or mirrors curved one way or the other, or is fully or partially reflected. Fourier optics forms much of the theory behind image processing techniques, as well as finding applications where information needs to be extracted from optical sources such as in quantum optics. To put it in a slightly more complex way, similar to the concept of *frequency* and *time* used in traditional Fourier transform theory, Fourier optics makes use of the spatial frequency domain (k_x, k_y) as the conjugate of the spatial (x, y) domain. Terms and concepts such as transform theory, spectrum, bandwidth, window functions and sampling from one-dimensional signal processing are commonly used.

Propagation of Light in Homogeneous, Source-free Media

Light can be described as a waveform propagating through free space (vacuum) or a material medium (such as air or glass). Mathematically, the (real valued) amplitude of one wave component is represented by a scalar wave function u that depends on both space and time:

$$u = u(\mathbf{r}, t)$$

where

$$\mathbf{r} = (x, y, z)$$

represents position in three dimensional space, and t represents time.

The wave equation

Fourier optics begins with the homogeneous, scalar wave equation (valid in source-free regions):

$$\left(\nabla^2 - \frac{1}{c^2} \frac{\partial^2}{\partial t^2} \right) u(\mathbf{r}, t) = 0.$$

where $u(r,t)$ is a real valued Cartesian component of an electromagnetic wave propagating through free space.

Sinusoidal Steady State

If light of a fixed frequency/wavelength/color (as from a laser) is assumed, then the time-harmonic form of the optical field is given as:

$$u(\mathbf{r}, t) = \text{Re}\left\{ \psi(\mathbf{r}) e^{j\omega t} \right\}.$$

where j is the imaginary unit,

$$\omega = 2\pi f$$

is the angular frequency (in radians per unit time) of the light waves, and

$$\psi(\mathbf{r}) = a(\mathbf{r}) e^{j\phi(\mathbf{r})}$$

is, in general, a complex quantity, with separate amplitude a and phase ϕ.

The Helmholtz equation

Substituting this expression into the wave equation yields the time-independent form of the wave equation, also known as the Helmholtz equation:

$$\left(\nabla^2 + k^2\right)\psi(\mathbf{r}) = 0$$

where

$$k = \frac{\omega}{c} = \frac{2\pi}{\lambda}$$

is the wave number, $\psi(\mathbf{r})$ is the time-independent, complex-valued component of the propagating wave. Note that the propagation constant, k, and the frequency, ω, are linearly related to one another, a typical characteristic of transverse electromagnetic (TEM) waves in homogeneous media.

Solving the Helmholtz Equation

Solutions to the Helmholtz equation may readily be found in rectangular coordinates via the principle of separation of variables for partial differential equations. This principle says that in separable orthogonal coordinates, an *elementary product solution* to this wave equation may be constructed of the following form:

$$\psi(x, y, z) = f_x(x) \times f_y(y) \times f_z(z)$$

i.e., as the product of a function of x, times a function of y, times a function of z. If this *elementary product solution* is substituted into the wave equation (2.0), using the scalar Laplacian in rectangular coordinates:

$$\nabla^2 \psi = \frac{\partial^2 \psi}{\partial x^2} + \frac{\partial^2 \psi}{\partial y^2} + \frac{\partial^2 \psi}{\partial z^2}$$

then the following equation for the 3 individual functions is obtained

$$f_x''(x)f_y(y)f_z(z) + f_x(x)f_y''(y)f_z(z) + f_x(x)f_y(y)f_z''(z) + k^2 f_x(x)f_y(y)f_z(z) = 0$$

which is readliy rearranged into the form:

$$\frac{f_x''(x)}{f_x(x)} + \frac{f_y''(y)}{f_y(y)} + \frac{f_z''(z)}{f_z(z)} + k^2 = 0$$

It may now be argued that each of the quotients in the equation above must, of necessity, be constant. For, say the first quotient is not constant, and is a function of x. None of the other terms in the equation has any dependence on the variable x. Therefore, the first term may not have any x-dependence either; it must be constant. The constant is denoted as $-k_x^2$. Reasoning in a similar way for the y and z quotients, three ordinary differential equations are obtained for the f_x, f_y and f_z, along with one *separation condition*:

$$\frac{d^2}{dx^2} f_x(x) + k_x^2 f_x(x) = 0$$

$$\frac{d^2}{dy^2} f_y(y) + k_y^2 f_y(y) = 0$$

$$\frac{d^2}{dz^2} f_z(z) + k_z^2 f_z(z) = 0$$

$$k_x^2 + k_y^2 + k_z^2 = k^2$$

Each of these 3 differential equations has the same solution: sines, cosines or complex exponentials. We'll go with the complex exponential for notational simplicity, compatibility with usual FT notation, and the fact that a two-sided integral of complex exponentials picks up both the sine and cosine contributions. As a result, the elementary product solution for E_u is:

$$\psi(x,y,z) = e^{j(k_x x + k_y y + k_z z)}$$

$$= e^{j(k_x x + k_y y)} e^{jk_z z}$$

$$= e^{j(k_x x + k_y y)} e^{\pm jz\sqrt{k^2 - k_x^2 - k_y^2}}$$

which represents a propagating or exponentially decaying uniform plane wave solution to the homogeneous wave equation. The - sign is used for a wave propagating/decaying in the +z direction and the + sign is used for a wave propagating/decaying in the -z direction (this follows the engineering time convention, which assumes an $e^{j\omega t}$ time dependence). This field represents a propagating plane wave when the quantity under the radical is positive, and an exponentially decaying wave when it is negative (in passive media, the root with a non-positive imaginary part must always be chosen, to represent uniform propagation or decay, but not amplification).

Product solutions to the Helmholtz equation are also readily obtained in cylindrical and spherical coordinates, yielding cylindrical and spherical harmonics (with the remaining separable coordinate systems being used much less frequently).

The complete solution: the superposition integral

A general solution to the homogeneous electromagnetic wave equation in rectangular coordinates may be formed as a weighted superposition of all possible elementary plane wave solutions as:

$$\psi(x,y,z) = \int_{-\infty}^{+\infty} \int_{-\infty}^{+\infty} \Psi_0(k_x, k_y) e^{j(k_x x + k_y y)} \, e^{\pm jz\sqrt{k^2 - k_x^2 - k_y^2}} \, dk_x dk_y \qquad (2.1)$$

Next, let

$$\psi_0(x,y) = \psi(x,y,z)|_{z=0} \,.$$

Then:

$$\psi_0(x,y) = \int_{-\infty}^{+\infty} \int_{-\infty}^{+\infty} \Psi_0(k_x, k_y) e^{j(k_x x + k_y y)} \, dk_x dk_y$$

This plane wave spectrum representation of the electromagnetic field is the basic foundation of Fourier optics (this point cannot be emphasized strongly enough), because when z=0, the equation above simply becomes a Fourier transform (FT) relationship between the field and its plane wave content (hence the name, "Fourier optics").

Thus:

$$\Psi_0(k_x, k_y) = \mathcal{F}\{\psi_0(x,y)\}$$

and

$$\psi_0(x,y) = \mathcal{F}^{-1}\{\Psi_0(k_x, k_y)\}$$

All spatial dependence of the individual plane wave components is described explicitly via the exponential functions. The coefficients of the exponentials are only functions of spatial wavenumber k_x, k_y, just as in ordinary Fourier analysis and Fourier transforms.

The Diffraction Limit

When

$$k_x^2 + k_y^2 > k^2$$

the plane waves are evanescent (decaying), so that any spatial frequency content in an object plane transparency which is finer than one wavelength will not be transferred over to the image plane, simply because the plane waves corresponding to that content cannot propagate. In connection with lithography of electronic components, this phenomenon is known as the diffraction limit and is the reason why light of progressively higher frequency (smaller wavelength, thus larger k) is required for etching progressively finer features in integrated circuits.

The Paraxial Approximation

Paraxial plane waves (Optic axis is assumed z-directed)

As shown above, an elementary product solution to the Helmholtz equation takes the form:

$$\psi(\mathbf{r}) = A(\mathbf{r})e^{-j\mathbf{k}\cdot\mathbf{r}}$$

where

$$\mathbf{k} = k_x\mathbf{x} + k_y\mathbf{y} + k_z\mathbf{z}$$

is the wave vector, and

$$k = \|\mathbf{k}\| = \sqrt{k_x^2 + k_y^2 + k_z^2} = \frac{\omega}{c}$$

is the wave number. Next, using the paraxial approximation, it is assumed that

$$k_x^2 + k_y^2 \ll k_z^2$$

or equivalently,

$$\sin\theta \approx \theta$$

where θ is the angle between the wave vector k and the z-axis.

As a result,

$$k_z = k\cos\theta \approx k(1 - \theta^2/2)$$

and

$$\psi(\mathbf{r}) \approx A(\mathbf{r})e^{-j(k_x x + k_y y)}e^{jkz\theta^2/2}e^{-jkz}$$

The Paraxial Wave Equation

Substituting this expression into the Helmholtz equation, the paraxial wave equation is derived:

$$\nabla_T^2 A - 2jk\frac{\partial A}{\partial z} = 0$$

where

$$\nabla_T^2 = \nabla^2 - \frac{\partial^2}{\partial z^2} = \frac{\partial^2}{\partial x^2} + \frac{\partial^2}{\partial y^2}$$

is the transverse Laplace operator, shown here in Cartesian coordinates.

The Far Field Approximation

The equation above may be evaluated asymptotically in the far field (using the stationary phase method) to show that the field at the distant point (x,y,z) is indeed due solely to the plane wave component (k_x, k_y, k_z) which propagates parallel to the vector (x,y,z), and whose plane is tangent to the phasefront at (x,y,z). The mathematical details of this process may be found in Scott or Scott . The result of performing a stationary phase integration on the expression above is the following expression,

$$E_u(r,\theta,\phi)=2\pi j \ (k \ \cos\theta)\frac{e^{-jkr}}{r}E_u(k \ \sin\theta \ \cos\phi, k \ \sin\theta \ \sin\phi) \qquad (2.2)$$

which clearly indicates that the field at (x,y,z) is directly proportional to the spectral component in the direction of (x,y,z), where,

$$x = r \ \sin\theta \ \cos\phi$$

$$y = r \ \sin\theta \ \sin\phi$$

$$z = r \ \cos\theta$$

and

$$k_x = k \ \sin\theta \ \cos\phi$$

$$k_y = k \ \sin\theta \ \sin\phi$$

$$k_z = k \ \cos\theta$$

Stated another way, the radiation pattern of any planar field distribution is the FT of that source distribution (see Huygens–Fresnel principle, wherein the same equation is developed using a

Green's function approach). Note that this is NOT a plane wave, as many might think. The $\frac{e^{-jkr}}{r}$ radial dependence is a spherical wave - both in magnitude and phase - whose local amplitude is the FT of the source plane distribution at that far field angle. The plane wave spectrum has nothing to do with saying that the field behaves something like a plane wave for far distances.

Spatial Versus Angular Bandwidth

Equation (2.2) above is critical to making the connection between *spatial bandwidth* (on the one hand) and *angular bandwidth* (on the other), in the far field. Note that the term "far field" usually means we're talking about a converging or diverging spherical wave with a pretty well defined phase center. The connection between spatial and angular bandwidth in the far field is essential in understanding the low pass filtering property of thin lenses.

Once the concept of angular bandwidth is understood, the optical scientist can "jump back and forth" between the spatial and spectral domains to quickly gain insights which would ordinarily not be so readily available just through spatial domain or ray optics considerations alone. For example, any source bandwidth which lies past the edge angle to the first lens (this edge angle sets the bandwidth of the optical system) will not be captured by the system to be processed.

As a side note, electromagnetics scientists have devised an alternative means for calculating the far zone electric field which does not involve stationary phase integration. They have devised a concept known as "fictitious magnetic currents" usually denoted by M, and defined as

$$\mathbf{M} = 2\mathbf{E}^{aper} \times \hat{\mathbf{z}}.$$

In this equation, it is assumed that the unit vector in the z-direction points into the half-space where the far field calculations will be made. These equivalent magnetic currents are obtained using equivalence principles which, in the case of an infinite planar interface, allow any electric currents, J to be "imaged away" while the fictitious magnetic currents are obtained from twice the aperture electric field. Then the radiated electric field is calculated from the magnetic currents using an equation similar to the equation for the magnetic field radiated by an electric current. In this way, a vector equation is obtained for the radiated electric field in terms of the aperture electric field and the derivation requires no use of stationary phase ideas.

The Plane Wave Spectrum: The Foundation of Fourier Optics

Fourier optics is somewhat different from ordinary ray optics typically used in the analysis and design of focused imaging systems such as cameras, telescopes and microscopes. Ray optics is the very first type of optics most of us encounter in our lives; it's simple to conceptualize and understand, and works very well in gaining a baseline understanding of common optical devices. Unfortunately, ray optics does not explain the operation of Fourier optical systems, which are in general not focused systems. Ray optics is a subset of wave optics (in the jargon, it is "the asymptotic zero-wavelength limit" of wave optics) and therefore has limited applicability. We have to know when it is valid and when it is not - and this is one of those times when it is not. For our current task, we must expand our understanding of optical phenomena to encompass wave optics, in which the optical field is seen as a solution to Maxwell's equations. This more general *wave optics* accurately explains the operation of Fourier optics devices.

In this section, we won't go all the way back to Maxwell's equations, but will start instead with the homogeneous Helmholtz equation (valid in source-free media), which is one level of refinement up from Maxwell's equations (Scott). From this equation, we'll show how infinite uniform plane waves comprise one field solution (out of many possible) in free space. These uniform plane waves form the basis for understanding Fourier optics.

The plane wave spectrum concept is the basic foundation of Fourier Optics. The plane wave spectrum is a continuous spectrum of *uniform* plane waves, and there is one plane wave component in the spectrum for every tangent point on the far-field phase front. The amplitude of that plane wave component would be the amplitude of the optical field at that tangent point. Again, this is true only in the far field, defined as: Range $= 2 D^2 / \lambda$ where D is the maximum linear extent of the optical sources and λ is the wavelength (Scott). The plane wave spectrum is often regarded as being discrete for certain types of periodic gratings, though in reality, the spectra from gratings are continuous as well, since no physical device can have the infinite extent required to produce a true line spectrum.

As in the case of electrical signals, bandwidth is a measure of how finely detailed an image is; the finer the detail, the greater the bandwidth required to represent it. A DC electrical signal is constant and has no oscillations; a plane wave propagating parallel to the optic axis has constant value in any x-y plane, and therefore is analogous to the (constant) DC component of an electrical signal. Bandwidth in electrical signals relates to the difference between the highest and lowest frequencies present in the spectrum of the signal. For *optical* systems, bandwidth also relates to spatial frequency content (spatial bandwidth), but it also has a secondary meaning. It also measures how far from the optic axis the corresponding plane waves are tilted, and so this type of bandwidth is often referred to also as angular bandwidth. It takes more frequency bandwidth to produce a short pulse in an electrical circuit, and more angular (or, spatial frequency) bandwidth to produce a sharp spot in an optical system.

The plane wave spectrum arises naturally as the eigenfunction or "natural mode" solution to the homogeneous electromagnetic wave equation in rectangular coordinates. In the frequency domain, with an assumed (engineering) time convention of $e^{j\omega t}$, the homogeneous electromagnetic wave equation is known as the Helmholtz equation and takes the form:

$$\nabla^2 E_u + k^2 E_u = 0 \qquad (2.0)$$

where $u = x, y, z$ and $k = 2\pi/\lambda$ is the wavenumber of the medium.

Eigenfunction (Natural Mode) Solutions: Background and Overview

In the case of differential equations, as in the case of matrix equations, whenever the right-hand side of an equation is zero (i.e., the forcing function / forcing vector is zero), the equation may still admit a non-trivial solution, known in applied mathematics as an eigenfunction solution, in physics as a "natural mode" solution and in electrical circuit theory as the "zero-input response." This is a concept that spans a wide range of physical disciplines. Common physical examples of *resonant* natural modes would include the resonant vibrational modes of stringed instruments (1D), percussion instruments (2D) or the former Tacoma Narrows Bridge (3D). Examples of *propagating* natural modes would include waveguide modes, optical fiber modes, solitons and Bloch waves. Infinite homogeneous media admit the rectangular, circular and spherical harmonic solutions to the Helmholtz equation, depending on the coordinate system under consideration. The propagating plane waves we'll study in this article are perhaps the simplest type of propagating waves found in any type of media.

There is a striking similarity between the Helmholtz equation (2.0) above, which may be written

$$(\nabla^2 + k^2)f = 0,$$

and the usual equation for the eigenvalues/eigenvectors of a square matrix, A,

$$(\mathbf{A} - \lambda \mathbf{I})\mathbf{x} = 0,$$

particularly since both the scalar Laplacian, ∇^2 and the matrix, A are linear operators on their respective function/vector spaces (the minus sign in the second equation is, for all intents and purposes, immaterial; the plus sign in the first equation however is significant). It is perhaps worthwhile to note that both the eigenfunction and eigenvector solutions to these two equations respectively, often yield an orthogonal set of functions/vectors which span (i.e., form a basis set for) the function/vector spaces under consideration. The interested reader may investigate other functional linear operators which give rise to different kinds of orthogonal eigenfunctions such as Legendre polynomials, Chebyshev polynomials and Hermite polynomials.

In the matrix case, eigenvalues λ may be found by setting the determinant of the matrix equal to zero, i.e. finding where the matrix has no inverse. Finite matrices have only a finite number of eigenvalues/eigenvectors, whereas linear operators can have a countably infinite number of eigenvalues/eigenfunctions (in confined regions) or uncountably infinite (continuous) spectra of solutions, as in unbounded regions.

In certain physics applications such as in the computation of bands in a periodic volume, it is often the case that the elements of a matrix will be very complicated functions of frequency and wavenumber, and the matrix will be non-singular for most combinations of frequency and wavenumber, but will also be singular for certain specific combinations. By finding which combinations of frequency and wavenumber drive the determinant of the matrix to zero, the propagation characteristics of the medium may be determined. Relations of this type, between frequency and wavenumber, are known as dispersion relations and some physical systems may admit many different kinds of dispersion relations. An example from electromagnetics is the ordinary waveguide, which may admit numerous dispersion relations, each associated with a unique mode of the waveguide. Each propagation mode of the waveguide is known as an eigenfunction solution (or eigenmode solution) to Maxwell's equations in the waveguide. Free space also admits eigenmode (natural mode) solutions (known more commonly as plane waves), but with the distinction that for any given frequency, free space admits a continuous modal spectrum, whereas waveguides have a discrete mode spectrum. In this case the dispersion relation is linear, as in section 1.2.

K-space

The separation condition,

$$k_x^2 + k_y^2 + k_z^2 = k^2$$

which is identical to the equation for the Euclidean metric in three-dimensional configuration space, suggests the notion of a k-vector in three-dimensional "k-space", defined (for propagating plane waves) in rectangular coordinates as:

$$\mathbf{k} = k_x \hat{\mathbf{x}} + k_y \hat{\mathbf{y}} + k_z \hat{\mathbf{z}}$$

and in the spherical coordinate system as

$$k_x = k \sin\theta \cos\phi$$

$$k_y = k \sin\theta \sin\phi$$

$$k_z = k \cos\theta$$

Use will be made of these spherical coordinate system relations in the next section.

The notion of k-space is central to many disciplines in engineering and physics, especially in the study of periodic volumes, such as in crystallography and the band theory of semiconductor materials.

The Two-dimensional Fourier Transform

Analysis Equation (calculating the spectrum of the function):

$$U(k_x,k_y) = \int_{-\infty}^{\infty}\int_{-\infty}^{\infty} u(x,y)e^{-j(k_x x+k_y y)}\,dxdy$$

Synthesis Equation (reconstructing the function from its spectrum):

$$u(x,y) = \frac{1}{(2\pi)^2}\int_{-\infty}^{\infty}\int_{-\infty}^{\infty} U(k_x,k_y)e^{j(k_x x+k_y y)}\,dk_x dk_y$$

Note: the normalizing factor of: $\frac{1}{(2\pi)^2}$ is present whenever angular frequency (radians) is used, but not when ordinary frequency (cycles) is used.

Optical Systems: General Overview and Analogy With Electrical Signal Processing Systems

An optical system consists of an input plane, and output plane, and a set of components that transforms the image *f* formed at the input into a different image *g* formed at the output. The output image is related to the input image by convolving the input image with the optical impulse response, *h* (known as the *point-spread function*, for focused optical systems). The impulse response uniquely defines the input-output behavior of the optical system. By convention, the optical axis of the system is taken as the *z*-axis. As a result, the two images and the impulse response are all functions of the transverse coordinates, *x* and *y*.

$$g(x,y) = h(x,y) * f(x,y)$$

The impulse response of an optical imaging system is the output plane field which is produced when an ideal mathematical point source of light is placed in the input plane (usually on-axis). In practice, it is not necessary to have an ideal point source in order to determine an exact impulse response. This is because any source bandwidth which lies outside the bandwidth of the system won't matter anyway (since it cannot even be captured by the optical system), so therefore it's not necessary in determining the impulse response. The source only needs to have at least as much (angular) bandwidth as the optical system.

Optical systems typically fall into one of two different categories. The first is the ordinary focused optical imaging system, wherein the input plane is called the object plane and the output plane is called the image plane. The field in the image plane is desired to be a high-quality reproduction of the field in the object plane. In this case, the impulse response of the optical system is desired to approximate a 2D delta function, at the same location (or a linearly scaled location) in the output plane corresponding to the location of the impulse in the input plane. The *actual* impulse response typically resembles an Airy function, whose radius is on the order of the wavelength of the light used. In this case, the impulse response is typically referred to as a point spread function, since the mathematical point of light in the object plane has been spread out into an Airy function in the image plane.

The second type is the optical image processing system, in which a significant feature in the input plane field is to be located and isolated. In this case, the impulse response of the system is desired to be a close replica (picture) of that feature which is being searched for in the input plane field, so that a convolution of the impulse response (an image of the desired feature) against the input plane field will produce a bright spot at the feature location in the output plane. It is this latter type of optical *image processing* system that is the subject of this section. Section 5.2 presents one hardware implementation of the optical image processing operations described in this section.

Input Plane

The input plane is defined as the locus of all points such that $z = 0$. The input image f is therefore

$$f(x, y) = U(x, y, z)\big|_{z=0}$$

Output Plane

The output plane is defined as the locus of all points such that $z = d$. The output image g is therefore

$$g(x, y) = U(x, y, z)\big|_{z=d}$$

The 2D convolution of input function against the impulse response function

$$g(x, y) = h(x, y) * f(x, y)$$

i.e.,

$$g(x, y) = \int_{-\infty}^{\infty} \int_{-\infty}^{\infty} h(x - x', y - y') f(x', y') dx' dy' \quad (4.1)$$

The alert reader will note that the integral above tacitly assumes that the impulse response is NOT a function of the position (x',y') of the impulse of light in the input plane (if this were not the case, this type of convolution would not be possible). This property is known as *shift invariance* (Scott). No optical system is perfectly shift invariant: as the ideal, mathematical point of light is scanned away from the optic axis, aberrations will eventually degrade the impulse response (known as a coma in focused imaging systems). However, high quality optical systems are often "shift invariant enough" over certain regions of the input plane that we may regard the impulse response as being a function of only the difference between input and output plane coordinates, and thereby use the equation above with impunity.

Also, this equation assumes unit magnification. If magnification is present, then eqn. (4.1) becomes

$$g(x, y) = \int_{-\infty}^{\infty} \int_{-\infty}^{\infty} h_M(x - Mx', y - My') f(x', y') dx' dy' \quad (4.2)$$

which basically translates the impulse response function, $h_M()$, from x' to x=Mx'. In (4.2), $h_M()$ will be a magnified version of the impulse response function $h()$ of a similar, unmagnified system, so that $h_M(x,y) = h(x/M, y/M)$.

Derivation of the Convolution Equation

The extension to two dimensions is trivial, except for the difference that causality exists in the time domain, but not in the spatial domain. Causality means that the impulse response $h(t - t')$ of an electrical system, due to an impulse applied at time t', must of necessity be zero for all times t such that $t - t' < 0$.

Obtaining the convolution representation of the system response requires representing the input signal as a weighted superposition over a train of impulse functions by using the *sifting property* of Dirac delta functions.

$$f(t) = \int_{-\infty}^{\infty} \delta(t-t') f(t') dt'$$

It is then presumed that the system under consideration is *linear*, that is to say that the output of the system due to two different inputs (possibly at two different times) is the sum of the individual outputs of the system to the two inputs, when introduced individually. Thus the optical system may contain no nonlinear materials nor active devices (except possibly, extremely linear active devices). The output of the system, for a single delta function input is defined as the *impulse response* of the system, h(t - t'). And, by our linearity assumption (i.e., that the output of system to a pulse train input is the sum of the outputs due to each individual pulse), we can now say that the general input function $f(t)$ produces the output:

$$g(t) = \int_{-\infty}^{\infty} h(t-t') f(t') dt'$$

where h(t - t') is the (impulse) response of the linear system to the delta function input δ(t - t'), applied at time t'. This is where the convolution equation above comes from. The convolution equation is useful because it is often much easier to find the response of a system to a delta function input - and then perform the convolution above to find the response to an arbitrary input - than it is to try to find the response to the arbitrary input directly. Also, the impulse response (in either time or frequency domains) usually yields insight to relevant figures of merit of the system. In the case of most lenses, the point spread function (PSF) is a pretty common figure of merit for evaluation purposes.

The same logic is used in connection with the Huygens–Fresnel principle, or Stratton-Chu formulation, wherein the "impulse response" is referred to as the Green's function of the system. So the spatial domain operation of a linear optical system is analogous in this way to the Huygens–Fresnel principle.

System transfer function

If the last equation above is Fourier transformed, it becomes:

$$G(\omega) = H(\omega) \cdot F(\omega)$$

where

> $G(\omega)$ is the spectrum of the output signal
>
> $H(\omega)$ is the system transfer function
>
> $F(\omega)$ is the spectrum of the input signal

In like fashion, (4.1) may be Fourier transformed to yield:

$$G(k_x, k_y) = H(k_x, k_y) \cdot F(k_x, k_y)$$

The system transfer function, $H(\omega)$. In optical imaging this function is better known as the optical transfer function *(Goodman)*.

Once again it may be noted from the discussion on the Abbe sine condition, that this equation assumes unit magnification.

This equation takes on its real meaning when the Fourier transform, $G(k_x, k_y)$ is associated with the coefficient of the plane wave whose transverse wavenumbers are (k_x, k_y).. Thus, the input-plane plane wave spectrum is transformed into the output-plane plane wave spectrum through the multiplicative action of the system transfer function. It is at this stage of understanding that the previous background on the plane wave spectrum becomes invaluable to the conceptualization of Fourier optical systems.

Applications of Fourier Optics Principles

Fourier optics is used in the field of optical information processing, the staple of which is the classical 4F processor.

The Fourier transform properties of a lens provide numerous applications in optical signal processing such as spatial filtering, optical correlation and computer generated holograms.

Fourier optical theory is used in interferometry, optical tweezers, atom traps, and quantum computing. Concepts of Fourier optics are used to reconstruct the phase of light intensity in the spatial frequency plane.

Fourier Transforming Property of Lenses

If a transmissive object is placed one focal length in front of a lens, then its Fourier transform will be formed one focal length behind the lens. Consider the figure to the right (click to enlarge)

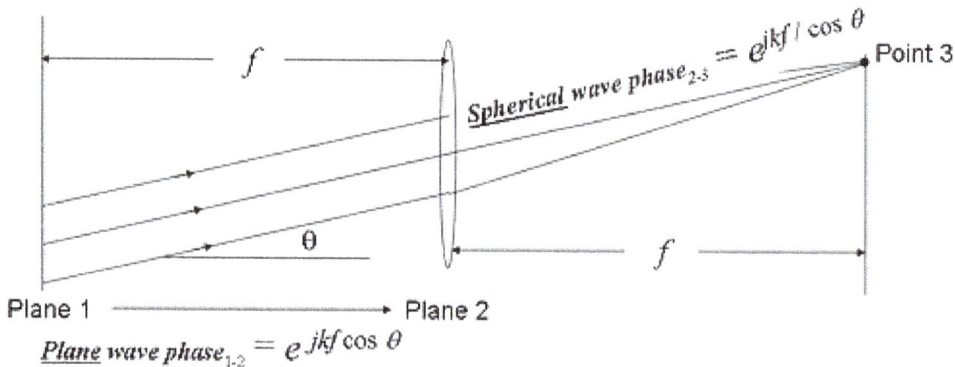

On the Fourier transforming property of lenses

In this figure, a plane wave incident from the left is assumed. The transmittance function in the front focal plane (i.e., Plane 1) *spatially modulates the incident plane wave* in magnitude and phase, *like on the left-hand side of eqn. (2.1)* (specified to z=0), and *in so doing, produces a spectrum of plane waves* corresponding to the FT of the transmittance function, *like on the right-hand side of eqn. (2.1)* (for z>0). The various plane wave components propagate at different tilt angles with respect to the optic axis of the lens (i.e., the horizontal axis). The finer the features in the transparency, the broader the angular bandwidth of the plane wave spectrum. We'll consider one such plane wave component, propagating at angle θ with respect to the optic axis. It is assumed that θ is small (paraxial approximation), so that

$$\frac{k_x}{k} = \sin\theta \simeq \theta$$

and

$$\frac{k_z}{k} = \cos\theta \simeq 1 - \frac{\theta^2}{2}$$

and

$$\frac{1}{\cos\theta} \simeq \frac{1}{1 - \frac{\theta^2}{2}} \simeq 1 + \frac{\theta^2}{2}$$

In the figure, the *plane wave* phase, moving horizontally from the front focal plane to the lens plane, is

$$e^{jkf\cos\theta}$$

and the *spherical wave* phase from the lens to the spot in the back focal plane is:

$$e^{jkf/\cos\theta}$$

and the sum of the two path lengths is $f(1 + \theta^2/2 + 1 - \theta^2/2) = 2f$ i.e., it is a constant value, independent of tilt angle, θ, for paraxial plane waves. Each paraxial plane wave component of the field in the front focal plane appears as a point spread function spot in the back focal plane, with an intensity and phase equal to the intensity and phase of the original plane wave component in the front focal plane. In other words, the field in the back focal plane is the Fourier transform of the field in the front focal plane.

All FT components are computed simultaneously - in parallel - at the speed of light. As an example, light travels at a speed of roughly 1 ft (0.30 m). / ns, so if a lens has a 1 ft (0.30 m). focal length, an entire 2D FT can be computed in about 2 ns (2 x 10⁻⁹ seconds). If the focal length is 1 in., then the time is under 200 ps. No electronic computer can compete with these kinds of numbers or perhaps ever hope to, although new supercomputers such as the petaflop IBM Roadrunner may actually prove faster than optics, as improbable as that may seem. However, their speed is obtained by combining numerous computers which, individually, are still slower than optics. The disadvantage of the optical FT is that, as the derivation shows, the FT relationship only holds for paraxial plane waves, so this FT "computer" is inherently bandlimited. On the other hand, since the wavelength of visible light is so minute in relation to even the smallest visible feature dimensions in the image i.e.,

$$k^2 \gg k_x^2 + k_y^2$$

(for all k_x, k_y within the spatial bandwidth of the image, so that k_z is nearly equal to k), the paraxial approximation is not terribly limiting in practice. And, of course, this is an analog - not a digital - computer, so precision is limited. Also, phase can be challenging to extract; often it is inferred interferometrically.

Optical processing is especially useful in real time applications where rapid processing of massive amounts of 2D data is required, particularly in relation to pattern recognition.

Object Truncation and Gibbs Phenomenon

The spatially modulated electric field, shown on the left-hand side of eqn. (2.1), typically only occupies a finite (usually rectangular) aperture in the x,y plane. The rectangular aperture function acts like a 2D square-top filter, where the field is assumed to be zero outside this 2D rectangle. The spatial domain integrals for calculating the FT coefficients on the right-hand side of eqn. (2.1) are truncated at the boundary of this aperture. This step truncation can introduce inaccuracies in both theoretical calculations and measured values of the plane wave coefficients on the RHS of eqn. (2.1).

Whenever a function is discontinuously truncated in one FT domain, broadening and rippling are introduced in the other FT domain. A perfect example from optics is in connection with the point spread function, which for on-axis plane wave illumination of a quadratic lens (with circular aperture), is an Airy function, $J_1(x)/x$. Literally, the point source has been "spread out" (with ripples added), to form the Airy point spread function (as the result of truncation of the plane wave spectrum by the finite aperture of the lens). This source of error is known as Gibbs phenomenon and it may be mitigated by simply ensuring that all significant content lies near the center of the transparency, or through the use of window functions which smoothly taper the field to zero at the frame boundaries. By the convolution theorem, the FT of an arbitrary transparency function - multiplied (or truncated) by an aperture function - is equal to the FT of the non-truncated transparency function convolved against the FT of the aperture function, which in this case becomes a type of "Greens function" or "impulse response function" in the spectral domain. Therefore, the image of a circular lens is equal to the object plane function convolved against the Airy function (the FT of a circular aperture function is $J_1(x)/x$ and the FT of a rectangular aperture function is a product of sinc functions, $\sin x/x$).

Fourier Analysis and Functional Decomposition

Even though the input transparency only occupies a finite portion of the x-y plane (Plane 1), the uniform plane waves comprising the plane wave spectrum occupy the entire x-y plane, which is why (for this purpose) only the longitudinal plane wave phase (in the z-direction, from Plane 1 to Plane 2) must be considered, and not the phase transverse to the z-direction. It is of course, very tempting to think that if a plane wave emanating from the finite aperture of the transparency is tilted too far from horizontal, it will somehow "miss" the lens altogether but again, since the uniform plane wave extends infinitely far in all directions in the transverse (x-y) plane, the planar wave components cannot miss the lens.

This issue brings up perhaps the predominant difficulty with Fourier analysis, namely that the input-plane function, defined over a finite support (i.e., over its own finite aperture), is being approximated with other functions (sinusiods) which have infinite support (*i.e.*, they are defined over the entire infinite x-y plane). This is unbelievably inefficient computationally, and is the principal reason why wavelets were conceived, that is to represent a function (defined on a finite interval or area) in terms of oscillatory functions which are also defined over finite intervals or areas. Thus, instead of getting the frequency content of the entire image all at once (along with the frequency content of the entire rest of the x-y plane, over which the image has zero value), the result is instead the frequency content of different parts of the image, which is usually much simpler. Unfortunately, wavelets in the x-y plane don't correspond to any known type of propagating wave function, in the same way that Fourier's sinusoids (in the x-y plane) correspond to plane wave functions in three dimensions. However, the FTs of most wavelets are well known and could possibly be shown to be equivalent to some useful type of propagating field.

On the other hand, Sinc functions and Airy functions - which are not only the point spread functions of rectangular and circular apertures, respectively, but are also cardinal functions commonly used for functional decomposition in interpolation/sampling theory [Scott 1990] - do correspond to converging or diverging spherical waves, and therefore could potentially be implemented as a whole new functional decomposition of the object plane function, thereby leading to another point of view similar in nature to Fourier optics. This would basically be the same as conventional ray optics, but with diffraction effects included. In this case, each point spread function would be a type of "smooth pixel," in much the same way that a soliton on a fiber is a "smooth pulse."

Perhaps a lens figure-of-merit in this "point spread function" viewpoint would be to ask how well a lens transforms an Airy function in the object plane into an Airy function in the image plane, as a function of radial distance from the optic axis, or as a function of the size of the object plane Airy function. This is somewhat like the point spread function, except now we're really looking at it as a kind of input-to-output plane transfer function (like MTF), and not so much in absolute terms, relative to a perfect point. Similarly, Gaussian wavelets, which would correspond to the waist of a propagating Gaussian beam, could also potentially be used in still another functional decomposition of the object plane field.

Far-field Range and the $2D^2 / \lambda$ Criterion

In the figure above, illustrating the Fourier transforming property of lenses, the lens is in the near field of the object plane transparency, therefore the object plane field at the lens may be regarded as a superposition of plane waves, each one of which propagates at some angle with respect to the z-axis. In this regard, the far-field criterion is loosely defined as: Range = $2 D^2 / \lambda$ where D is the maximum linear extent of the optical sources and λ is the wavelength (Scott). The D of the transparency is on the order of cm (10^{-2} m) and the wavelength of light is on the order of 10^{-6} m, therefore D/λ for the whole transparency is on the order of 10^4. This times D is on the order of 10^2 m, or hundreds of meters. On the other hand, the far field distance from a PSF spot is on the order of λ. This is because D for the spot is on the order of λ, so that D/λ is on the order of unity; this times D (i.e., λ) is on the order of λ (10^{-6} m).

Since the lens is in the far field of any PSF spot, the field incident on the lens from the spot may be regarded as being a spherical wave, as in eqn. (2.2), not as a plane wave spectrum, as in eqn. (2.1). On the other hand, the lens is in the near field of the entire input plane transparency, therefore eqn. (2.1) - the full plane wave spectrum - accurately represents the field incident on the lens from that larger, extended source.

Lens as a Low-pass Filter

A lens is basically a low-pass plane wave filter. Consider a "small" light source located on-axis in the object plane of the lens. It is assumed that the source is small enough that, by the far-field criterion, the lens is in the far field of the "small" source. Then, the field radiated by the small source is a spherical wave which is modulated by the FT of the source distribution, as in eqn. (2.2), Then, the lens passes - from the object plane over onto the image plane - only that por-tion of the radiated spherical wave which lies inside the edge angle of the lens. In this far-field case, truncation of the radiated spherical wave is equivalent to truncation of the plane wave spectrum of the small source. So, the plane wave components in this far-field spherical wave, which lie beyond the edge angle of the lens, are not captured by the lens and are not transferred over to the image plane. Note: this logic is valid only for small sources, such that the lens is in the far field region of the source, according to the 2 D^2 / λ criterion mentioned previously. If an object plane transparency is imagined as a summation over small sources (as in the Whittaker–Shannon interpolation formula, Scott), each of which has its spectrum truncated in this fashion, then every point of the entire object plane transparency suffers the same effects of this low pass filtering.

Loss of the high (spatial) frequency content causes blurring and loss of sharpness. Bandwidth truncation causes a (fictitious, mathematical, ideal) point source in the object plane to be blurred (or, spread out) in the image plane, giving rise to the term, "point spread function." Whenever bandwidth is expanded or contracted, image size is typi-cally contracted or expanded accordingly, in such a way that the space-bandwidth product remains constant, by Heisenberg's principle (Scott and Abbe sine condition).

Coherence and Fourier Transforming

While working in the frequency domain, with an assumed e$^{j\omega t}$ (engineering) time dependence, coherent (laser) light is implicitly assumed, which has a delta function dependence in the frequency domain. Light at different (delta function) frequencies will "spray" the plane wave spectrum out at different angles, and as a result these plane wave components will be focused at different places in the output plane. The Fourier transforming property of lenses works best with coherent light, unless there is some special reason to combine light of different frequencies, to achieve some special purpose.

Hardware Implementation of the System Transfer Function: The 4F Correlator

The theory on optical transfer functions presented in section 4 is somewhat abstract. However, there is one very well known device which implements the system transfer function H in hardware using only 2 identical lenses and a transparency plate - the 4F correlator. Although one important application of this device would certainly be to implement the mathematical operations of cross-correlation and convolution, this device - 4 focal lengths long - actually serves a wide variety of image processing operations that go well beyond what its name implies. A diagram of a typical 4F correlator is shown in the figure below (click to enlarge). This device may be readily understood by combining the plane wave spectrum representation of the electric field (*section 2*) with the Fourier transforming property of quadratic lenses (*section 5.1*) to yield the optical image processing operations described in section 4.

The 4F correlator is based on the convolution theorem from Fourier transform theory, which states that convolution in the spatial (x,y) domain is equivalent to direct multiplication in the spatial frequency (k_x, k_y) domain (aka: *spectral domain*). Once again, a plane wave is assumed incident from the left and a transparency containing one 2D function, $f(x,y)$, is placed in the input plane of the correlator, located one focal length in front of the first lens. The transparency spatially modulates the incident plane wave in magnitude and phase, like on the left-hand side of eqn. (2.1), and in so doing, produces a spectrum of plane waves corresponding to the FT of the transmittance function, like on the right-hand side of eqn. (2.1). That spectrum is then formed as an "image" one focal length behind the first lens, as shown. A transmission mask containing the FT of the second function, $g(x,y)$, is placed in this same plane, one focal length behind the first lens, causing the transmission through the mask to be equal to the product, $F(k_x,k_y)$ x $G(k_x,k_y)$. This product now lies in the "input plane" of the second lens (one focal length in front), so that the FT of this product (i.e., the convolution of $f(x,y)$ and $g(x,y)$), is formed in the back focal plane of the second lens.

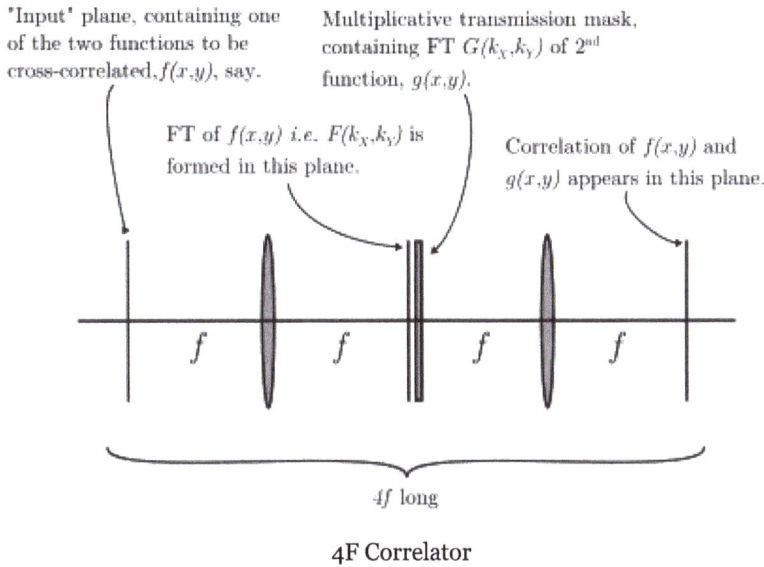

"Input" plane, containing one of the two functions to be cross-correlated, $f(x,y)$, say.

Multiplicative transmission mask, containing FT $G(k_x,k_y)$ of 2nd function, $g(x,y)$.

FT of $f(x,y)$ i.e. $F(k_x,k_y)$ is formed in this plane.

Correlation of $f(x,y)$ and $g(x,y)$ appears in this plane.

f f f f

4f long

4F Correlator

If an ideal, mathematical point source of light is placed on-axis in the input plane of the first lens, then there will be a uniform, collimated field produced in the output plane of the first lens. When this uniform, collimated field is multiplied by the FT plane mask, and then Fourier transformed by the second lens, the output plane field (which in this case is the *impulse response* of the correlator) is just our correlating function, $g(x,y)$. In practical applications, $g(x,y)$ will be some type of feature which must be identified and located within the input plane field. In military applications, this feature may be a tank, ship or airplane which must be quickly identified within some more complex scene.

The 4F correlator is an excellent device for illustrating the "systems" aspects of optical instruments, alluded to in *section 4* above. The FT plane mask function, $G(k_x,k_y)$ is the system transfer function of the correlator, which we'd in general denote as $H(k_x,k_y)$, and it is the FT of the impulse response function of the correlator, $h(x,y)$ which is just our correlating function $g(x,y)$. And, as mentioned above, the impulse response of the correlator is just a picture of the feature we're trying to find in the input image. In the 4F correlator, the system transfer function $H(k_x,k_y)$ is directly multiplied against the spectrum $F(k_x,k_y)$ of the input function, to produce the spectrum of the output function. This is how electrical signal processing systems operate on 1D temporal signals.

Afterword: Plane wave spectrum within the broader context of functional decomposition

Electrical fields can be represented mathematically in many different ways. In the Huygens–Fresnel or Stratton-Chu viewpoints, the electric field is represented as a superposition of point sources, each one of which gives rise to a Green's function field. The total field is then the weighted sum of all of the individual Green's function fields. That seems to be the most natural way of viewing the electric field for most people - no doubt because most of us have, at one time or another, drawn out the circles with protractor and paper, much the same way Thomas Young did in his classic paper on the double-slit experiment. However, it is by no means the only way to represent the electric field, which may also be represented as a spectrum of sinusoidally varying plane waves. In addition, Frits Zernike proposed still another functional decomposition based on his Zernike polynomials, defined on the unit disc. The third-order (and lower) Zernike polynomials correspond to the normal lens aberrations. And still another functional decomposition could be made in terms of Sinc functions and Airy functions, as in the Whittaker–Shannon interpolation formula and the Nyquist–Shannon sampling theorem. All of these functional decompositions have utility in different circumstances. The optical scientist having access to these various representational forms has available a richer insight to the nature of these marvelous fields and their properties. These different ways of looking at the field are not conflicting or contradictory, rather, by exploring their connections, one can often gain deeper insight into the nature of wave fields.

Functional Decomposition and Eigenfunctions

The twin subjects of eigenfunction expansions and functional decomposition, are not completely independent. The eigenfunction expansions to certain linear operators defined over a given domain, will often yield a countably infinite set of orthogonal functions which will span that domain. Depending on the operator and the dimensionality (and shape, and boundary conditions) of its domain, many different types of functional decompositions are, in principle, possible.

Fermat's Principle

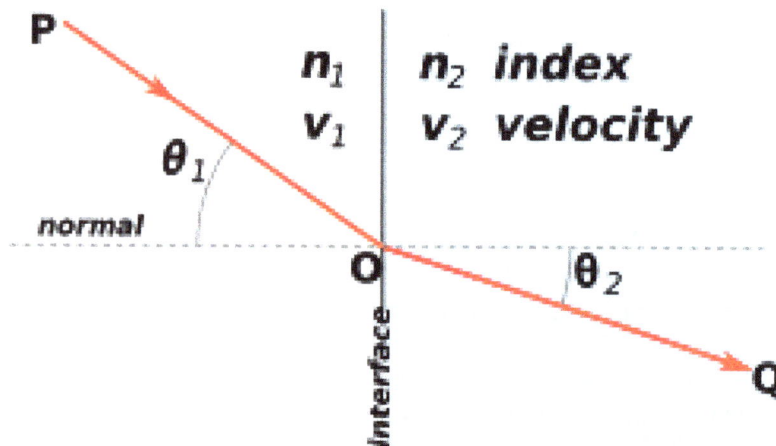

Fermat's principle leads to Snell's law; when the sines of the angles in the different media are in the same proportion as the propagation velocities, the time to get from P to Q is minimized.

In optics, Fermat's principle or the principle of least time, named after French mathematician Pierre de Fermat, is the principle that the path taken between two points by a ray of light is the path that can be traversed in the least time. This principle is sometimes taken as the definition of a ray of light. However, this version of the principle is not general; a more modern statement of the principle is that rays of light traverse the path of stationary optical length with respect to variations of the path. In other words, a ray of light prefers the path such that there are other paths, arbitrarily nearby on either side, along which the ray would take almost exactly the same time to traverse.

Fermat's principle can be used to describe the properties of light rays reflected off mirrors, refracted through different media, or undergoing total internal reflection. It follows mathematically from Huygens' principle (at the limit of small wavelength). Fermat's text *Analyse des réfractions* exploits the technique of adequality to derive Snell's law of refraction and the law of reflection.

Fermat's principle has the same form as Hamilton's principle and it is the basis of Hamiltonian optics.

Modern Version

The time T a point of the electromagnetic wave needs to cover a path between the points A and B is given by:

$$T = \int_{t_0}^{t_1} dt = \frac{1}{c}\int_{t_0}^{t_1} \frac{c}{v}\frac{ds}{dt}dt = \frac{1}{c}\int_A^B nds$$

c is the speed of light in vacuum, ds an infinitesimal displacement along the ray, $v = ds/dt$ the speed of light in a medium and $n = c/v$ the refractive index of that medium, t_0 is the starting time (the wave front is in A), t_1 is the arrival time at B. The optical path length of a ray from a point A to a point B is defined by:

$$S = \int_A^B nds$$

and it is related to the travel time by $S = cT$. The optical path length is a purely geometrical quantity since time is not considered in its calculation. An extremum in the light travel time between two points A and B is equivalent to an extremum of the optical path length between those two points. The historical form proposed by Fermat is incomplete. A complete modern statement of the variational Fermat principle is that

the optical length of the path followed by light between two fixed points, A and B, is an extremum. The optical length is defined as the physical length multiplied by the refractive index of the material."

In the context of calculus of variations this can be written as

$$\delta S = \delta \int_A^B nds = 0$$

In general, the refractive index is a scalar field of position in space, that is, $n = n(x_1, x_2, x_3)$ in 3D euclidean space. Assuming now that light has a component that travels along the x_3 axis, the path of a light ray may be parametrized as $s = (x_1(x_3), x_2(x_3), x_3)$ and

$$nds = n\frac{\sqrt{dx_1^2 + dx_2^2 + dx_3^2}}{dx_3}dx_3 = n\sqrt{1 + \dot{x}_1^2 + \dot{x}_2^2}\,dx_3$$

where $\dot{x}_k = dx_k/dx_3$. The principle of Fermat can now be written as

$$\delta S = \delta \int_{x_{3A}}^{x_{3B}} n(x_1, x_2, x_3)\sqrt{1 + \dot{x}_1^2 + \dot{x}_2^2}\,dx_3$$

$$= \delta \int_{x_{3A}}^{x_{3B}} L(x_1(x_3), x_2(x_3), \dot{x}_1(x_3), \dot{x}_2(x_3), x_3)dx_3 = 0$$

which has the same form as Hamilton's principle but in which x_3 takes the role of time in classical mechanics. Function $L(x_1, x_2, \dot{x}_1, \dot{x}_2, x_3)$ is the optical Lagrangian from which the Lagrangian and Hamiltonian (as in Hamiltonian mechanics) formulations of geometrical optics may be derived.

Derivation

Classically, Fermat's principle can be considered as a mathematical consequence of Huygens' principle. Indeed, of all secondary waves (along all possible paths) the waves with the extrema (stationary) paths contribute most due to constructive interference. Suppose that light waves propagate from A to B by all possible routes AB_j, unrestricted initially by rules of geometrical or physical optics. The various optical paths AB_j will vary by amounts greatly in excess of one wavelength, and so the waves arriving at B will have a large range of phases and will tend to interfere destructively. But if there is a shortest route AB_0, and the optical path varies smoothly through it, then a considerable number of neighboring routes close to AB_0 will have optical paths differing from AB_0 by second-order amounts only and will therefore interfere constructively. Waves along and close to this shortest route will thus dominate and AB_0 will be the route along which the light is seen to travel.

Fermat's principle is the main principle of quantum electrodynamics which states that any particle (e.g. a photon or an electron) propagates over all available, unobstructed paths and that the interference, or superposition, of its wavefunction over all those paths at the point of observer gives the probability of detecting the particle at this point. Thus, because the extremal paths (shortest, longest, or stationary) cannot be completely canceled out, they contribute most to this interference. In humans, for example, Fermat's principle can be demonstrated in a situation when a lifeguard has to find the fastest way to traverse both beach and water in order to reach a drowning swimmer. The principle has been tested in studies with ants, in which the ants' nest is on one end of a container and food is on the opposite end, but the ants choose to follow the path of least time, rather than the most direct path.

In the classic mechanics of waves, Fermat's principle follows from the extremum principle of mechanics.

History

Euclid, c. 320 BCE in his Catoptrics (on mirrors, including spherical mirrors) and Optics puts the foundations for reflection, which is repeated by Ptolemy and then in his more detailed books that have surviced Hero of Alexandria (Heron) (c. 60) described the principle of reflection, which stated that a ray of light that goes from point A to point B, suffering any number of reflections on flat mirrors, in the same medium, has a smaller path length than any nearby path. In fact the very basic principle is already in the myth of the Allegory of the Cave by Plato.

Pierre de Fermat

Ibn al-Haytham (Alhacen), in his *Book of Optics* (1021), expanded the principle to both reflection and refraction, and expressed an early version of the principle of least time. His experiments were based on earlier works on refraction carried out by the Greek scientist Ptolemy

The generalized principle of least time in its modern form was stated by Fermat in a letter dated January 1, 1662, to Cureau de la Chambre. It was met with objections made in May 1662 by Claude Clerselier, an expert in optics and leading spokesman for the Cartesians at that time. Amongst his objections, Clerselier states:

... The principle which you take as the basis for your proof, namely that Nature always acts by using the simplest and shortest paths, is merely a moral, and not a physical one. It is not, and cannot be, the cause of any effect in Nature

The original French, from Mahoney, is as follows:

Le principe que vous prenez pour fondement de votre démonstration, à savoir que la nature agit toujours par les voies les plus courtes et les plus simples, n'est qu'un principe moral et non point physique, qui n'est point et qui ne peut être la cause d'aucun effet de la nature.

Indeed, Fermat's principle does not hold standing alone, we now know it can be derived from earlier principles such as Huygens' principle.

Historically, Fermat's principle has served as a guiding principle in the formulation of physical laws with the use of variational calculus.

References

- Hsien-Che Lee (2005). Introduction to Color Imaging Science. Cambridge University Press. p. 57. ISBN 978-0-521-84388-1.

- Geoffrey G. Attridge (2000). "Sensitometry". In Ralph E. Jacobson; Sidney F. Ray; Geoffrey G. Attridge; Norman R. Axford. The Manual of Photography: Photographic and Digital Imaging (9th ed.). Oxford: Focal Press. pp. 218–223. ISBN 0-240-51574-9.

- Rob Sheppard (2010). Digital Photography: Top 100 Simplified Tips & Tricks (4th ed.). John Wiley and Sons. p. 40. ISBN 978-0-470-59710-1.

- Barbara A. Lynch-Johnt & Michelle Perkins (2008). Illustrated Dictionary of Photography. Amherst Media. p. 15. ISBN 978-1-58428-222-8.

- David K. Lynch, William Charles Livingston (2001). Color and light in nature. Cambridge University Press. p. 31. ISBN 978-0-521-77504-5. Retrieved 2011-04-02.

- Yu Timofeev and A. V. Vasilev (2008-05-01). Theoretical Fundamentals of Atmospheric Optics. Cambridge International Science Publishing. p. 174. ISBN 978-1-904602-25-5. Retrieved 2012-02-23.

- Craig F. Bohren and Eugene Edmund Clothiaux (2006). Fundamentals of atmospheric radiation: an introduction with 400 problems. Wiley-VCH. p. 427. ISBN 978-3-527-40503-9.

- Lee M. Grenci and Jon M. Nese (2001). A world of weather: fundamentals of meteorology: a text/ laboratory manual. Kendall Hunt. p. 330. ISBN 978-0-7872-7716-1. Retrieved 2011-04-12.

- JaapJan Zeeberg (2001). Climate and glacial history of the Novaya Zemlya archipelago, Russian Arctic: with notes on the region's history of exploration. JaapJan Zeeberg. p. 149. ISBN 978-90-5170-563-8. Retrieved 2011-04-02.

- John A. Day (2005). The Book of Clouds. Sterling Publishing Company, Inc. pp. 124–127. ISBN 978-1-4027-2813-6. Retrieved 2010-10-09.

- Camilleri, K. (2009). Heisenberg and the Interpretation of Quantum Mechanics: the Physicist as Philosopher, Cambridge University Press, Cambridge UK, ISBN 978-0-521-88484-6.

- Buchwald, Jed (1989). The Rise of the Wave Theory of Light: Optical Theory and Experiment in the Early Nineteenth Century. Chicago: University of Chicago Press. ISBN 0-226-07886-8. OCLC 18069573.

- Penrose, Roger (2007). The Road to Reality: A Complete Guide to the Laws of the Universe. Vintage. p. 521, §21.10. ISBN 978-0-679-77631-4.

- Walter Schumann (2009). Gemstones of the World: Newly Revised & Expanded Fourth Edition. Sterling Publishing Company, Inc. pp. 41–2. ISBN 978-1-4027-6829-3. Retrieved 31 December 2011.

- Ariel Lipson, Stephen G. Lipson, Henry Lipson, Optical Physics 4th Edition, Cambridge University Press, ISBN 978-0-521-49345-1.

Applications of Optics

Optics is used for scientific and commercial use. Fiber- optic communication is the technique of conveying information from one place to another. This is done by sending pulses of light through an optical fiber. Alternatively the other uses of optics are electromagnetically induced transparency, superlens, optical fiber, spectrometer and optical computing. This section has been carefully written to provide an easy understanding of the theory of optics.

Fiber-optic Communication

Fiber-optic communication is a method of transmitting information from one place to another by sending pulses of light through an optical fiber. The light forms an electromagnetic carrier wave that is modulated to carry information. First developed in the 1970s, fiber-optics have revolutionized the telecommunications industry and have played a major role in the advent of the Information Age. Because of its advantages over electrical transmission, optical fibers have largely replaced copper wire communications in core networks in the developed world. Optical fiber is used by many telecommunications companies to transmit telephone signals, Internet communication, and cable television signals. Researchers at Bell Labs have reached internet speeds of over 100 petabit×kilometer per second using fiber-optic communication.

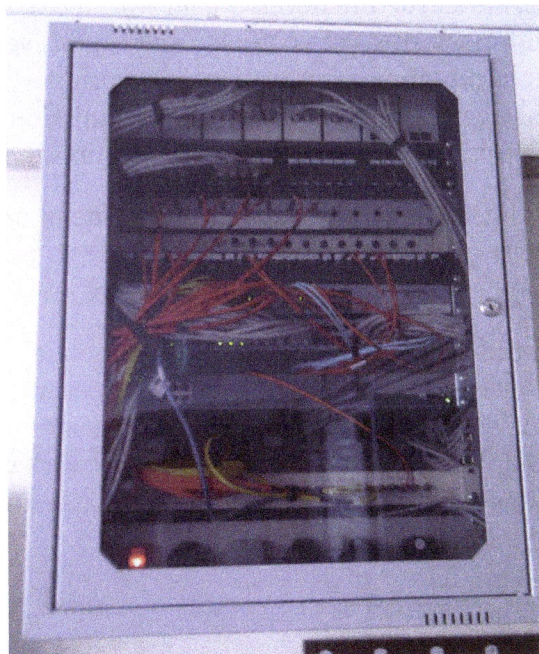

An optical fiber junction box. The yellow cables are single mode fibers; the orange and blue cables are multi-mode fibers: 62.5/125 μm OM1 and 50/125 μm OM3 fibers, respectively.

Stealth installing a 432-count dark fibre cable underneath the streets of Midtown Manhattan, New York City

The process of communicating using fiber-optics involves the following basic steps: Creating the optical signal involving the use of a transmitter, relaying the signal along the fiber, ensuring that the signal does not become too distorted or weak, receiving the optical signal, and converting it into an electrical signal.

Applications

Optical fiber is used by many telecommunications companies to transmit telephone signals, Internet communication, and cable television signals. Due to much lower attenuation and interference, optical fiber has large advantages over existing copper wire in long-distance and high-demand applications. However, infrastructure development within cities was relatively difficult and time-consuming, and fiber-optic systems were complex and expensive to install and operate. Due to these difficulties, fiber-optic communication systems have primarily been installed in long-distance applications, where they can be used to their full transmission capacity, offsetting the increased cost. Since 2000, the prices for fiber-optic communications have dropped considerably.

The price for rolling out fiber to the home has currently become more cost-effective than that of rolling out a copper based network. Prices have dropped to $850 per subscriber in the US and lower in countries like The Netherlands, where digging costs are low and housing density is high.

Since 1990, when optical-amplification systems became commercially available, the telecommunications industry has laid a vast network of intercity and transoceanic fiber communication lines. By 2002, an intercontinental network of 250,000 km of submarine communications cable with a capacity of 2.56 Tb/s was completed, and although specific network capacities are privileged information, telecommunications investment reports indicate that network capacity has increased dramatically since 2004.

History

In 1880 Alexander Graham Bell and his assistant Charles Sumner Tainter created a very early precursor to fiber-optic communications, the Photophone, at Bell's newly established Volta Laboratory in Washington, D.C. Bell considered it his most important invention. The device allowed for the transmission of sound on a beam of light. On June 3, 1880, Bell conducted the world's first wireless telephone transmission between two buildings, some 213 meters apart. Due to its use of an atmospheric transmission medium, the Photophone would not prove practical until advances in laser and optical fiber technologies permitted the secure transport of light. The Photophone's first practical use came in military communication systems many decades later.

In 1954 Harold Hopkins and Narinder Singh Kapany showed that rolled fiber glass allowed light to be transmitted. Initially it was considered that the light can traverse in only straight medium.

In 1966 Charles K. Kao and George Hockham proposed optical fibers at STC Laboratories (STL) at Harlow, England, when they showed that the losses of 1,000 dB/km in existing glass (compared to 5-10 dB/km in coaxial cable) was due to contaminants, which could potentially be removed.

Optical fiber was successfully developed in 1970 by Corning Glass Works, with attenuation low enough for communication purposes (about 20dB/km), and at the same time GaAs semiconductor lasers were developed that were compact and therefore suitable for transmitting light through fiber optic cables for long distances.

After a period of research starting from 1975, the first commercial fiber-optic communications system was developed, which operated at a wavelength around 0.8 μm and used GaAs semiconductor lasers. This first-generation system operated at a bit rate of 45 Mbps with repeater spacing of up to 10 km. Soon on 22 April 1977, General Telephone and Electronics sent the first live telephone traffic through fiber optics at a 6 Mbit/s throughput in Long Beach, California.

In October 1973, Corning Glass signed a development contract with CSELT and Pirelli aimed to test fiber optics in an urban environment: in September 1977, the second cable in this test series, named COS-2, was experimentally deployed in two lines (9 km) in Turin, for the first time in a big city, at a speed of 140 Mbit/s.

The second generation of fiber-optic communication was developed for commercial use in the early 1980s, operated at 1.3 μm, and used InGaAsP semiconductor lasers. These early systems were initially limited by multi mode fiber dispersion, and in 1981 the single-mode fiber was revealed to greatly improve system performance, however practical connectors capable of working with single mode fiber proved difficult to develop. In 1984, they had already developed a fiber optic cable that would help further their progress toward making fiber optic cables that would circle the globe. Canadian service provider SaskTel had completed construction of what was then the world's longest commercial fiberoptic network, which covered 3,268 km and linked 52 communities. By 1987, these systems were operating at bit rates of up to 1.7 Gb/s with repeater spacing up to 50 km.

The first transatlantic telephone cable to use optical fiber was TAT-8, based on Desurvire optimized laser amplification technology. It went into operation in 1988.

Third-generation fiber-optic systems operated at 1.55 μm and had losses of about 0.2 dB/km. This development was spurred by the discovery of Indium gallium arsenide and the development of the Indium Gallium Arsenide photodiode by Pearsall. Engineers overcame earlier difficulties with pulse-spreading at that wavelength using conventional InGaAsP semiconductor lasers. Scientists overcame this difficulty by using dispersion-shifted fibers designed to have minimal dispersion at 1.55 μm or by limiting the laser spectrum to a single longitudinal mode. These developments eventually allowed third-generation systems to operate commercially at 2.5 Gbit/s with repeater spacing in excess of 100 km.

The fourth generation of fiber-optic communication systems used optical amplification to reduce the need for repeaters and wavelength-division multiplexing to increase data capacity. These two improvements caused a revolution that resulted in the doubling of system capacity every six months starting in 1992 until a bit rate of 10 Tb/s was reached by 2001. In 2006 a bit-rate of 14 Tbit/s was reached over a single 160 km line using optical amplifiers.

The focus of development for the fifth generation of fiber-optic communications is on extending the wavelength range over which a WDM system can operate. The conventional wavelength window, known as the C band, covers the wavelength range 1.53-1.57 μm, and *dry fiber* has a low-loss window promising an extension of that range to 1.30-1.65 μm. Other developments include the concept of "optical solitons, " pulses that preserve their shape by counteracting the effects of dispersion with the nonlinear effects of the fiber by using pulses of a specific shape.

In the late 1990s through 2000, industry promoters, and research companies such as KMI, and RHK predicted massive increases in demand for communications bandwidth due to increased use of the Internet, and commercialization of various bandwidth-intensive consumer services, such as video on demand. Internet protocol data traffic was increasing exponentially, at a faster rate than integrated circuit complexity had increased under Moore's Law. From the bust of the dot-com bubble through 2006, however, the main trend in the industry has been consolidation of firms and offshoring of manufacturing to reduce costs. Companies such as Verizon and AT&T have taken advantage of fiber-optic communications to deliver a variety of high-throughput data and broadband services to consumers' homes.

Technology

Modern fiber-optic communication systems generally include an optical transmitter to convert an electrical signal into an optical signal to send into the optical fiber, a cable containing bundles of multiple optical fibers that is routed through underground conduits and buildings, multiple kinds of amplifiers, and an optical receiver to recover the signal as an electrical signal. The information transmitted is typically digital information generated by computers, telephone systems, and cable television companies.

Transmitters

The most commonly used optical transmitters are semiconductor devices such as light-emitting diodes (LEDs) and laser diodes. The difference between LEDs and laser diodes is that LEDs produce incoherent light, while laser diodes produce coherent light. For use in optical communications, semiconductor optical transmitters must be designed to be compact, efficient, and reliable, while operating in an optimal wavelength range, and directly modulated at high frequencies.

In its simplest form, a LED is a forward-biased p-n junction, emitting light through spontaneous emission, a phenomenon referred to as electroluminescence. The emitted light is incoherent with a relatively wide spectral width of 30-60 nm. LED light transmission is also inefficient, with only about 1% of input power, or about 100 microwatts, eventually converted into launched power which has been coupled into the optical fiber. However, due to their relatively simple design, LEDs are very useful for low-cost applications.

A GBIC module (shown here with its cover removed), is an optical and electrical transceiver. The electrical connector is at top right, and the optical connectors are at bottom left

Communications LEDs are most commonly made from Indium gallium arsenide phosphide (In-GaAsP) or gallium arsenide (GaAs). Because InGaAsP LEDs operate at a longer wavelength than GaAs LEDs (1.3 micrometers vs. 0.81-0.87 micrometers), their output spectrum, while equivalent in energy is wider in wavelength terms by a factor of about 1.7. The large spectrum width of LEDs is subject to higher fiber dispersion, considerably limiting their bit rate-distance product (a common measure of usefulness). LEDs are suitable primarily for local-area-network applications with bit rates of 10-100 Mbit/s and transmission distances of a few kilometers. LEDs have also been developed that use several quantum wells to emit light at different wavelengths over a broad spectrum, and are currently in use for local-area WDM (Wavelength-Division Multiplexing) networks.

Today, LEDs have been largely superseded by VCSEL (Vertical Cavity Surface Emitting Laser) devices, which offer improved speed, power and spectral properties, at a similar cost. Common VCSEL devices couple well to multi mode fiber.

A semiconductor laser emits light through stimulated emission rather than spontaneous emission, which results in high output power (~100 mW) as well as other benefits related to the nature of coherent light. The output of a laser is relatively directional, allowing high coupling efficiency (~50 %) into single-mode fiber. The narrow spectral width also allows for high bit rates since it reduces the effect of chromatic dispersion. Furthermore, semiconductor lasers can be modulated directly at high frequencies because of short recombination time.

Commonly used classes of semiconductor laser transmitters used in fiber optics include VCSEL (Vertical-Cavity Surface-Emitting Laser), Fabry–Pérot and DFB (Distributed Feed Back).

Laser diodes are often directly modulated, that is the light output is controlled by a current applied directly to the device. For very high data rates or very long distance *links*, a laser source may be operated continuous wave, and the light modulated by an external device such as an electro-absorption modulator or Mach–Zehnder interferometer. External modulation increases the achievable link distance by eliminating laser chirp, which broadens the linewidth of directly modulated lasers, increasing the chromatic dispersion in the fiber.

A transceiver is a device combining a transmitter and a receiver in a single housing.

Receivers

The main component of an optical receiver is a photodetector, which converts light into electricity using the photoelectric effect. The primary photodetectors for telecommunications are made from Indium gallium arsenide The photodetector is typically a semiconductor-based photodiode. Several types of photodiodes include p-n photodiodes, p-i-n photodiodes, and avalanche photodiodes. Metal-semiconductor-metal (MSM) photodetectors are also used due to their suitability for circuit integration in regenerators and wavelength-division multiplexers.

Optical-electrical converters are typically coupled with a transimpedance amplifier and a limiting amplifier to produce a digital signal in the electrical domain from the incoming optical signal, which may be attenuated and distorted while passing through the channel. Further signal processing such as clock recovery from data (CDR) performed by a phase-locked loop may also be applied before the data is passed on.

Fiber Cable Types

A cable reel trailer with conduit that can carry optical fiber

An optical fiber cable consists of a core, cladding, and a buffer (a protective outer coating), in which the cladding guides the light along the core by using the method of total internal reflection. The core and the cladding (which has a lower-refractive-index) are usually made of high-quality silica glass, although they can both be made of plastic as well. Connecting two optical fibers is done by fusion splicing or mechanical splicing and requires special skills and interconnection technology due to the microscopic precision required to align the fiber cores.

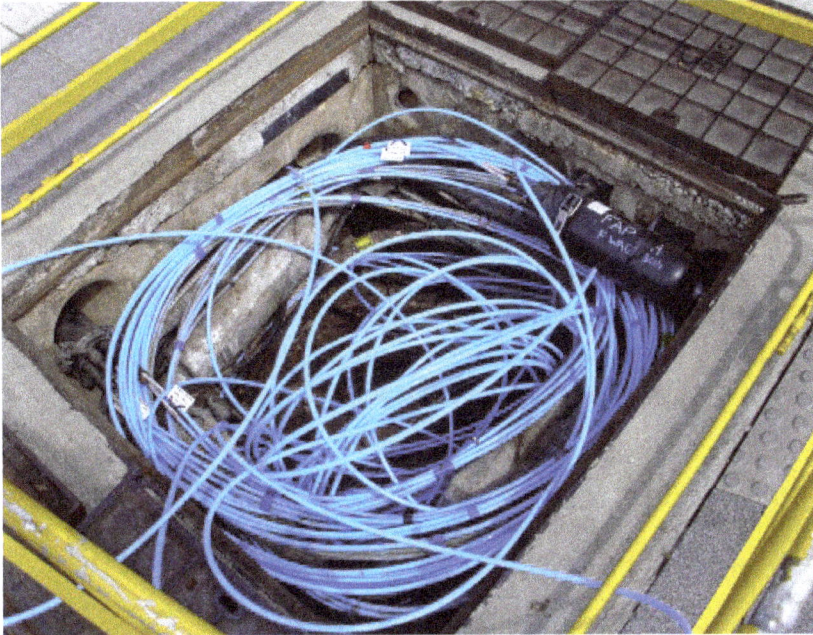

Multi-mode optical fiber in an underground service pit

Two main types of optical fiber used in optic communications include multi-mode optical fibers and single-mode optical fibers. A multi-mode optical fiber has a larger core (≥ 50 micrometers), allowing less precise, cheaper transmitters and receivers to connect to it as well as cheaper connectors. However, a multi-mode fiber introduces multimode distortion, which often limits the bandwidth and length of the link. Furthermore, because of its higher dopant content, multi-mode fibers are usually expensive and exhibit higher attenuation. The core of a single-mode fiber is smaller (<10 micrometers) and requires more expensive components and interconnection methods, but allows much longer, higher-performance links.

In order to package fiber into a commercially viable product, it typically is protectively coated by using ultraviolet (UV), light-cured acrylate polymers, then terminated with optical fiber connectors, and finally assembled into a cable. After that, it can be laid in the ground and then run through the walls of a building and deployed aerially in a manner similar to copper cables. These fibers require less maintenance than common twisted pair wires, once they are deployed.

Specialized cables are used for long distance subsea data transmission, e.g. transatlantic communications cable. New (2011–2013) cables operated by commercial enterprises (Emerald Atlantis, Hibernia Atlantic) typically have four strands of fiber and cross the Atlantic (NYC-London) in 60-70ms. Cost of each such cable was about $300M in 2011. *source: The Chronicle Herald.*

Another common practice is to bundle many fiber optic strands within long-distance power transmission cable. This exploits power transmission rights of way effectively, ensures a power company can own and control the fiber required to monitor its own devices and lines, is effectively immune to tampering, and simplifies the deployment of smart grid technology.

Amplifier

The transmission distance of a fiber-optic communication system has traditionally been limited by fiber attenuation and by fiber distortion. By using opto-electronic repeaters, these problems have been eliminated. These repeaters convert the signal into an electrical signal, and then use a transmitter to send the signal again at a higher intensity than was received, thus counteracting the loss incurred in the previous segment. Because of the high complexity with modern wavelength-division multiplexed signals (including the fact that they had to be installed about once every 20 km), the cost of these repeaters is very high.

An alternative approach is to use an optical amplifier, which amplifies the optical signal directly without having to convert the signal into the electrical domain. It is made by doping a length of fiber with the rare-earth mineral erbium, and *pumping* it with light from a laser with a shorter wavelength than the communications signal (typically 980 nm). Amplifiers have largely replaced repeaters in new installations.

Wavelength-division Multiplexing

Wavelength-division multiplexing (WDM) is the practice of multiplying the available capacity of optical fibers through use of parallel channels, each channel on a dedicated wavelength of light. This requires a wavelength division multiplexer in the transmitting equipment and a demultiplexer (essentially a spectrometer) in the receiving equipment. Arrayed waveguide gratings are commonly used for multiplexing and demultiplexing in WDM. Using WDM technology now commercially available, the bandwidth of a fiber can be divided into as many as 160 channels to support a combined bit rate in the range of 1.6 Tbit/s.

Parameters

Bandwidth–distance Product

Because the effect of dispersion increases with the length of the fiber, a fiber transmission system is often characterized by its bandwidth–distance product, usually expressed in units of MHz·km. This value is a product of bandwidth and distance because there is a trade off between the bandwidth of the signal and the distance it can be carried. For example, a common multi-mode fiber with bandwidth–distance product of 500 MHz·km could carry a 500 MHz signal for 1 km or a 1000 MHz signal for 0.5 km.

Engineers are always looking at current limitations in order to improve fiber-optic communication, and several of these restrictions are currently being researched.

Record Speeds

Each fiber can carry many independent channels, each using a different wavelength of light (wavelength-division multiplexing). The net data rate (data rate without overhead bytes) per fiber is the per-channel data rate reduced by the FEC overhead, multiplied by the number of channels (usually up to eighty in commercial dense WDM systems as of 2008).

Year	Organization	Effective speed	WDM channels	Per channel speed	Distance
2009	Alcatel-Lucent	15 Tbit/s	155	100 Gbit/s	90 km
2010	NTT	69.1 Tbit/s	432	171 Gbit/s	240 km
2011	KIT	26 Tbit/s	1	26 Tbit/s	50 km
2011	NEC	101 Tbit/s	370	273 Gbit/s	165 km
2012	NEC, Corning	1.05 Petabit/s	12 core fiber		52.4 km

While the physical limitations of electrical cable prevent speeds in excess of 10 Gigabits per second, the physical limitations of fiber optics have not yet been reached

In 2013, *New Scientist* reported that a team at the University of Southampton had achieved a throughput of 73.7 Tbit per second, with the signal traveling at 99.7% the speed of light through a hollow-core photonic crystal fiber.

Dispersion

For modern glass optical fiber, the maximum transmission distance is limited not by direct material absorption but by several types of dispersion, or spreading of optical pulses as they travel along the fiber. Dispersion in optical fibers is caused by a variety of factors. Intermodal dispersion, caused by the different axial speeds of different transverse modes, limits the performance of multi-mode fiber. Because single-mode fiber supports only one transverse mode, intermodal dispersion is eliminated.

In single-mode fiber performance is primarily limited by chromatic dispersion (also called group velocity dispersion), which occurs because the index of the glass varies slightly depending on the wavelength of the light, and light from real optical transmitters necessarily has nonzero spectral width (due to modulation). Polarization mode dispersion, another source of limitation, occurs because although the single-mode fiber can sustain only one transverse mode, it can carry this mode with two different polarizations, and slight imperfections or distortions in a fiber can alter the propagation velocities for the two polarizations. This phenomenon is called fiber birefringence and can be counteracted by polarization-maintaining optical fiber. Dispersion limits the bandwidth of the fiber because the spreading optical pulse limits the rate that pulses can follow one another on the fiber and still be distinguishable at the receiver.

Some dispersion, notably chromatic dispersion, can be removed by a 'dispersion compensator'. This works by using a specially prepared length of fiber that has the opposite dispersion to that induced by the transmission fiber, and this sharpens the pulse so that it can be correctly decoded by the electronics.

Attenuation

Fiber attenuation, which necessitates the use of amplification systems, is caused by a combination of material absorption, Rayleigh scattering, Mie scattering, and connection losses. Although material absorption for pure silica is only around 0.03 dB/km (modern fiber has attenuation around 0.3 dB/km), impurities in the original optical fibers caused attenuation of about 1000 dB/km. Other forms of attenuation are caused by physical stresses to the fiber, microscopic fluctuations in density, and imperfect splicing techniques.

Transmission Windows

Each effect that contributes to attenuation and dispersion depends on the optical wavelength. There are wavelength bands (or windows) where these effects are weakest, and these are the most favorable for transmission. These windows have been standardized, and the currently defined bands are the following:

Band	Description	Wavelength Range
O band	original	1260 to 1360 nm
E band	extended	1360 to 1460 nm
S band	short wavelengths	1460 to 1530 nm
C band	conventional ("erbium window")	1530 to 1565 nm
L band	long wavelengths	1565 to 1625 nm
U band	ultralong wavelengths	1625 to 1675 nm

Note that this table shows that current technology has managed to bridge the second and third windows that were originally disjoint.

Historically, there was a window used below the O band, called the first window, at 800-900 nm; however, losses are high in this region so this window is used primarily for short-distance communications. The current lower windows (O and E) around 1300 nm have much lower losses. This region has zero dispersion. The middle windows (S and C) around 1500 nm are the most widely used. This region has the lowest attenuation losses and achieves the longest range. It does have some dispersion, so dispersion compensator devices are used to remove this.

Regeneration

When a communications link must span a larger distance than existing fiber-optic technology is capable of, the signal must be *regenerated* at intermediate points in the link by optical communications repeaters. Repeaters add substantial cost to a communication system, and so system designers attempt to minimize their use.

Recent advances in fiber and optical communications technology have reduced signal degradation so far that *regeneration* of the optical signal is only needed over distances of hundreds of kilometers. This has greatly reduced the cost of optical networking, particularly over undersea spans where the cost and reliability of repeaters is one of the key factors determining the performance of the whole cable system. The main advances contributing to these performance improvements are dispersion management, which seeks to balance the effects of dispersion against non-linearity; and solitons, which use nonlinear effects in the fiber to enable dispersion-free propagation over long distances.

Last Mile

Although fiber-optic systems excel in high-bandwidth applications, optical fiber has been slow to achieve its goal of fiber to the premises or to solve the last mile problem. However, as bandwidth demand increases, more and more progress towards this goal can be observed. In Japan, for instance EPON has largely replaced DSL as a broadband Internet source. South Korea's KT also provides a service called FTTH (Fiber To The Home), which provides fiber-optic connections to the subscriber's home. The largest FTTH deployments are in Japan, South Korea, and China. Singapore started implementation of their all-fiber Next Generation Nationwide Broadband Network (Next Gen NBN), which is slated for completion in 2012 and is being installed by OpenNet. Since they began rolling out services in September 2010, Network coverage in Singapore has reached 85% nationwide.

In the US, Verizon Communications provides a FTTH service called FiOS to select high-ARPU (Average Revenue Per User) markets within its existing territory. The other major surviving ILEC (or Incumbent Local Exchange Carrier), AT&T, uses a FTTN (Fiber To The Node) service called U-verse with twisted-pair to the home. Their MSO competitors employ FTTN with coax using HFC. All of the major access networks use fiber for the bulk of the distance from the service provider's network to the customer.

Also in the US, Wilson Utilities, located in Wilson, North Carolina, has implemented FTTH and has successfully achieved 1 gigabit fiber to the home. This was implemented in late 2013. Wilson Utilities first rolled out their FTTH (Fiber to the Home) in 2012 with speeds offerings of 20/40/60/100 megabits per second. Their service is referred to as GreenLight.

Some other small cities in the US, such as Morristown, TN, have had their local utility company, Morristown Utility Systems in this case, deploy FTTH, offering symmetric gigabit speeds to each subscriber (though most are 50/50 or 100/100 mbit). It's called MUS Fibernet. AT&T and others have aggressively sought legislation at the state level to prevent further competition from municipalities, despite their low investment in rural areas.

The globally dominant access network technology is EPON (Ethernet Passive Optical Network). In Europe, and among telcos in the United States, BPON (ATM-based Broadband PON) and GPON (Gigabit PON) had roots in the FSAN (Full Service Access Network) and ITU-T standards organizations under their control.

Comparison With Electrical Transmission

A mobile fiber optic splice lab used to access and splice underground cables

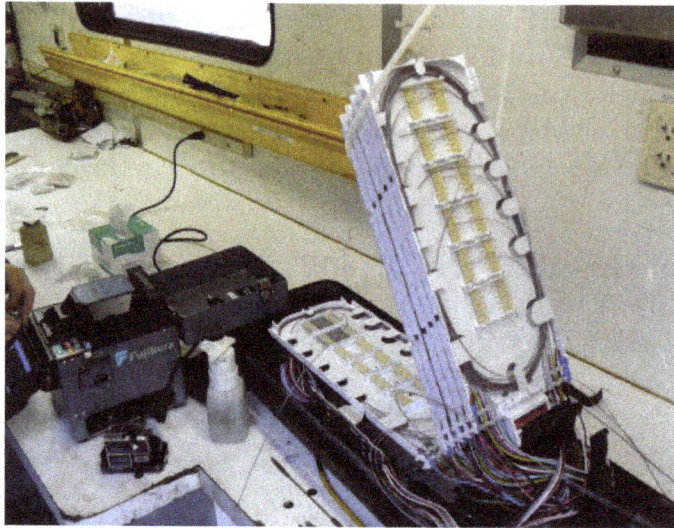

An underground fiber optic splice enclosure opened up

The choice between optical fiber and electrical (or copper) transmission for a particular system is made based on a number of trade-offs. Optical fiber is generally chosen for systems requiring higher bandwidth or spanning longer distances than electrical cabling can accommodate.

The main benefits of fiber are its exceptionally low loss (allowing long distances between amplifiers/repeaters), its absence of ground currents and other parasite signal and power issues common to long parallel electric conductor runs (due to its reliance on light rather than electricity for transmission, and the dielectric nature of fiber optic), and its inherently high data-carrying capacity. Thousands of electrical links would be required to replace a single high bandwidth fiber cable. Another benefit of fibers is that even when run alongside each other for long distances, fiber cables experience effectively no crosstalk, in contrast to some types of electrical transmission lines. Fiber can be installed in areas with high electromagnetic interference (EMI), such as alongside utility lines, power lines, and railroad tracks. Nonmetallic all-dielectric cables are also ideal for areas of high lightning-strike incidence.

For comparison, while single-line, voice-grade copper systems longer than a couple of kilometers require in-line signal repeaters for satisfactory performance; it is not unusual for optical systems to go over 100 kilometers (62 mi), with no active or passive processing. Single-mode fiber cables are commonly available in 12 km lengths, minimizing the number of splices required over a long cable run. Multi-mode fiber is available in lengths up to 4 km, although industrial standards only mandate 2 km unbroken runs.

In short distance and relatively low bandwidth applications, electrical transmission is often preferred because of its

- Lower material cost, where large quantities are not required

- Lower cost of transmitters and receivers

- Capability to carry electrical power as well as signals (in appropriately designed cables)

- Ease of operating transducers in linear mode.

- Crosstalk from nearby cables and other parasitical unwanted signals increase profits from replacement and mitigation devices.

Optical fibers are more difficult and expensive to splice than electrical conductors. And at higher powers, optical fibers are susceptible to fiber fuse, resulting in catastrophic destruction of the fiber core and damage to transmission components.

Because of these benefits of electrical transmission, optical communication is not common in short box-to-box, backplane, or chip-to-chip applications; however, optical systems on those scales have been demonstrated in the laboratory.

In certain situations fiber may be used even for short distance or low bandwidth applications, due to other important features:

- Immunity to electromagnetic interference, including nuclear electromagnetic pulses.

- High electrical resistance, making it safe to use near high-voltage equipment or between areas with different earth potentials.

- Lighter weight—important, for example, in aircraft.

- No sparks—important in flammable or explosive gas environments.

- Not electromagnetically radiating, and difficult to tap without disrupting the signal—important in high-security environments.

- Much smaller cable size—important where pathway is limited, such as networking an existing building, where smaller channels can be drilled and space can be saved in existing cable ducts and trays.

- Resistance to corrosion due to non-metallic transmission medium

Optical fiber cables can be installed in buildings with the same equipment that is used to install copper and coaxial cables, with some modifications due to the small size and limited pull tension and bend radius of optical cables. Optical cables can typically be installed in duct systems in spans of 6000 meters or more depending on the duct's condition, layout of the duct system, and installation technique. Longer cables can be coiled at an intermediate point and pulled farther into the duct system as necessary.

Governing Standards

In order for various manufacturers to be able to develop components that function compatibly in fiber optic communication systems, a number of standards have been developed. The International Telecommunications Union publishes several standards related to the characteristics and performance of fibers themselves, including

- ITU-T G.651, "Characteristics of a 50/125 μm multimode graded index optical fibre cable"

- ITU-T G.652, "Characteristics of a single-mode optical fibre cable"

Other standards specify performance criteria for fiber, transmitters, and receivers to be used together in conforming systems. Some of these standards are:

- 100 Gigabit Ethernet

- 10 Gigabit Ethernet

- Fibre Channel

- Gigabit Ethernet

- HIPPI

- Synchronous Digital Hierarchy

- Synchronous Optical Networking

- Optical Transport Network (OTN)

TOSLINK is the most common format for digital audio cable using plastic optical fiber to connect digital sources to digital receivers.

Electromagnetically Induced Transparency

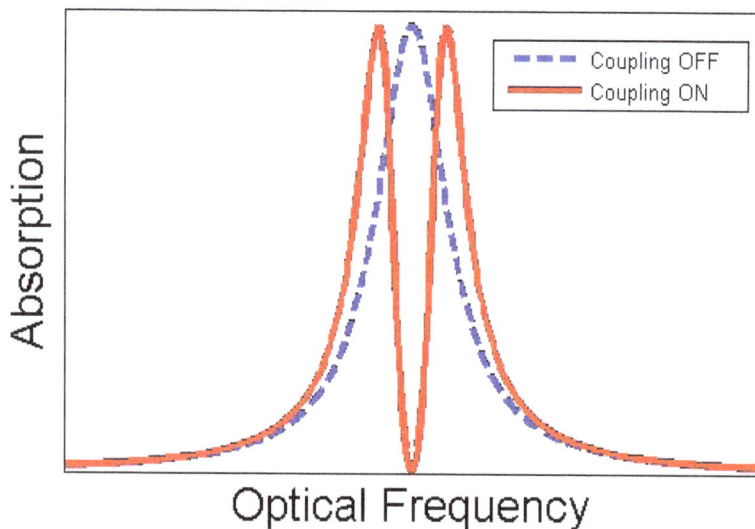

The effect of EIT on a typical absorption line. A weak probe normally experiences absorption shown in blue. A second coupling beam induces EIT and creates a "window" in the absorption region (red). This plot is a computer simulation of EIT in an InAs/GaAs quantum dot

Electromagnetically induced transparency (EIT) is a coherent optical nonlinearity which renders a medium transparent window over a narrow spectral range within an absorption line. Extreme dispersion is also created within this transparency "window" which leads to "slow light", described below. It is in essence a quantum interference effect that permits the propagation of light through an otherwise opaque atomic medium.

Observation of EIT involves two optical fields (highly coherent light sources, such as lasers) which are tuned to interact with three quantum states of a material. The "probe" field is tuned near resonance between two of the states and measures the absorption spectrum of the transition. A much stronger "coupling" field is tuned near resonance at a different transition. If the states are selected properly, the presence of the coupling field will create a spectral "window" of transparency which will be detected by the probe. The coupling laser is sometimes referred to as the "control" or "pump", the latter in analogy to incoherent optical nonlinearities such as spectral hole burning or saturation.

EIT is based on the destructive interference of the transition probability amplitude between atomic states. Closely related to EIT are coherent population trapping (CPT) phenomena.

The quantum interference in EIT can be exploited to laser cool atomic particles, even down to the quantum mechanical ground state of motion. This has recently been used to directly image individual atoms trapped in an optical lattice.

Medium Requirements

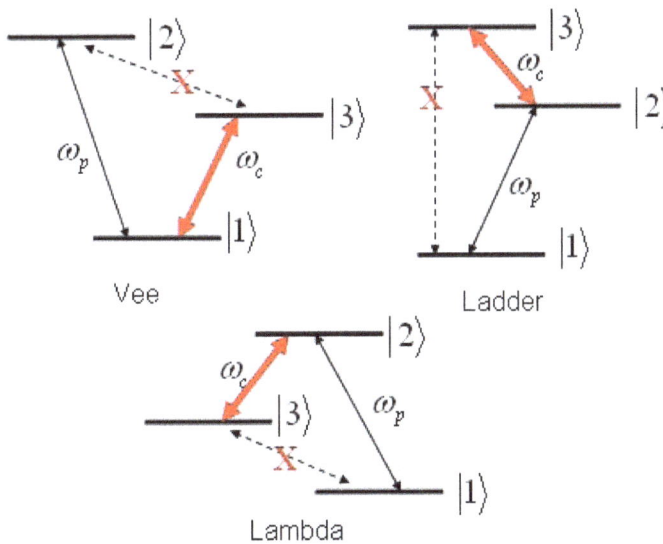

EIT level schemes can be sorted into three categories; vee, ladder, and lambda.

There are specific restrictions on the configuration of the three states. Two of the three possible transitions between the states must be "dipole allowed", i.e. the transitions can be induced by an oscillating electric field. The third transition must be "dipole forbidden." One of the three states is connected to the other two by the two optical fields. The three types of EIT schemes are differentiated by the energy differences between this state and the other two. The schemes are the ladder, vee, and lambda. Any real material system may contain many triplets of states which could theoretically support EIT, but there are several practical limitations on which levels can actually be used.

Also important are the dephasing rates of the individual states. In any real system at finite temperature there are processes which cause a scrambling of the phase of the quantum states. In the gas phase, this means usually collisions. In solids, dephasing is due to interaction of the electronic states with the host lattice. The dephasing of state $|3\rangle$ is especially important; ideally $|3\rangle$ should be a robust, metastable state.

Current EIT research uses atomic systems in dilute gases, solid solutions, or more exotic states such as Bose–Einstein condensate. EIT has been demonstrated in electromechanical and optomechanical systems, where it is known as optomechanically induced transparency. Work is also being done in semiconductor nanostructures such as quantum wells, quantum wires and quantum dots

Theory

EIT was first proposed theoretically by professor Jakob Khanin and graduate student Olga Kocharovskaya at Gorky State University (present: Nizhny Novgorod), Russia; there are now several different approaches to a theoretical treatment of EIT. One approach is to extend the density matrix treatment used to derive Rabi oscillation of a two-state, single field system. In this picture the probability amplitude for the system to transfer between states can interfere destructively, preventing absorption. In this context, "interference" refers to interference between *quantum events* (transitions) and not optical interference of any kind. As a specific example, consider the lambda scheme shown above. Absorption of the probe is defined by transition from $|1\rangle$ to $|2\rangle$. The fields can drive population from $|1\rangle$-$|2\rangle$ directly or from $|1\rangle$-$|2\rangle$-$|3\rangle$-$|2\rangle$. The probability amplitudes for the different paths interfere destructively. If $|3\rangle$ has a comparatively long lifetime, then the result will be a transparent window completely inside of the $|1\rangle$-$|2\rangle$ absorption line.

Another approach is the "dressed state" picture, wherein the system + coupling field Hamiltonian is diagonalized and the effect on the probe is calculated in the new basis. In this picture EIT resembles a combination of Autler-Townes splitting and Fano interference between the dressed states. Between the doublet peaks, in the center of the transparency window, the quantum probability amplitudes for the probe to cause a transition to either state cancel.

A polariton picture is particularly important in describing stopped light schemes. Here, the photons of the probe are coherently "transformed" into "dark state polaritons" which are excitations of the medium. These excitations exist (or can be "stored") for a length of time dependent only on the dephasing rates.

Slow Light and Stopped Light

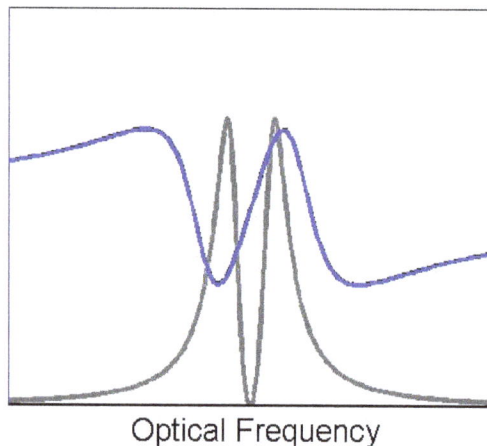

Optical Frequency

Rapid change of index of refraction (blue) in a region of rapidly changing absorption (gray) associated with EIT. The steep and positive linear region of the refractive index in the center of the transparency window gives rise to slow light

It is important to realize that EIT is only one of many diverse mechanisms which can produce slow light. The Kramers–Kronig relations dictate that a change in absorption (or gain) over a narrow spectral range must be accompanied by a change in refractive index over a similarly narrow region. This rapid and *positive* change in refractive index produces an extremely low group velocity. The first experimental observation of the low group velocity produced by EIT was by Boller, İmamoğlu, and Harris at Stanford University in 1991 in strontium. In 1999 Lene Hau reported slowing light in a medium of ultracold sodium atoms, achieving this by using quantum interference effects responsible for electromagnetically induced transparency (EIT). Her group performed copious research regarding EIT with Stephen E. Harris. "Using detailed numerical simulations, and analytical theory, we study properties of micro-cavities which incorporate materials that exhibit Electro-magnetically Induced Transparency (EIT) or Ultra Slow Light (USL). We find that such systems, while being miniature in size (order wavelength), and integrable, can have some outstanding properties. In particular, they could have lifetimes orders of magnitude longer than other existing systems, and could exhibit non-linear all-optical switching at single photon power levels. Potential applications include miniature atomic clocks, and all-optical quantum information processing." The current record for slow light in an EIT medium is held by Budker, Kimball, Rochester, and Yashchuk at U.C. Berkeley in 1999. Group velocities as low as 8 m/s were measured in a warm thermal rubidium vapor.

Stopped light, in the context of an EIT medium, refers to the *coherent* transfer of photons to the quantum system and back again. In principle, this involves switching *off* the coupling beam in an adiabatic fashion while the probe pulse is still inside of the EIT medium. There is experimental evidence of trapped pulses in EIT medium. In authors created a stationary light pulse inside the atomic coherent media. In 2009 researchers from Harvard University and MIT demonstrated a few-photon optical switch for quantum optics based on the slow light ideas. Lene Hau and a team from Harvard University were the first to demonstrate stopped light.

Superlens

A practical superlens, or super lens, is a lens which uses metamaterials to go beyond the diffraction limit. The diffraction limit is a feature of conventional lenses and microscopes that limits the fineness of their resolution. Many lens designs have been proposed that go beyond the diffraction limit in some way, but there are constraints and obstacles involved in realizing each of them.

Development of the Concept

As Ernst Abbe reported in 1873, the lens of a camera or microscope is incapable of capturing some very fine details of any given image. The super lens, on the other hand, is intended to capture these fine details. Consequently, conventional lens limitation has inhibited progress in certain areas of the biological sciences. This is because a virus or DNA molecule is out of visual range with the highest powered microscopes. Also, this limitation inhibits seeing the minute processes of cellular proteins moving alongside microtubules of a living cell in their natural environments. Additionally, computer chips and the interrelated microelectronics are manufactured to smaller and smaller scales. This requires specialized optical equipment, which is also limited because these use the conventional lens. Hence, the principles governing a super lens show that it has potential for imaging a DNA molecule and cellular protein processes, or aiding in the manufacture of even smaller computer chips and microelectronics.

Furthermore, conventional lenses capture only the propagating light waves. These are waves that travel from a light source or an object to a lens, or the human eye. This can alternatively be studied as the far field. In contrast, a superlens captures propagating light waves and waves that stay on top of the surface of an object, which, alternatively, can be studied as both the far field and the near field.

In the early 20th century the term "superlens" was used by Dennis Gabor to describe something quite different: a compound lenslet array system.

Theory

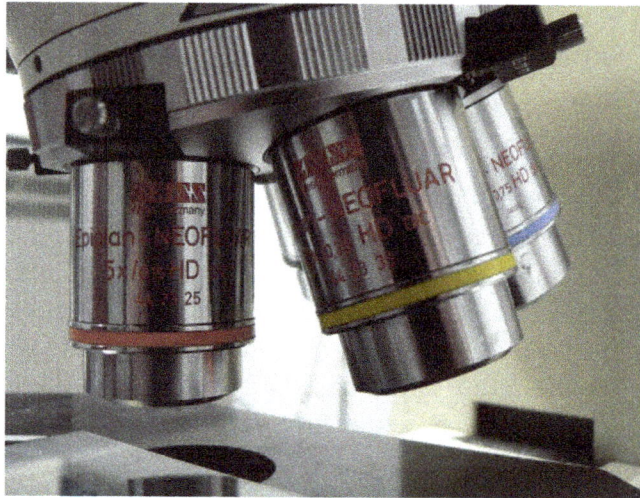

The Binocular microscope is a conventional optical system. Spatial resolution is confined by a diffraction limit that is a little above 200 nanometers.

Image Formation

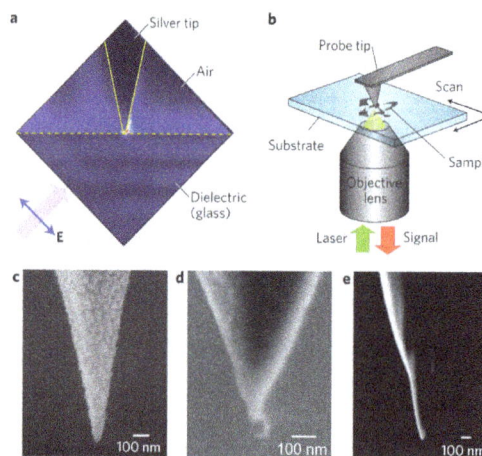

Schematic depictions and images of commonly used metallic nanoprobes that can be used to see a sample in nanometer resolution. Notice that the tips of the three nanoprobes are 100 nanometers.

An image of an object can be defined as a tangible or visible representation of the features of that object. A requirement for image formation is interaction with fields of electromagnetic radiation. Furthermore, the level of feature detail, or image resolution, is limited to a length of a wave of radiation. For example, with optical microscopy, image production and resolution depends on the length of a wave of visible light. However, with a superlens, this limitation may be removed, and a new class of image generated.

Electron beam lithography can overcome this resolution limit. Optical microscopy, on the other hand cannot, being limited to some value just above 200 nanometers. However, new technologies combined with optical microscopy are beginning to allow for increased feature resolution.

One definition of being constrained by the resolution barrier, is a resolution cut off at half the wavelength of light. The visible spectrum has a range that extends from 390 nanometers to 750 nanometers. Green light, half way in between, is around 500 nanometers. Microscopy takes into account parameters such as lens aperture, distance from the object to the lens, and the refractive index of the observed material. This combination defines the resolution cutoff, or Microscopy's optical limit, which tabulates to 200 nanometers. Therefore, conventional lenses, which literally construct an image of an object by using "ordinary" light waves, discard information that produce very fine, and minuscule details of the object that are contained in evanescent waves. These dimensions are less than 200 nanometers. For this reason, conventional optical systems, such as microscopes, have been unable to accurately image very small, nanometer-sized structures or nanometer-sized organisms in vivo, such as individual viruses, or DNA molecules.

The limitations of standard optical microscopy (bright field microscopy) lie in three areas:

- The technique can only image dark or strongly refracting objects effectively.

- Diffraction limits the object, or cell's, resolution to approximately 200 nanometers.

- Out of focus light from points outside the focal plane reduces image clarity.

Live biological cells in particular generally lack sufficient contrast to be studied successfully, because the internal structures of the cell are mostly colorless and transparent. The most common way to increase contrast is to stain the different structures with selective dyes, but often this involves killing and fixing the sample. Staining may also introduce artifacts, apparent structural details that are caused by the processing of the specimen and are thus not a legitimate feature of the specimen.

Conventional Lens

The conventional glass lens is pervasive throughout our society and in the sciences. It is one of the fundamental tools of optics simply because it interacts with various wavelengths of light. At the same time, the wavelength of light can be analogous to the width of a pencil used to draw the ordinary images. The limit becomes noticeable, for example, when the laser used in a digital video system can only detect and deliver details from a DVD based on the wavelength of light. The image cannot be rendered any sharper beyond this limitation.

DVD – A DVD is a large, fast compact digital disk that can hold video, audio, and/or computer data. A laser is employed for data transfer.

Thus, when an object emits or reflects light there are two types of electromagnetic radiation associated with this phenomenon. These are the near field radiation and the far field radiation. As implied by its description, the far field escapes beyond the object. It is then easily captured and manipulated by a conventional glass lens. However, useful (nanometer-sized) resolution details are not observed, because they are hidden in the near field. They remain localized, staying much closer to the light emitting object, unable to travel, and unable to be captured by the conventional lens. Controlling the near field radiation, for high resolution, can be accomplished with a new class of materials not easily obtained in nature. These are unlike familiar solids, such as crystals, which derive their properties from atomic and molecular units. The new material class, termed metamaterials, obtains its properties from its artificially larger structure. This has resulted in novel properties, and novel responses, which allow for details of images that surpass the limitations imposed by the wavelength of light.

Subwavelength Imaging

This has led to the desire to view live biological cell interactions in a real time, natural environment, and the need for *subwavelength imaging*. Subwavelength imaging can be defined as optical microscopy with the ability to see details of an object or organism below the wavelength of visible light. In other words, to have the capability to observe, in real time, below 200 nanometers. Optical microscopy is a non-invasive technique and technology because everyday light is the transmission medium. Imaging below the optical limit in optical microscopy (subwavelength) can be engineered for the cellular level, and nanometer level in principle.

For example, in 2007 a technique was demonstrated where a metamaterials-based lens coupled with a conventional optical lens could manipulate visible light to see (nanoscale) patterns that were too small to be observed with an ordinary optical microscope. This has potential applications not only for observing a whole living cell, or for observing cellular processes, such as how proteins and fats move in and out of cells. In the technology domain, it could be used to improve the first steps of photolithography and nanolithography, essential for manufacturing ever smaller computer chips.

The "Electrocomposeur" was an electron-beam lithography machine (Electron microscope) designed
for mask writing. It was developed in the early 1970s and deployed in the mid 1970s

Focusing at subwavelength has become a unique imaging technique which allows visualization of features on the viewed object which are smaller than the wavelength of the photons in use. A photon is the minimum unit of light. While previously thought to be physically impossible, subwavelength imaging has been made possible through the development of metamaterials. This is generally accomplished using a layer of metal such as gold or silver a few atoms thick, which acts as a superlens, or by means of 1D and 2D photonic crystals. There is a subtle interplay between propagating waves, evanescent waves, near field imaging and far field imaging discussed in the sections below.

Early Subwavelength Imaging

Metamaterial lenses (*Superlens*) are able to reconstruct nanometer sized images by producing a negative refractive index in each instance. This compensates for the swiftly decaying evanescent waves. Prior to metamaterials, numerous other techniques had been proposed and even demonstrated for creating super-resolution microscopy. As far back as 1928, Edward Hutchinson Synge, a scientist, is given credit for conceiving and developing the idea for what would ultimately become near-field scanning optical microscopy.

In 1974 proposals for two-dimensional fabrication techniques were presented. These proposals included contact imaging to create a pattern in relief, photolithography, electron lithography, X-ray lithography, or ion bombardment, on an appropriate planar substrate. The shared technological goals of the metamaterial lens and the variety of lithography aim to optically resolve features having dimensions much smaller than that of the vacuum wavelength of the exposing light. In 1981 two different techniques of contact imaging of planar (flat) submicroscopic metal patterns with blue light (400 nm) were demonstrated. One demonstration resulted in an image resolution of 100 nm and the other a resolution of 50 to 70 nm.

Since at least 1998 near field optical lithography was designed to create nanometer-scale features. Research on this technology continued as the first experimentally demonstrated negative index metamaterial came into existence in 2000–2001. The effectiveness of electron-beam lithography was also being researched at the beginning of the new millennium for nanometer-scale applications. Imprint lithography was shown to have desirable advantages for nanometer-scaled research and technology.

Advanced deep UV photolithography can now offer sub-100 nm resolution, yet the minimum feature size and spacing between patterns are determined by the diffraction limit of light. Its derivative technologies such as evanescent near-field lithography, near-field interference lithography, and phase-shifting mask lithography were developed to overcome the diffraction limit.

In the year 2000, John Pendry proposed using a metamaterial lens to achieve nanometer-scaled imaging for focusing below the wavelength of light.

Analysis of the Diffraction Limit

The original problem of the perfect lens: The general expansion of an EM field emanating from a source consists of both propagating waves and near-field or evanescent waves. An example of a 2-D line source with an electric field which has S-polarization will have plane waves consisting of propagating and evanescent components, which advance parallel to the interface. As both the propagating and the smaller evanescent waves advance in a direction parallel to the medium interface, evanescent waves decay in the direction of propagation. Ordinary (positive index) optical elements can refocus the propagating components, but the exponentially decaying inhomogeneous components are always lost, leading to the diffraction limit for focusing to an image.

A superlens is a lens which is capable of subwavelength imaging, allowing for magnification of near field rays. Conventional lenses have a resolution on the order of one wavelength due to the so-called diffraction limit. This limit hinders imaging very small objects, such as individual atoms, which are much smaller than the wavelength of visible light. A superlens is able to beat the diffraction limit. An example is the initial lens described by Pendry, which uses a slab of material with a negative index of refraction as a flat lens. In theory, a perfect lens would be capable of perfect focus — meaning that it could perfectly reproduce the electromagnetic field of the source plane at the image plane.

The Diffraction Limit as Restriction on Resolution

The performance limitation of conventional lenses is due to the diffraction limit. Following Pendry (2000), the diffraction limit can be understood as follows. Consider an object and a lens placed along the z-axis so the rays from the object are traveling in the +z direction. The field emanating from the object can be written in terms of its angular spectrum method, as a superposition of plane waves:

$$E(x,y,z,t) = \sum_{k_x,k_y} A(k_x,k_y)e^{i(k_z z + k_y y + k_x x - \omega t)}$$

where k_z is a function of k_x, k_y as:

$$k_z = \sqrt{\frac{\omega^2}{c^2} - \left(k_x^2 + k_y^2\right)}$$

Only the positive square root is taken as the energy is going in the +z direction. All of the components of the angular spectrum of the image for which is real are transmitted and re-focused by an ordinary lens. However, if

$$k_x^2 + k_y^2 > \frac{\omega^2}{c^2}$$

then k_z becomes imaginary, and the wave is an evanescent wave whose amplitude decays as the wave propagates along the z-axis. This results in the loss of the high angular frequency components of the wave, which contain information about the high frequency (small scale) features of the object being imaged. The highest resolution that can be obtained can be expressed in terms of the wavelength:

$$k_{max} \approx \frac{\omega}{c} = \frac{2\pi}{\lambda}$$

$$\Delta x_{min} \approx \lambda$$

A superlens overcomes the limit. A Pendry-type superlens has an index of n = −1 (ε = −1, μ = −1), and in such a material, transport of energy in the +z direction requires the z-component of the wave vector to have opposite sign:

$$k_z' = -\sqrt{\frac{\omega^2}{c^2} - \left(k_x^2 + k_y^2\right)}$$

For large angular frequencies, the evanescent wave now grows, so with proper lens thickness, all components of the angular spectrum can be transmitted through the lens undistorted. There are no problems with conservation of energy, as evanescent waves carry none in the direction of growth: the Poynting vector is oriented perpendicularly to the direction of growth. For traveling waves inside a perfect lens, the Poynting vector points in direction opposite to the phase velocity.

Effects of Negative Index of Refraction

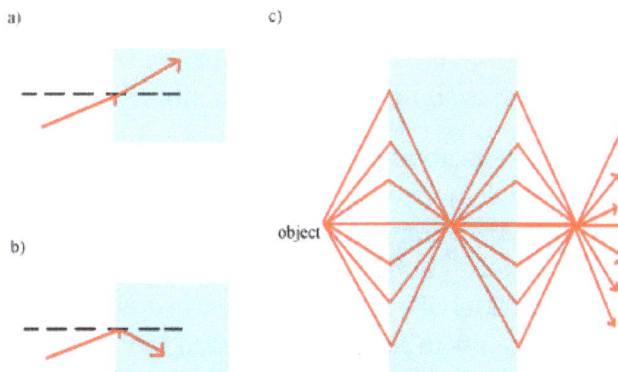

When a wave strikes a positive refraction index material from a vacuum. b) When a wave strikes a negative refraction index material from a vacuum. c) When an object is placed in front of an object with n = −1, light from it is refracted so it focuses once inside the lens and once outside. This allows subwavelength imaging.

Normally when a wave passes through the interface of two materials, the wave appears on the opposite side of the normal. However, if the interface is between a material with a positive index of refraction and another material with a negative index of refraction, the wave will appear on the same side of the normal. Pendry's idea of a perfect lens is a flat material where n = −1. Such a lens allows for near field rays—which normally decay due to the diffraction limit—to focus once within the lens and once outside the lens, allowing for subwavelength imaging.

Superlens Development and Construction

Superlens construction was at one time thought to be impossible. In 2000, Pendry claimed that a simple slab of left-handed material would do the job. The experimental realization of such a lens took, however, some more time, because it is not that easy to fabricate metamaterials with both negative permittivity and permeability. Indeed, no such material exists naturally and construction of the required metamaterials is non-trivial. Furthermore, it was shown that the parameters of the material are extremely sensitive (the index must equal −1); small deviations make the sub-wavelength resolution unobservable. Due to the resonant nature of metamaterials, on which many (proposed) implementations of superlenses depend, metamaterials are highly dispersive. The sensitive nature of the superlens to the material parameters causes superlenses based on metamaterials to have a limited usable frequency range. This initial theoretical superlens design consisted of a metamaterial that compensated for wave decay and reconstructs images in the near field. Both propagating and evanescent waves could contribute to the resolution of the image.

Pendry also suggested that a lens having only one negative parameter would form an approximate superlens, provided that the distances involved are also very small and provided that the source polarization is appropriate. For visible light this is a useful substitute, since engineering metamaterials with a negative permeability at the frequency of visible light is difficult. Metals are then a good alternative as they have negative permittivity (but not negative permeability). Pendry suggested using silver due to its relatively low loss at the predicted wavelength of operation (356 nm). In 2003 Pendry's theory was first experimentally demonstrated by Parimi et al. at RF/microwave frequencies. In 2005, two independent groups verified Pendry's lens at UV range, both using thin layers of silver illuminated with UV light to produce "photographs" of objects smaller than the wavelength. Negative refraction of visible light was experimentally verified in an yttrium orthovanadate (YVO_4) bicrystal in 2003.

It was discovered that a simple superlens design for microwaves could use an array of parallel conducting wires. This structure was shown to be able to improve the resolution of MRI imaging.

In 2004, the first superlens with a negative refractive index provided resolution three times better than the diffraction limit and was demonstrated at microwave frequencies. In 2005, the first near field superlens was demonstrated by N.Fang *et al.*, but the lens did not rely on negative refraction. Instead, a thin silver film was used to enhance the evanescent modes through surface plasmon coupling. Almost at the same time Melville and Blaikie succeeded with a near field superlens. Other groups followed. Two developments in superlens research were reported in 2008. In the second case, a metamaterial was formed from silver nanowires which were electrochemically deposited in porous aluminium oxide. The material exhibited negative refraction.

The superlens has not yet been demonstrated at visible or near-infrared frequencies (Nielsen, R.

B.; 2010). Furthermore, as dispersive materials, these are limited to functioning at a single wavelength. Proposed solutions are metal–dielectric composites (MDCs) and multilayer lens structures. The multi-layer superlens appears to have better subwavelength resolution than the single layer superlens. Losses are less of a concern with the multi-layer system, but so far it appears to be impractical because of impedance mis-match.

Perfect Lenses

When the world is observed through conventional lenses, the sharpness of the image is determined by and limited to the wavelength of light. Around the year 2000, a slab of negative index metamaterial was theorized to create a lens with capabilities beyond conventional (positive index) lenses. Pendry proposed that a thin slab of negative refractive metamaterial might overcome known problems with common lenses to achieve a "perfect" lens that would focus the entire spectrum, both the propagating as well as the evanescent spectra.

A slab of silver was proposed as the metamaterial. More specifically, such silver thin film can be regarded as a metasurface. As light moves away (propagates) from the source, it acquires an arbitrary phase. Through a conventional lens the phase remains consistent, but the evanescent waves decay exponentially. In the flat metamaterial DNG slab, normally decaying evanescent waves are contrarily amplified. Furthermore, as the evanescent waves are now amplified, the phase is reversed.

Therefore, a type of lens was proposed, consisting of a metal film metamaterial. When illuminated near its plasma frequency, the lens could be used for superresolution imaging that compensates for wave decay and reconstructs images in the near-field. In addition, both propagating and evanescent waves contribute to the resolution of the image.

Pendry suggested that left-handed slabs allow "perfect imaging" if they are completely lossless, impedance matched, and their refractive index is −1 relative to the surrounding medium. Theoretically, this would be a breakthrough in that the optical version resolves objects as minuscule as nanometers across. Pendry predicted that Double negative metamaterials (DNG) with a refractive index of $n = -1$, can act, at least in principle, as a "perfect lens" allowing imaging resolution which is limited not by the wavelength, but rather by material quality.

Other Studies Concerning the Perfect Lens

Further research demonstrated that Pendry's theory behind the perfect lens was not exactly correct. The analysis of the focusing of the evanescent spectrum (equations 13–21 in reference) was flawed. In addition, this applies to only one (theoretical) instance, and that is one particular medium that is lossless, nondispersive and the constituent parameters are defined as:

$$\varepsilon(\omega) / \varepsilon_0 = \mu(\omega) / \mu_0 = -1, \text{ which in turn results in a negative refraction of } n = -1$$

However, the final intuitive result of this theory that both the propagating and evanescent waves are focused, resulting in a converging focal point within the slab and another convergence (focal point) beyond the slab turned out to be correct.

If the DNG metamaterial medium has a large negative index or becomes lossy or dispersive, Pendry's perfect lens effect cannot be realized. As a result, the perfect lens effect does not exist in general. According to FDTD simulations at the time (2001), the DNG slab acts like a converter from a pulsed cylindrical wave to a pulsed beam. Furthermore, in reality (in practice), a DNG medium must be and is dispersive and lossy, which can have either desirable or undesirable effects, depending on the research or application. Consequently, Pendry's perfect lens effect is inaccessible with any metamaterial designed to be a DNG medium.

Another analysis, in 2002, of the perfect lens concept showed it to be in error while using the lossless, dispersionless DNG as the subject. This analysis mathematically demonstrated that subtleties of evanescent waves, restriction to a finite slab and absorption had led to inconsistencies and divergencies that contradict the basic mathematical properties of scattered wave fields. For example, this analysis stated that absorption, which is linked to dispersion, is always present in practice, and absorption tends to transform amplified waves into decaying ones inside this medium (DNG).

A third analysis of Pendry's perfect lens concept, published in 2003, used the recent demonstration of negative refraction at microwave frequencies as confirming the viability of the fundamental concept of the perfect lens. In addition, this demonstration was thought to be experimental evidence that a planar DNG metamaterial would refocus the far field radiation of a point source. However, the perfect lens would require significantly different values for permittivity, permeability, and spatial periodicity than the demonstrated negative refractive sample.

This study agrees that any deviation from conditions where $\varepsilon = \mu = -1$ results in the normal, conventional, imperfect image that degrades exponentially i.e., the diffraction limit. The perfect lens solution in the absence of losses is again, not practical, and can lead to paradoxical interpretations.

It was determined that although resonant surface plasmons are undesirable for imaging, these turn out to be essential for recovery of decaying evanescent waves. This analysis discovered that metamaterial periodicity has a significant effect on the recovery of types of evanescent components. In addition, achieving subwavelength resolution is possible with current technologies. Negative refractive indices have been demonstrated in structured metamaterials. Such materials can be engineered to have tunable material parameters, and so achieve the optimal conditions. Losses can be minimized in structures utilizing superconducting elements. Furthermore, consideration of alternate structures may lead to configurations of left-handed materials that can achieve subwavelength focusing. Such structures were being studied at the time.

Near-field Imaging With Magnetic Wires

Pendry's theoretical lens was designed to focus both propagating waves and the near-field evanescent waves. From permittivity «ε» and magnetic permeability «μ» an index of refraction «n» is derived. The index of refraction determines how light is bent on traversing from one material to another. In 2003, it was suggested that a metamaterial constructed with alternating, parallel, layers of $n = -1$ materials and $n = +1$ materials, would be a more effective design for a metamaterial lens. It is an effective medium made up of a multi-layer stack, which exhibits birefringence, $n_z = \infty$, $n_x = 0$. The effective refractive indices are then perpendicular and parallel, respectively.

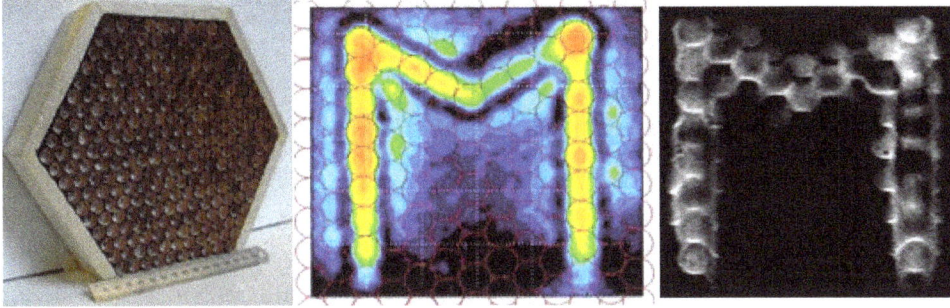

A prism composed of high performance Swiss rolls which behaves as a magnetic faceplate, transferring a magnetic field distribution faithfully from the input to the output face.

Like a conventional lens, the z-direction is along the axis of the roll. The resonant frequency (w_o) – close to 21.3 MHz – is determined by the construction of the roll. Damping is achieved by the inherent resistance of the layers and the lossy part of permittivity.

Simply put, as the field pattern is transferred from the input to the output face of a slab, so the image information is transported across each layer. This was experimentally demonstrated. To test the two-dimensional imaging performance of the material, an antenna was constructed from a pair of anti-parallel wires in the shape of the letter M. This generated a line of magnetic flux, so providing a characteristic field pattern for imaging. It was placed horizontally, and the material, consisting of 271 Swiss rolls tuned to 21.5 MHz, was positioned on top of it. The material does indeed act as an image transfer device for the magnetic field. The shape of the antenna is faithfully reproduced in the output plane, both in the distribution of the peak intensity, and in the "valleys" that bound the M.

A consistent characteristic of the very near (evanescent) field is that the electric and magnetic fields are largely decoupled. This allows for nearly independent manipulation of the electric field with the permittivity and the magnetic field with the permeability.

Furthermore, this is highly anisotropic system. Therefore, the transverse (perpendicular) components of the EM field which radiate the material, that is the wavevector components k_x and k_y, are decoupled from the longitudinal component k_z. So, the field pattern should be transferred from the input to the output face of a slab of material without degradation of the image information.

Optical Super Lens with Silver Metamaterial

In 2003, a group of researchers showed that optical evanescent waves would be enhanced as they passed through a silver metamaterial lens. This was referred to as a diffraction-free lens. Although a coherent, high-resolution, image was not intended, nor achieved, regeneration of the evanescent field was experimentally demonstrated.

By 2003 it was known for decades that evanescent waves could be enhanced by producing excited states at the interface surfaces. However, the use of surface plasmons to reconstruct evanescent components was not tried until Pendry's recent proposal. By studying films of varying thickness it has been noted that a rapidly growing transmission coefficient occurs, under the appropriate conditions. This demonstration provided direct evidence that the foundation of superlensing is solid, and suggested the path that will enable the observation of superlensing at optical wavelengths.

In 2005, a coherent, high-resolution, image was produced (based on the 2003 results). A thinner slab of silver (35 nm) was better for sub–diffraction-limited imaging, which results in one-sixth of the illumination wavelength. This type of lens was used to compensate for wave decay and reconstruct images in the near-field. Prior attempts to create a working superlens used a slab of silver that was too thick.

Objects were imaged as small as 40 nm across. In 2005 the imaging resolution limit for optical microscopes was at about one tenth the diameter of a red blood cell. With the silver superlens this results in a resolution of one hundredth of the diameter of a red blood cell.

Conventional lenses, whether man-made or natural, create images by capturing the propagating light waves all objects emit and then bending them. The angle of the bend is determined by the index of refraction and has always been positive until the fabrication of artificial negative index materials. Objects also emit evanescent waves that carry details of the object, but are unobtainable with conventional optics. Such evanescent waves decay exponentially and thus never become part of the image resolution, an optics threshold known as the diffraction limit. Breaking this diffraction limit, and capturing evanescent waves are critical to the creation of a 100-percent perfect representation of an object.

In addition, conventional optical materials suffer a diffraction limit because only the propagating components are transmitted (by the optical material) from a light source. The non-propagating components, the evanescent waves, are not transmitted. Moreover, lenses that improve image resolution by increasing the index of refraction are limited by the availability of high-index materials, and point by point subwavelength imaging of electron microscopy also has limitations when compared to the potential of a working superlens. Scanning electron and atomic force microscopes are now used to capture detail down to a few nanometers. However, such microscopes create images by scanning objects point by point, which means they are typically limited to non-living samples, and image capture times can take up to several minutes.

With current optical microscopes, scientists can only make out relatively large structures within a cell, such as its nucleus and mitochondria. With a superlens, optical microscopes could one day reveal the movements of individual proteins traveling along the microtubules that make up a cell's skeleton, the researchers said. Optical microscopes can capture an entire frame with a single snapshot in a fraction of a second. With superlenses this opens up nanoscale imaging to living materials, which can help biologists better understand cell structure and function in real time.

Advances of magnetic coupling in the THz and infrared regime provided the realization of a possible metamaterial superlens. However, in the near field, the electric and magnetic responses of materials are decoupled. Therefore, for transverse magnetic (TM) waves, only the permittivity needed to be considered. Noble metals, then become natural selections for superlensing because negative permittivity is easily achieved.

By designing the thin metal slab so that the surface current oscillations (the surface plasmons) match the evanescent waves from the object, the superlens is able to substantially enhance the amplitude of the field. Superlensing results from the enhancement of evanescent waves by surface plasmons.

The key to the superlens is its ability to significantly enhance and recover the evanescent waves that carry information at very small scales. This enables imaging well below the diffraction limit. No lens is yet able to completely reconstitute all the evanescent waves emitted by an object, so the goal of a 100-percent perfect image will persist. However, many scientists believe that a true perfect lens is not possible because there will always be some energy absorption loss as the waves pass through any known material. In comparison, the superlens image is substantially better than the one created without the silver superlens.

50-nm Flat Silver Layer

In February 2004, an electromagnetic radiation focusing system, based on a negative index metamaterial plate, accomplished subwavelength imaging in the microwave domain. This showed that obtaining separated images at much less than the wavelength of light is possible. Also, in 2004, a silver layer was used for sub-micrometre near-field imaging. Super high resolution was not achieved, but this was intended. The silver layer was too thick to allow significant enhancements of evanescent field components.

In early 2005, feature resolution was achieved with a different silver layer. Though this was not an actual image, it was intended. Dense feature resolution down to 250 nm was produced in a 50 nm thick photoresist using illumination from a mercury lamp. Using simulations (FDTD), the study noted that resolution improvements could be expected for imaging through silver lenses, rather than another method of near field imaging.

Building on this prior research, super resolution was achieved at optical frequencies using a 50 nm flat silver layer. The capability of resolving an image beyond the diffraction limit, for far-field imaging, is defined here as superresolution.

The image fidelity is much improved over earlier results of the previous experimental lens stack. Imaging of sub-micrometre features has been greatly improved by using thinner silver and spacer layers, and by reducing the surface roughness of the lens stack. The ability of the silver lenses to image the gratings has been used as the ultimate resolution test, as there is a concrete limit for the ability of a conventional (far field) lens to image a periodic object – in this case the image is a diffraction grating. For normal-incidence illumination the minimum spatial period that can be resolved with wavelength λ through a medium with refractive index n is λ/n. Zero contrast would therefore be expected in any (conventional) far-field image below this limit, no matter how good the imaging resist might be.

The (super) lens stack here results in a computational result of a diffraction-limited resolution of 243 nm. Gratings with periods from 500 nm down to 170 nm are imaged, with the depth of the modulation in the resist reducing as the grating period reduces. All of the gratings with periods above the diffraction limit (243 nm) are well resolved. The key results of this experiment are super-imaging of the sub-diffraction limit for 200 nm and 170 nm periods. In both cases the gratings are resolved, even though the contrast is diminished, but this gives experimental confirmation of Pendry's superlensing proposal.

Negative Index GRIN Lenses

Gradient Index (GRIN) – The larger range of material response available in metamaterials should lead to improved GRIN lens design. In particular, since the permittivity and permeability of a metamaterial can be adjusted independently, metamaterial GRIN lenses can presumably be better matched to free space. The GRIN lens is constructed by using a slab of NIM with a variable index of refraction in the y direction, perpendicular to the direction of propagation z.

Far-field Superlens

In 2005, a group proposed a theoretical way to overcome the near-field limitation using a new device termed a far-field superlens (FSL), which is a properly designed periodically corrugated metallic slab-based superlens.

Imaging was experimentally demonstrated in the far field, taking the next step after near-field experiments. The key element is termed as a far-field superlens (FSL) which consists of a conventional superlens and a nanoscale coupler.

Focusing beyond the Diffraction Limit with Far-field Time Reversal

An approach is presented for subwavelength focusing of microwaves using both a time-reversal mirror placed in the far field and a random distribution of scatterers placed in the near field of the focusing point.

Hyperlens

Once capability for near-field imaging was demonstrated, the next step was to project a near-field image into the far-field. This concept, including technique and materials, is dubbed "hyperlens".,

In May 2012, an ultraviolet (1200-1400 THz) hyperlens was created using alternating layers of boron nitride and graphene.

The capability of a metamaterial-hyperlens for sub-diffraction-limited imaging is shown below.

Sub-diffraction Imaging in the Far Field

With conventional optical lenses, the far field is a limit that is too distant for evanescent waves to arrive intact. When imaging an object, this limits the optical resolution of lenses to the order of the wavelength of light These non-propagating waves carry detailed information in the form of high spatial resolution, and overcome limitations. Therefore, projecting image details, normally limited by diffraction into the far field does require recovery of the evanescent waves.

In essence steps leading up to this investigation and demonstration was the employment of an anisotropic metamaterial with a hyperbolic dispersion. The effect was such that ordinary evanescent waves propagate along the radial direction of the layered metamaterial. On a microscopic level the large spatial frequency waves propagate through coupled surface plasmon excitations between the metallic layers.

In 2007, just such an anisotropic metamaterial was employed as a magnifying optical hyperlens. The hyperlens consisted of a curved periodic stack of thin silver and alumina (at 35 nanometers thick) deposited on a half-cylindrical cavity, and fabricated on a quartz substrate. The radial and tangential permittivities have different signs.

Upon illumination, the scattered evanescent field from the object enters the anisotropic medium and propagates along the radial direction. Combined with another effect of the metamaterial, a magnified image at the outer diffraction limit-boundary of the hyperlens occurs. Once the magnified feature is larger than (beyond) the diffraction limit, it can then be imaged with a conventional optical microscope, thus demonstrating magnification and projection of a sub-diffraction-limited image into the far field.

The hyperlens magnifies the object by transforming the scattered evanescent waves into propagating waves in the anisotropic medium, projecting a spatial resolution high-resolution image into the far field. This type of metamaterials-based lens, paired with a conventional optical lens is therefore able to reveal patterns too small to be discerned with an ordinary optical microscope. In one experiment, the lens was able to distinguish two 35-nanometer lines etched 150 nanometers apart. Without the metamaterials, the microscope showed only one thick line.

In a control experiment, the line pair object was imaged without the hyperlens. The line pair could not be resolved because of the diffraction limit of the (optical) aperture was limited to 260 nm. Because the hyperlens supports the propagation of a very broad spectrum of wave vectors, it can magnify arbitrary objects with sub-diffraction-limited resolution.

Although this work appears to be limited by being only a cylindrical hyperlens, the next step is to design a spherical lens. That lens will exhibit three-dimensional capability. Near-field optical microscopy uses a tip to scan an object. In contrast, this optical hyperlens magnifies an image that is sub-diffraction-limited. The magnified sub-diffraction image is then projected into the far field.

The optical hyperlens shows a notable potential for applications, such as real-time biomolecular imaging and nanolithography. Such a lens could be used to watch cellular processes that have been impossible to see. Conversely, it could be used to project an image with extremely fine features onto a photoresist as a first step in photolithography, a process used to make computer chips. The hyperlens also has applications for DVD technology.

In 2010, spherical hyperlens for two dimensional imaging at visible frequencies is demonstrated experimentally. The spherical hyperlens based on silver and titanium oxide alternating layers has strong anisotropic hyperbolic dispersion allowing super-resolution with visible spectrum. The resolution is 160 nm at visible spectrum. It will enable biological imaging such as cell and DNA with a strong benefit of magnifying sub-diffraction resolution into far-field.

Plasmon-assisted Microscopy

Super-imaging in the Visible Frequency Range

In 2007 researchers demonstrated super imaging using materials, which create negative refractive index and lensing is achieved in the visible range.

Continual improvements in optical microscopy are needed to keep up with the progress in nanotechnology and microbiology. Advancement in spatial resolution is key. Conventional optical microscopy is limited by a diffraction limit which is on the order of 200 nanometers (wavelength). This means that viruses, proteins, DNA molecules and many other samples are hard to observe with a regular (optical) microscope. The lens previously demonstrated with negative refractive index material, a thin planar superlens, does not provide magnification beyond the diffraction limit of conventional microscopes. Therefore, images smaller than the conventional diffraction limit will still be unavailable.

Another approach achieving super-resolution at visible wavelength is recently developed spherical hyperlens based on silver and titanium oxide alternating layers. It has strong anisotropic hyperbolic dispersion allowing super-resolution with converting evanescent waves into propagating waves. This method is non-fluorescence based super-resolution imaging, which results in real-time imaging without any reconstruction of images and information.

Super Resolution far-field Microscopy Techniques

By 2008 the diffraction limit has been surpassed and lateral imaging resolutions of 20 to 50 nm have been achieved by several "super-resolution" far-field microscopy techniques, including stimulated emission depletion (STED) and its related RESOLFT (reversible saturable optically linear fluorescent transitions) microscopy; saturated structured illumination microscopy (SSIM) ; stochastic optical reconstruction microscopy (STORM); photoactivated localization microscopy (PALM); and other methods using similar principles.

Cylindrical Superlens Via Coordinate Transformation

This began with a proposal by Pendry, in 2003. Magnifying the image required a new design concept in which the surface of the negatively refracting lens is curved. One cylinder touches another cylinder, resulting in a curved cylindrical lens which reproduced the contents of the smaller cylinder in magnified but undistorted form outside the larger cylinder. Coordinate transformations are required to curve the original perfect lens into the cylindrical, lens structure.

This was followed by a 36-page conceptual and mathematical proof in 2005, that the cylindrical superlens works in the quasistatic regime. The debate over the perfect lens is discussed first.

In 2007, a superlens utilizing coordinate transformation was again the subject. However, in addition to image transfer other useful operations were discussed; translation, rotation, mirroring and inversion as well as the superlens effect. Furthermore, elements that perform magnification are described, which are free from geometric aberrations, on both the input and output sides while utilizing free space sourcing (rather than waveguide). These magnifying elements also operate in the near and far field, transferring the image from near field to far field.

The cylindrical magnifying superlens was experimentally demonstrated in 2007 by two groups, Liu et al. and Smolyaninov et al.

Nano-optics With Metamaterials

Nanohole Array as a Lens

Work in 2007 demonstrated that a quasi-periodic array of nanoholes, in a metal screen, were able to focus the optical energy of a plane wave to form subwavelength spots (hot spots). The distances for the spots was a few tens of wavelengths on the other side of the array, or, in other words, opposite the side of the incident plane wave. The quasi-periodic array of nanoholes functioned as a light concentrator.

In June 2008, this was followed by the demonstrated capability of an array of quasi-crystal nanoholes in a metal screen. More than concentrating hot spots, an image of the point source is displayed a few tens of wavelengths from the array, on the other side of the array (the image plane). Also this type of array exhibited a 1 to 1 linear displacement, – from the location of the point source to its respective, parallel, location on the image plane. In other words, from x to x + δx. For example, other point sources were similarly displaced from x' to x' + δx', from x^ to x^ + δx^, and from x^^ to x^^ + δx^^, and so on. Instead of functioning as a light concentrator, this performs the function of conventional lens imaging with a 1 to 1 correspondence, albeit with a point source.

However, resolution of more complicated structures can be achieved as constructions of multiple point sources. The fine details, and brighter image, that are normally associated with the high numerical apertures of conventional lenses can be reliably produced. Notable applications for this technology arise when conventional optics is not suitable for the task at hand. For example, this technology is better suited for X-ray imaging, or nano-optical circuits, and so forth.

Nanolens

In 2010, a nano-wire array prototype, described as a three-dimensional (3D) metamaterial-nanolens, consisting of bulk nanowires deposited in a dielectric substrate was fabricated and tested.

The metamaterial nanolens was constructed of millions of nanowires at 20 nanometers in diameter. These were precisely aligned and a packaged configuration was applied. The lens is able to depict a clear, high-resolution image of nano-sized objects because it uses both normal propagating EM radiation, and evanescent waves to construct the image. Super-resolution imaging was demonstrated over a distance of 6 times the wavelength (λ), in the far-field, with a resolution of at least $\lambda/4$. This is a significant improvement over previous research and demonstration of other near field and far field imaging, including nanohole arrays discussed below.

Light Transmission Properties of Holey Metal Films

2009-12. The light transmission properties of holey metal films in the metamaterial limit, where the unit length of the periodic structures is much smaller than the operating wavelength, are analyzed theoretically.

Transporting an Image Through a Subwavelength Hole

Theoretically it appears possible to transport a complex electromagnetic image through a tiny subwavelength hole with diameter considerably smaller than the diameter of the image, without losing the subwavelength details.

Nanoparticle Imaging – quantum Dots

When observing the complex processes in a living cell, significant processes (changes) or details are easy to overlook. This can more easily occur when watching changes that take a long time to unfold and require high-spatial-resolution imaging. However, recent research offers a solution to scrutinize activities that occur over hours or even days inside cells, potentially solving many of the mysteries associated with molecular-scale events occurring in these tiny organisms.

A joint research team, working at the National Institute of Standards and Technology (NIST) and the National Institute of Allergy and Infectious Diseases (NIAID), has discovered a method of using nanoparticles to illuminate the cellular interior to reveal these slow processes. Nanoparticles, thousands of times smaller than a cell, have a variety of applications. One type of nanoparticle called a quantum dot glows when exposed to light. These semiconductor particles can be coated with organic materials, which are tailored to be attracted to specific proteins within the part of a cell a scientist wishes to examine.

Notably, quantum dots last longer than many organic dyes and fluorescent proteins that were previously used to illuminate the interiors of cells. They also have the advantage of monitoring changes in cellular processes while most high-resolution techniques like electron microscopy only provide images of cellular processes frozen at one moment. Using quantum dots, cellular processes involving the dynamic motions of proteins, are observable (elucidated).

The research focused primarily on characterizing quantum dot properties, contrasting them with other imaging techniques. In one example, quantum dots were designed to target a specific type of human red blood cell protein that forms part of a network structure in the cell's inner membrane. When these proteins cluster together in a healthy cell, the network provides mechanical flexibility to the cell so it can squeeze through narrow capillaries and other tight spaces. But when the cell gets infected with the malaria parasite, the structure of the network protein changes.

Because the clustering mechanism is not well understood, it was decided to examine it with the quantum dots. If a technique could be developed to visualize the clustering, then the progress of a malaria infection could be understood, which has several distinct developmental stages.

Research efforts revealed that as the membrane proteins bunch up, the quantum dots attached to them are induced to cluster themselves and glow more brightly, permitting real time observation as the clustering of proteins progresses. More broadly, the research discovered that when quantum dots attach themselves to other nanomaterials, the dots' optical properties change in unique ways in each case. Furthermore, evidence was discovered that quantum dot optical properties are altered as the nanoscale environment changes, offering greater possibility of using quantum dots to sense the local biochemical environment inside cells.

Some concerns remain over toxicity and other properties. However, the overall findings indicate that quantum dots could be a valuable tool to investigate dynamic cellular processes.

The abstract from the related published research paper states (in part): Results are presented regarding the dynamic fluorescence properties of bioconjugated nanocrystals or quantum dots (QDs) in different chemical and physical environments. A variety of QD samples was prepared and compared: isolated individual QDs, QD aggregates, and QDs conjugated to other nanoscale materials...

Optical Fiber

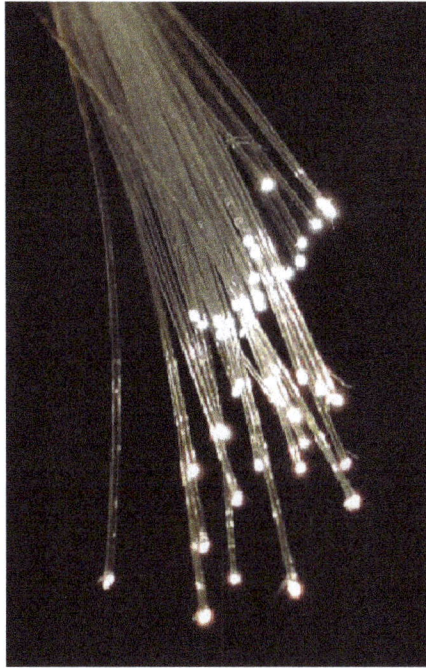

A bundle of optical fibers

Fiber crew installing a 432-count fiber cable underneath the streets of Midtown Manhattan, New York City

An optical fiber (or optical fibre) is a flexible, transparent fiber made by drawing glass (silica) or plastic to a diameter slightly thicker than that of a human hair. Optical fibers are used most often as a means to transmit light between the two ends of the fiber and find wide usage in fiber-optic communications, where they permit transmission over longer distances and at higher bandwidths (data rates) than wire cables. Fibers are used instead of metal wires because signals travel along them with lesser amounts of loss; in addition, fibers are also immune to electromagnetic interference, a problem from which metal wires suffer excessively. Fibers are also used for illumination, and are wrapped in bundles so that they may be used to carry images, thus allowing viewing in confined spaces, as in the case of a fiberscope. Specially designed fibers are also used for a variety of other applications, some of them being fiber optic sensors and fiber lasers.

Optical fibers typically include a transparent core surrounded by a transparent cladding material with a lower index of refraction. Light is kept in the core by the phenomenon of total internal reflection which causes the fiber to act as a waveguide. Fibers that support many propagation paths or transverse modes are called multi-mode fibers (MMF), while those that support a single mode are called single-mode fibers (SMF). Multi-mode fibers generally have a wider core diameter and are used for short-distance communication links and for applications where high power must be transmitted Single-mode fibers are used for most communication links longer than 1,000 meters (3,300 ft)

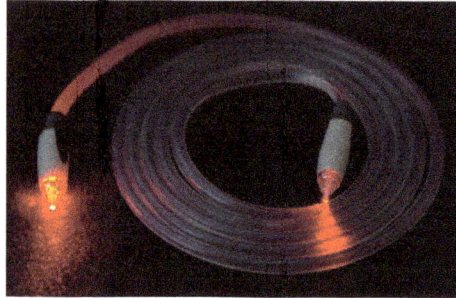

A TOSLINK fiber optic audio cable with red light being shone in one end transmits the light to the other end

An important aspect of a fiber optic communication is that of extension of the fiber optic cables such that the losses brought about by joining two different cables is kept to a minimum. Joining lengths of optical fiber often proves to be more complex than joining electrical wire or cable and involves careful cleaving of the fibers, perfect alignment of the fiber cores, and the splicing of these aligned fiber cores. For applications that demand a permanent connection a mechanical splice which holds the ends of the fibers together mechanically could be used or a fusion splice that uses heat to fuse the ends of the fibers together could be used. Temporary or semi-permanent connections are made by means of specialized optical fiber connectors.

A wall-mount cabinet containing optical fiber interconnects. The yellow cables are single mode fibers; the orange and aqua cables are multi-mode fibers: 50/125 µm OM2 and 50/125 µm OM3 fibers respectively.

The field of applied science and engineering concerned with the design and application of optical fibers is known as fiber optics.

History

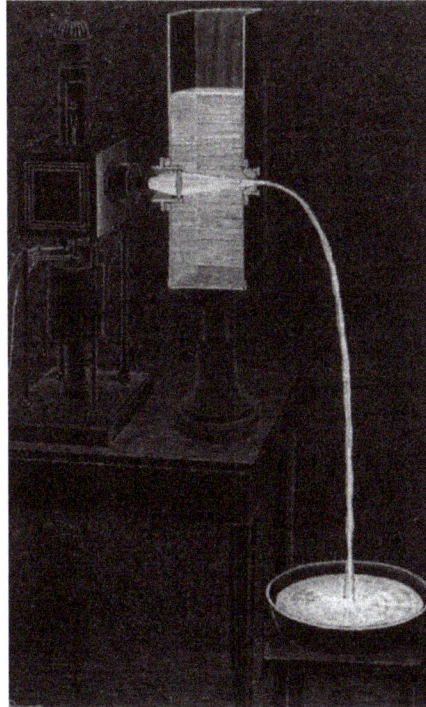

Daniel Colladon first described this "light fountain" or "light pipe" in an 1842 article titled *On the reflections of a ray of light inside a parabolic liquid stream*. This particular illustration comes from a later article by Colladon, in 1884.

Guiding of light by refraction, the principle that makes fiber optics possible, was first demonstrated by Daniel Colladon and Jacques Babinet in Paris in the early 1840s. John Tyndall included a demonstration of it in his public lectures in London, 12 years later. Tyndall also wrote about the property of total internal reflection in an introductory book about the nature of light in 1870:

When the light passes from air into water, the refracted ray is bent *towards* the perpendicular... When the ray passes from water to air it is bent *from* the perpendicular... If the angle which the ray in water encloses with the perpendicular to the surface be greater than 48 degrees, the ray will not quit the water at all: it will be *totally reflected* at the surface.... The angle which marks the limit where total reflection begins is called the limiting angle of the medium. For water this angle is 48°27′, for flint glass it is 38°41′, while for diamond it is 23°42′.

Unpigmented human hairs have also been shown to act as an optical fiber.

Practical applications, such as close internal illumination during dentistry, appeared early in the twentieth century. Image transmission through tubes was demonstrated independently by the radio experimenter Clarence Hansell and the television pioneer John Logie Baird in the 1920s. The principle was first used for internal medical examinations by Heinrich Lamm in the following decade. Modern optical fibers, where the glass fiber is coated with a transparent cladding to offer a more suitable refractive index, appeared later in the decade. Development then focused on fiber bundles for image transmission. Harold Hopkins and Narinder Singh Kapany at Imperial College in London achieved low-loss light transmission through a 75 cm long bundle which combined several thousand fibers. Their article titled "A flexible fibrescope, using static scanning" was published in the journal *Nature* in 1954. The first fiber optic semi-flexible gastroscope was patented by Basil Hirschowitz, C. Wilbur Peters, and Lawrence E. Curtiss, researchers at the University of Michigan, in 1956. In the process of developing the gastroscope, Curtiss produced the first glass-clad fibers; previous optical fibers had relied on air or impractical oils and waxes as the low-index cladding material.

A variety of other image transmission applications soon followed.

In 1880 Alexander Graham Bell and Sumner Tainter invented the Photophone at the Volta Laboratory in Washington, D.C., to transmit voice signals over an optical beam. It was an advanced form of telecommunications, but subject to atmospheric interferences and impractical until the secure transport of light that would be offered by fiber-optical systems. In the late 19th and early 20th centuries, light was guided through bent glass rods to illuminate body cavities. Jun-ichi Nishizawa, a Japanese scientist at Tohoku University, also proposed the use of optical fibers for communications in 1963, as stated in his book published in 2004 in India. Nishizawa invented other technologies that contributed to the development of optical fiber communications, such as the graded-index optical fiber as a channel for transmitting light from semiconductor lasers. The first working fiber-optical data transmission system was demonstrated by German physicist Manfred Börner at Telefunken Research Labs in Ulm in 1965, which was followed by the first patent application for this technology in 1966. NASA used fiber optics in the television cameras that were sent to the moon. At the time, the use in the cameras was classified *confidential*, and employees handling the cameras had to be supervised by someone with an appropriate security clearance.

Charles K. Kao and George A. Hockham of the British company Standard Telephones and Cables (STC) were the first to promote the idea that the attenuation in optical fibers could be reduced below 20 decibels per kilometer (dB/km), making fibers a practical communication medium. They proposed that the attenuation in fibers available at the time was caused by impurities that could be removed, rather than by fundamental physical effects such as scattering. They correctly and systematically theorized the light-loss properties for optical fiber, and pointed out the right material to use for such fibers — silica glass with high purity. This discovery earned Kao the Nobel Prize in Physics in 2009.

The crucial attenuation limit of 20 dB/km was first achieved in 1970, by researchers Robert D. Maurer, Donald Keck, Peter C. Schultz, and Frank Zimar working for American glass maker Corning Glass Works, now Corning Incorporated. They demonstrated a fiber with 17 dB/km attenuation by doping silica glass with titanium. A few years later they produced a fiber with only 4 dB/km attenuation using germanium dioxide as the core dopant. Such low attenuation ushered in the era of optical fiber telecommunication. In 1981, General Electric produced fused quartz ingots that could be drawn into strands 25 miles (40 km) long.

The Italian research center CSELT worked with Corning to develop practical optical fiber cables, resulting in the first metropolitan fiber optic cable being deployed in Torino in 1977. CSELT also developed an early technique for splicing optical fibers, called Springroove.

Attenuation in modern optical cables is far less than in electrical copper cables, leading to long-haul fiber connections with repeater distances of 70–150 kilometers (43–93 mi). The erbium-doped fiber amplifier, which reduced the cost of long-distance fiber systems by reducing or eliminating optical-electrical-optical repeaters, was co-developed by teams led by David N. Payne of the University of Southampton and Emmanuel Desurvire at Bell Labs in 1986. Robust modern optical fiber uses glass for both core and sheath, and is therefore less prone to aging. It was invented by Gerhard Bernsee of Schott Glass in Germany in 1973.

The emerging field of photonic crystals led to the development in 1991 of photonic-crystal fiber, which guides light by diffraction from a periodic structure, rather than by total internal reflection. The first photonic crystal fibers became commercially available in 2000. Photonic crystal fibers can carry higher power than conventional fibers and their wavelength-dependent properties can be manipulated to improve performance.

Uses

Communication

Optical fiber can be used as a medium for telecommunication and computer networking because it is flexible and can be bundled as cables. It is especially advantageous for long-distance communications, because light propagates through the fiber with little attenuation compared to electrical cables. This allows long distances to be spanned with few repeaters.

The per-channel light signals propagating in the fiber have been modulated at rates as high as 111 gigabits per second (Gbit/s) by NTT, although 10 or 40 Gbit/s is typical in deployed systems. In June 2013, researchers demonstrated transmission of 400 Gbit/s over a single channel using 4-mode orbital angular momentum multiplexing.

Each fiber can carry many independent channels, each using a different wavelength of light (wavelength-division multiplexing (WDM)). The net data rate (data rate without overhead bytes) per fiber is the per-channel data rate reduced by the FEC overhead, multiplied by the number of channels (usually up to eighty in commercial dense WDM systems as of 2008). As of 2011 the record for bandwidth on a single core was 101 Tbit/s (370 channels at 273 Gbit/s each). The record for a multi-core fiber as of January 2013 was 1.05 petabits per second. In 2009, Bell Labs broke the 100 (petabit per second)×kilometer barrier (15.5 Tbit/s over a single 7,000 km fiber).

For short distance application, such as a network in an office building, fiber-optic cabling can save space in cable ducts. This is because a single fiber can carry much more data than electrical cables such as standard category 5 Ethernet cabling, which typically runs at 100 Mbit/s or 1 Gbit/s speeds. Fiber is also immune to electrical interference; there is no cross-talk between signals in different cables, and no pickup of environmental noise. Non-armored fiber cables do not conduct electricity, which makes fiber a good solution for protecting communications equipment in high voltage environments, such as power generation facilities, or metal communication structures prone to lightning strikes. They can also be used in environments where explosive fumes are present, without danger of ignition. Wiretapping (in this case, fiber tapping) is more difficult compared to electrical connections, and there are concentric dual-core fibers that are said to be tap-proof.

Fibers are often also used for short-distance connections between devices. For example, most high-definition televisions offer a digital audio optical connection. This allows the streaming of audio over light, using the TOSLINK protocol.

Advantages Over Copper Wiring

The advantages of optical fiber communication with respect to copper wire systems are:

Broad Bandwidth

A single optical fiber can carry over 3,000,000 full-duplex voice calls or 90,000 TV channels.

Immunity to Electromagnetic Interference

Light transmission through optical fibers is unaffected by other electromagnetic radiation nearby. The optical fiber is electrically non-conductive, so it does not act as an antenna to pick up electromagnetic signals. Information traveling inside the optical fiber is immune to electromagnetic interference, even electromagnetic pulses generated by nuclear devices.

Low Attenuation Loss Over Long Distances

Attenuation loss can be as low as 0.2 dB/km in optical fiber cables, allowing transmission over long distances without the need for repeaters.

Electrical Insulator

Optical fibers do not conduct electricity, preventing problems with ground loops and conduction of lightning. Optical fibers can be strung on poles alongside high voltage power cables.

Material Cost and Theft Prevention

Conventional cable systems use large amounts of copper. Global copper prices experienced a boom in the 2000s, and copper has been a target of metal theft.

Security of Information Passed Down the Cable

Copper can be tapped with very little chance of detection.

Sensors

Fibers have many uses in remote sensing. In some applications, the sensor is itself an optical fiber. In other cases, fiber is used to connect a non-fiberoptic sensor to a measurement system. Depending on the application, fiber may be used because of its small size, or the fact that no electrical power is needed at the remote location, or because many sensors can be multiplexed along the length of a fiber by using different wavelengths of light for each sensor, or by sensing the time delay as light passes along the fiber through each sensor. Time delay can be determined using a device such as an *optical time-domain reflectometer*.

Optical fibers can be used as sensors to measure strain, temperature, pressure and other quantities by modifying a fiber so that the property to measure modulates the intensity, phase, polarization, wavelength, or transit time of light in the fiber. Sensors that vary the intensity of light are the simplest, since only a simple source and detector are required. A particularly useful feature of such fiber optic sensors is that they can, if required, provide distributed sensing over distances of up to one meter. In contrast, highly localized measurements can be provided by integrating miniaturized sensing elements with the tip of the fiber. These can be implemented by various micro- and nanofabrication technologies, such that they do not exceed the microscopic boundary of the fiber tip, allowing such applications as insertion into blood vessels via hypodermic needle.

Extrinsic fiber optic sensors use an optical fiber cable, normally a multi-mode one, to transmit modulated light from either a non-fiber optical sensor—or an electronic sensor connected to an optical transmitter. A major benefit of extrinsic sensors is their ability to reach otherwise inaccessible places. An example is the measurement of temperature inside aircraft jet engines by using a fiber to transmit radiation into a radiation pyrometer outside the engine. Extrinsic sensors can be used in the same way to measure the internal temperature of electrical transformers, where the extreme electromagnetic fields present make other measurement techniques impossible. Extrinsic sensors measure vibration, rotation, displacement, velocity, acceleration, torque, and twisting. A solid state version of the gyroscope, using the interference of light, has been developed. The *fiber optic gyroscope (FOG)* has no moving parts, and exploits the *Sagnac effect* to detect mechanical rotation.

Common uses for fiber optic sensors includes advanced intrusion detection security systems. The light is transmitted along a fiber optic sensor cable placed on a fence, pipeline, or communication cabling, and the returned signal is monitored and analyzed for disturbances. This return signal is digitally processed to detect disturbances and trip an alarm if an intrusion has occurred.

Optical fibers are widely used as components of optical chemical sensors and optical biosensors.

Power Transmission

Optical fiber can be used to transmit power using a photovoltaic cell to convert the light into electricity. While this method of power transmission is not as efficient as conventional ones, it is especially useful in situations where it is desirable not to have a metallic conductor as in the case of use near MRI machines, which produce strong magnetic fields. Other examples are for powering electronics in high-powered antenna elements and measurement devices used in high-voltage transmission equipment.

Other Uses

A frisbee illuminated by fiber optics

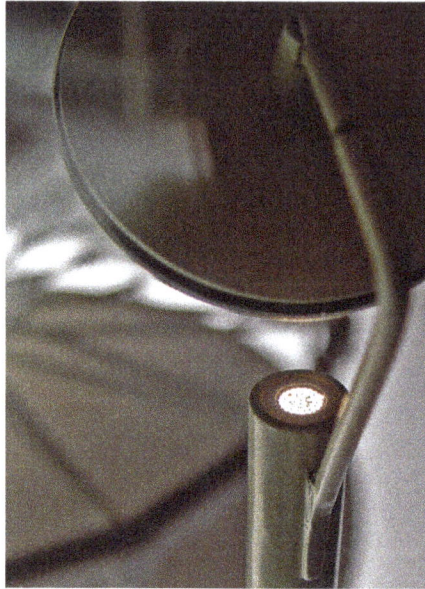

Light reflected from optical fiber illuminates exhibited model

Optical fibers have a wide number of applications. They are used as light guides in medical and other applications where bright light needs to be shone on a target without a clear line-of-sight path. In some buildings, optical fibers route sunlight from the roof to other parts of the building. Optical fiber lamps are used for illumination in decorative applications, including signs, art, toys and artificial Christmas trees. Swarovski boutiques use optical fibers to illuminate their crystal showcases from many different angles while only employing one light source. Optical fiber is an intrinsic part of the light-transmitting concrete building product, LiTraCon.

Use of optical fiber in a decorative lamp or nightlight.

Optical fiber is also used in imaging optics. A coherent bundle of fibers is used, sometimes along with lenses, for a long, thin imaging device called an endoscope, which is used to view objects through a small hole. Medical endoscopes are used for minimally invasive exploratory or surgical procedures. Industrial endoscopes are used for inspecting anything hard to reach, such as jet engine interiors. Many microscopes use fiber-optic light sources to provide intense illumination of samples being studied.

In spectroscopy, optical fiber bundles transmit light from a spectrometer to a substance that cannot be placed inside the spectrometer itself, in order to analyze its composition. A spectrometer analyzes substances by bouncing light off and through them. By using fibers, a spectrometer can be used to study objects remotely.

An optical fiber doped with certain rare earth elements such as erbium can be used as the gain medium of a laser or optical amplifier. Rare-earth-doped optical fibers can be used to provide signal amplification by splicing a short section of doped fiber into a regular (undoped) optical fiber line. The doped fiber is optically pumped with a second laser wavelength that is coupled into the line in addition to the signal wave. Both wavelengths of light are transmitted through the doped fiber, which transfers energy from the second pump wavelength to the signal wave. The process that causes the amplification is stimulated emission.

Optical fiber is also widely exploited as a nonlinear medium. The glass medium supports a host of nonlinear optical interactions, and the long interaction lengths possible in fiber facilitate a variety of phenomena, which are harnessed for applications and fundamental investigation. Conversely, fiber nonlinearity can have deleterious effects on optical signals, and measures are often required to minimize such unwanted effects.

Optical fibers doped with a wavelength shifter collect scintillation light in physics experiments.

Fiber optic sights for handguns, rifles, and shotguns use pieces of optical fiber to improve visibility of markings on the sight.

Principle of Operation

An overview of the operating principles of the optical fiber

An optical fiber is a cylindrical dielectric waveguide (nonconducting waveguide) that transmits light along its axis, by the process of total internal reflection. The fiber consists of a *core* surrounded by a *cladding* layer, both of which are made of dielectric materials. To confine the optical signal in the core, the refractive index of the core must be greater than that of the cladding. The boundary between the core and cladding may either be abrupt, in *step-index fiber*, or gradual, in *graded-index fiber*.

Index of Refraction

The index of refraction (or refractive index) is a way of measuring the speed of light in a material. Light travels fastest in a vacuum, such as in outer space. The speed of light in a vacuum is about 300,000 kilometers (186,000 miles) per second. The refractive index of a medium is calculated by dividing the speed of light in a vacuum by the speed of light in that medium. The refractive index of a vacuum is therefore 1, by definition. A typical singlemode fiber used for telecommunications has a cladding made of pure silica, with an index of 1.444 at 1,500 nm, and a core of doped silica with an index around 1.4475. The larger the index of refraction, the slower light travels in that medium. From this information, a simple rule of thumb is that a signal using optical fiber for communication will travel at around 200,000 kilometers per second. To put it another way, the signal will take 5 milliseconds to travel 1,000 kilometers in fiber. Thus a phone call carried by fiber between Sydney and New York, a 16,000-kilometer distance, means that there is a minimum delay of 80 millisec onds (about $\frac{1}{12}$ of a second) between when one caller speaks and the other hears. (The fiber in this case will probably travel a longer route, and there will be additional delays due to communication equipment switching and the process of encoding and decoding the voice onto the fiber).

Total Internal Reflection

When light traveling in an optically dense medium hits a boundary at a steep angle (larger than the critical angle for the boundary), the light is completely reflected. This is called total internal reflection. This effect is used in optical fibers to confine light in the core. Light travels through the fiber core, bouncing back and forth off the boundary between the core and cladding. Because the light must strike the boundary with an angle greater than the critical angle, only light that enters the fiber within a certain range of angles can travel down the fiber without leaking out. This range of angles is called the acceptance cone of the fiber. The size of this acceptance cone is a function of the refractive index difference between the fiber's core and cladding.

In simpler terms, there is a maximum angle from the fiber axis at which light may enter the fiber so that it will propagate, or travel, in the core of the fiber. The sine of this maximum angle is the numerical aperture (NA) of the fiber. Fiber with a larger NA requires less precision to splice and work with than fiber with a smaller NA. Single-mode fiber has a small NA.

Multi-mode Fiber

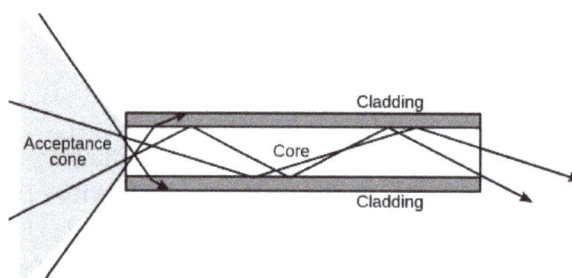

The propagation of light through a multi-mode optical fiber.

Fiber with large core diameter (greater than 10 micrometers) may be analyzed by geometrical optics. Such fiber is called *multi-mode iber*, from the electromagnetic analysis. In a step-index multi-mode fiber, rays of light are guided along the fiber core by total internal reflection. Rays that meet the core-cladding boundary at a high angle (measured relative to a line normal to the boundary), greater than the critical angle for this boundary, are completely reflected. The critical angle (minimum angle for total internal reflection) is determined by the difference in index of refraction between the core and cladding materials. Rays that meet the boundary at a low angle are refracted from the core into the cladding, and do not convey light and hence information along the fiber. The critical angle determines the acceptance angle of the fiber, often reported as a numerical aperture. A high numerical aperture allows light to propagate down the fiber in rays both close to the axis and at various angles, allowing efficient coupling of light into the fiber. However, this high numerical aperture increases the amount of dispersion as rays at different angles have different path lengths and therefore take different times to traverse the fiber.

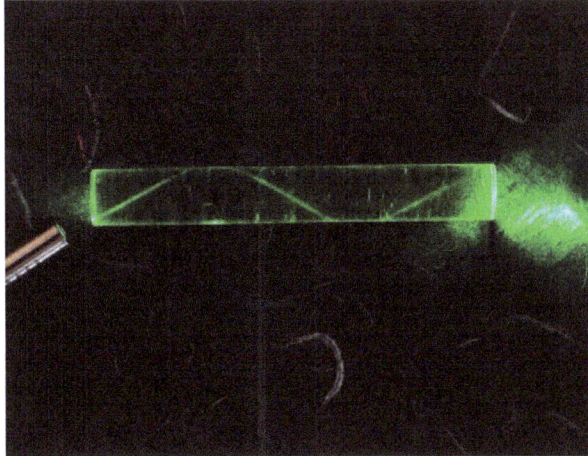

A laser bouncing down an acrylic rod, illustrating the total internal reflection of light in a multi-mode optical fiber.

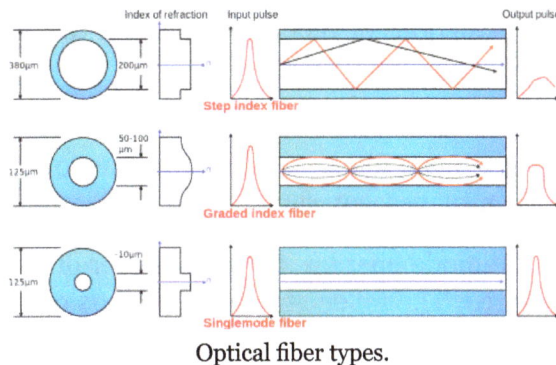

Optical fiber types.

In graded-index fiber, the index of refraction in the core decreases continuously between the axis and the cladding. This causes light rays to bend smoothly as they approach the cladding, rather than reflecting abruptly from the core-cladding boundary. The resulting curved paths reduce multi-path dispersion because high angle rays pass more through the lower-index periphery of the core, rather than the high-index center. The index profile is chosen to minimize the difference in axial propagation speeds of the various rays in the fiber. This ideal index profile is very close to a parabolic relationship between the index and the distance from the axis.

Single-mode Fiber

The structure of a typical single-mode fiber.
1. Core: 8 μm diameter
2. Cladding: 125 μm dia.
3. Buffer: 250 μm dia.
4. Jacket: 400 μm dia.

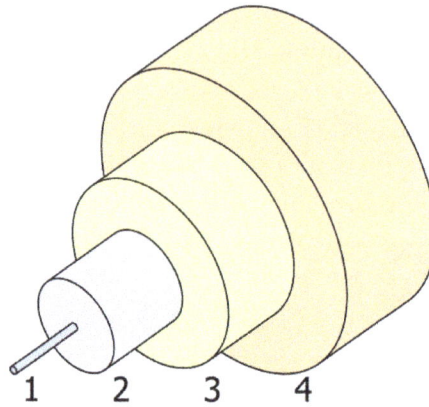

Fiber with a core diameter less than about ten times the wavelength of the propagating light cannot be modeled using geometric optics. Instead, it must be analyzed as an electromagnetic structure, by solution of Maxwell's equations as reduced to the electromagnetic wave equation. The electromagnetic analysis may also be required to understand behaviors such as speckle that occur when coherent light propagates in multi-mode fiber. As an optical waveguide, the fiber supports one or more confined transverse modes by which light can propagate along the fiber. Fiber supporting only one mode is called *single-mode* or *mono-mode fiber*. The behavior of larger-core multi-mode fiber can also be modeled using the wave equation, which shows that such fiber supports more than one mode of propagation (hence the name). The results of such modeling of multi-mode fiber approximately agree with the predictions of geometric optics, if the fiber core is large enough to support more than a few modes.

The waveguide analysis shows that the light energy in the fiber is not completely confined in the core. Instead, especially in single-mode fibers, a significant fraction of the energy in the bound mode travels in the cladding as an evanescent wave.

The most common type of single-mode fiber has a core diameter of 8–10 micrometers and is designed for use in the near infrared. The mode structure depends on the wavelength of the light used, so that this fiber actually supports a small number of additional modes at visible wavelengths. Multi-mode fiber, by comparison, is manufactured with core diameters as small as 50 micrometers and as large as hundreds of micrometers. The normalized frequency V for this fiber should be less than the first zero of the Bessel function J_0 (approximately 2.405).

Special-purpose Fiber

Some special-purpose optical fiber is constructed with a non-cylindrical core and/or cladding layer, usually with an elliptical or rectangular cross-section. These include polarization-maintaining fiber and fiber designed to suppress whispering gallery mode propagation. Polarization-maintaining fiber is a unique type of fiber that is commonly used in fiber optic sensors due to its ability to maintain the polarization of the light inserted into it.

Photonic-crystal fiber is made with a regular pattern of index variation (often in the form of cylindrical holes that run along the length of the fiber). Such fiber uses diffraction effects instead of or in addition to total internal reflection, to confine light to the fiber's core. The properties of the fiber can be tailored to a wide variety of applications.

Mechanisms of Attenuation

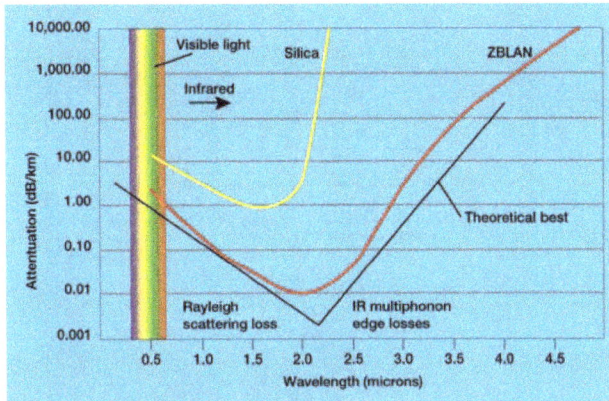

Light attenuation by ZBLAN and silica fibers

Attenuation in fiber optics, also known as transmission loss, is the reduction in intensity of the light beam (or signal) as it travels through the transmission medium. Attenuation coefficients in fiber optics usually use units of dB/km through the medium due to the relatively high quality of transparency of modern optical transmission media. The medium is usually a fiber of silica glass that confines the incident light beam to the inside. Attenuation is an important factor limiting the transmission of a digital signal across large distances. Thus, much research has gone into both limiting the attenuation and maximizing the amplification of the optical signal. Empirical research has shown that attenuation in optical fiber is caused primarily by both scattering and absorption. Single-mode optical fibers can be made with extremely low loss. Corning's SMF-28 fiber, a standard single-mode fiber for telecommunications wavelengths, has a loss of 0.17 dB/km at 1550 nm. For example, an 8 km length of SMF-28 transmits nearly 75% of light at 1,550 nm. It has been noted that if ocean water was as clear as fiber, one could see all the way to the bottom even of the Marianas Trench in the Pacific Ocean, a depth of 36,000 feet.

Light Scattering

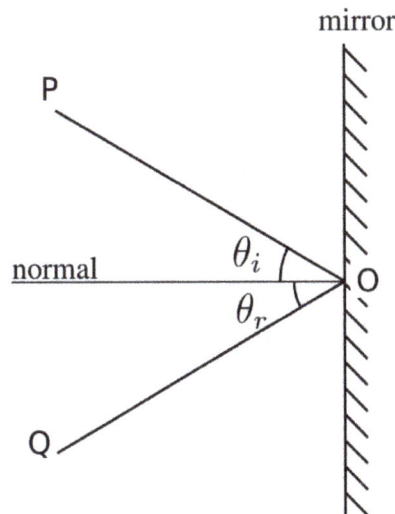

Specular reflection

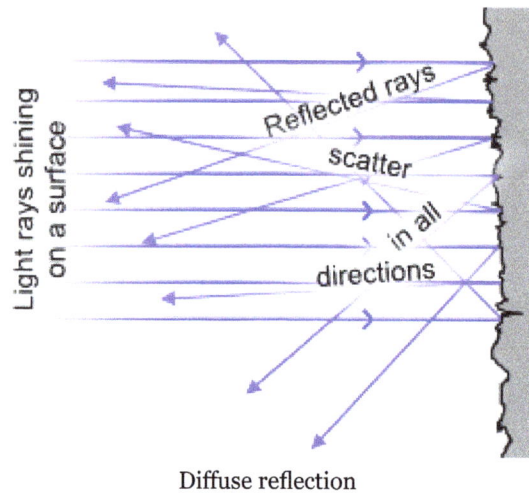

Diffuse reflection

The propagation of light through the core of an optical fiber is based on total internal reflection of the lightwave. Rough and irregular surfaces, even at the molecular level, can cause light rays to be reflected in random directions. This is called diffuse reflection or scattering, and it is typically characterized by wide variety of reflection angles.

Light scattering depends on the wavelength of the light being scattered. Thus, limits to spatial scales of visibility arise, depending on the frequency of the incident light-wave and the physical dimension (or spatial scale) of the scattering center, which is typically in the form of some specific micro-structural feature. Since visible light has a wavelength of the order of one micrometer (one millionth of a meter) scattering centers will have dimensions on a similar spatial scale.

Thus, attenuation results from the incoherent scattering of light at internal surfaces and interfaces. In (poly)crystalline materials such as metals and ceramics, in addition to pores, most of the internal surfaces or interfaces are in the form of grain boundaries that separate tiny regions of crystalline order. It has recently been shown that when the size of the scattering center (or grain boundary) is reduced below the size of the wavelength of the light being scattered, the scattering no longer occurs to any significant extent. This phenomenon has given rise to the production of transparent ceramic materials.

Similarly, the scattering of light in optical quality glass fiber is caused by molecular level irregularities (compositional fluctuations) in the glass structure. Indeed, one emerging school of thought is that a glass is simply the limiting case of a polycrystalline solid. Within this framework, "domains" exhibiting various degrees of short-range order become the building blocks of both metals and alloys, as well as glasses and ceramics. Distributed both between and within these domains are micro-structural defects that provide the most ideal locations for light scattering. This same phenomenon is seen as one of the limiting factors in the transparency of IR missile domes.

At high optical powers, scattering can also be caused by nonlinear optical processes in the fiber.

UV-Vis-IR Absorption

In addition to light scattering, attenuation or signal loss can also occur due to selective absorption of specific wavelengths, in a manner similar to that responsible for the appearance of color. Primary material considerations include both electrons and molecules as follows:

- At the electronic level, it depends on whether the electron orbitals are spaced (or "quantized") such that they can absorb a quantum of light (or photon) of a specific wavelength or frequency in the ultraviolet (UV) or visible ranges. This is what gives rise to color.

- At the atomic or molecular level, it depends on the frequencies of atomic or molecular vibrations or chemical bonds, how close-packed its atoms or molecules are, and whether or not the atoms or molecules exhibit long-range order. These factors will determine the capacity of the material transmitting longer wavelengths in the infrared (IR), far IR, radio and microwave ranges.

The design of any optically transparent device requires the selection of materials based upon knowledge of its properties and limitations. The Lattice absorption characteristics observed at the lower frequency regions (mid IR to far-infrared wavelength range) define the long-wavelength transparency limit of the material. They are the result of the interactive coupling between the motions of thermally induced vibrations of the constituent atoms and molecules of the solid lattice and the incident light wave radiation. Hence, all materials are bounded by limiting regions of absorption caused by atomic and molecular vibrations (bond-stretching)in the far-infrared (>10 μm).

Thus, multi-phonon absorption occurs when two or more phonons simultaneously interact to produce electric dipole moments with which the incident radiation may couple. These dipoles can absorb energy from the incident radiation, reaching a maximum coupling with the radiation when the frequency is equal to the fundamental vibrational mode of the molecular dipole (e.g. Si-O bond) in the far-infrared, or one of its harmonics.

The selective absorption of infrared (IR) light by a particular material occurs because the selected frequency of the light wave matches the frequency (or an integer multiple of the frequency) at which the particles of that material vibrate. Since different atoms and molecules have different natural frequencies of vibration, they will selectively absorb different frequencies (or portions of the spectrum) of infrared (IR) light.

Reflection and transmission of light waves occur because the frequencies of the light waves do not match the natural resonant frequencies of vibration of the objects. When IR light of these frequencies strikes an object, the energy is either reflected or transmitted.

Manufacturing

Materials

Glass optical fibers are almost always made from silica, but some other materials, such as fluorozirconate, fluoroaluminate, and chalcogenide glasses as well as crystalline materials like sapphire, are used for longer-wavelength infrared or other specialized applications. Silica and fluoride glasses usually have refractive indices of about 1.5, but some materials such as the chalcogenides can have indices as high as 3. Typically the index difference between core and cladding is less than one percent.

Plastic optical fibers (POF) are commonly step-index multi-mode fibers with a core diameter of 0.5

millimeters or larger. POF typically have higher attenuation coefficients than glass fibers, 1 dB/m or higher, and this high attenuation limits the range of POF-based systems.

Silica

Silica exhibits fairly good optical transmission over a wide range of wavelengths. In the near-infrared (near IR) portion of the spectrum, particularly around 1.5 μm, silica can have extremely low absorption and scattering losses of the order of 0.2 dB/km. Such remarkably low losses are possible only because ultra-pure silicon is available, it being essential for manufacturing integrated circuits and discrete transistors. A high transparency in the 1.4-μm region is achieved by maintaining a low concentration of hydroxyl groups (OH). Alternatively, a high OH concentration is better for transmission in the ultraviolet (UV) region.

Silica can be drawn into fibers at reasonably high temperatures, and has a fairly broad glass transformation range. One other advantage is that fusion splicing and cleaving of silica fibers is relatively effective. Silica fiber also has high mechanical strength against both pulling and even bending, provided that the fiber is not too thick and that the surfaces have been well prepared during processing. Even simple cleaving (breaking) of the ends of the fiber can provide nicely flat surfaces with acceptable optical quality. Silica is also relatively chemically inert. In particular, it is not hygroscopic (does not absorb water).

Silica glass can be doped with various materials. One purpose of doping is to raise the refractive index (e.g. with germanium dioxide (GeO_2) or aluminium oxide (Al_2O_3)) or to lower it (e.g. with fluorine or boron trioxide (B_2O_3)). Doping is also possible with laser-active ions (for example, rare earth-doped fibers) in order to obtain active fibers to be used, for example, in fiber amplifiers or laser applications. Both the fiber core and cladding are typically doped, so that the entire assembly (core and cladding) is effectively the same compound (e.g. an aluminosilicate, germanosilicate, phosphosilicate or borosilicate glass).

Particularly for active fibers, pure silica is usually not a very suitable host glass, because it exhibits a low solubility for rare earth ions. This can lead to quenching effects due to clustering of dopant ions. Aluminosilicates are much more effective in this respect.

Silica fiber also exhibits a high threshold for optical damage. This property ensures a low tendency for laser-induced breakdown. This is important for fiber amplifiers when utilized for the amplification of short pulses.

Because of these properties silica fibers are the material of choice in many optical applications, such as communications (except for very short distances with plastic optical fiber), fiber lasers, fiber amplifiers, and fiber-optic sensors. Large efforts put forth in the development of various types of silica fibers have further increased the performance of such fibers over other materials.

Fluoride Glass

Fluoride glass is a class of non-oxide optical quality glasses composed of fluorides of various metals. Because of their low viscosity, it is very difficult to completely avoid crystallization while processing it through the glass transition (or drawing the fiber from the melt). Thus, although heavy metal fluoride glasses (HMFG) exhibit very low optical attenuation, they are not only difficult to manufacture, but are quite fragile, and have poor resistance to moisture and other environmental attacks. Their best attribute is that they lack the absorption band associated with the hydroxyl (OH) group (3,200–3,600 cm^{-1}; i.e., 2,777–3,125 nm or 2.78–3.13 μm), which is present in nearly all oxide-based glasses.

An example of a heavy metal fluoride glass is the ZBLAN glass group, composed of zirconium, barium, lanthanum, aluminium, and sodium fluorides. Their main technological application is as optical waveguides in both planar and fiber form. They are advantageous especially in the mid-infrared (2,000–5,000 nm) range.

HMFGs were initially slated for optical fiber applications, because the intrinsic losses of a mid-IR fiber could in principle be lower than those of silica fibers, which are transparent only up to about 2 μm. However, such low losses were never realized in practice, and the fragility and high cost of fluoride fibers made them less than ideal as primary candidates. Later, the utility of fluoride fibers for various other applications was discovered. These include mid-IR spectroscopy, fiber optic sensors, thermometry, and imaging. Also, fluoride fibers can be used for guided lightwave transmission in media such as YAG (yttrium aluminium garnet) lasers at 2.9 μm, as required for medical applications (e.g. ophthalmology and dentistry).

Phosphate Glass

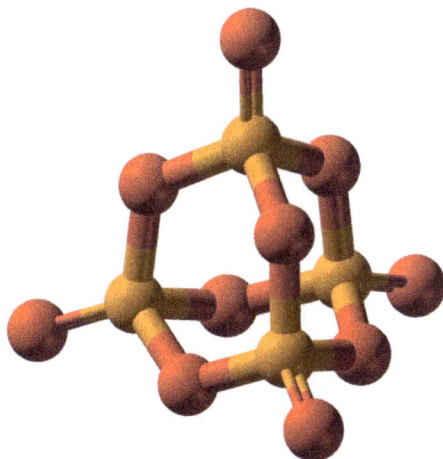

The P_4O_{10} cagelike structure—the basic building block for phosphate glass.

Phosphate glass constitutes a class of optical glasses composed of metaphosphates of various metals. Instead of the SiO_4 tetrahedra observed in silicate glasses, the building block for this glass former is phosphorus pentoxide (P_2O_5), which crystallizes in at least four different forms. The most familiar polymorph (see figure) comprises molecules of P_4O_{10}.

Phosphate glasses can be advantageous over silica glasses for optical fibers with a high concentration of doping rare earth ions. A mix of fluoride glass and phosphate glass is fluorophosphate glass.

Chalcogenide Glass

The chalcogens—the elements in group 16 of the periodic table—particularly sulfur (S), selenium

(Se) and tellurium (Te)—react with more electropositive elements, such as silver, to form chalco-genides. These are extremely versatile compounds, in that they can be crystalline or amorphous, metallic or semiconducting, and conductors of ions or electrons. Glass containing chalcogenides can be used to make fibers for far infrared transmission

Process

Preform

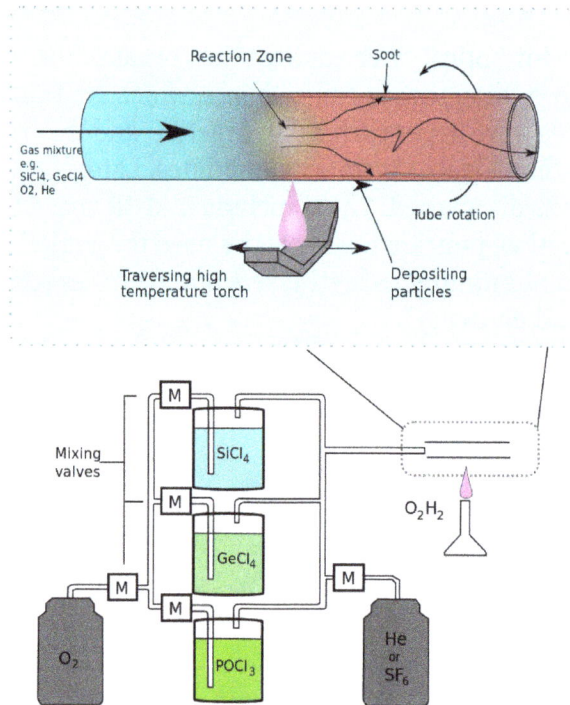

Illustration of the modified chemical vapor deposition (inside) process

Standard optical fibers are made by first constructing a large-diameter "preform" with a carefully controlled refractive index profile, and then "pulling" the preform to form the long, thin optical fiber. The preform is commonly made by three chemical vapor deposition methods: *inside vapor deposition*, *outside vapor deposition*, and *vapor axial deposition*.

With *inside vapor deposition*, the preform starts as a hollow glass tube approximately 40 centime-ters (16 in) long, which is placed horizontally and rotated slowly on a lathe. Gases such as silicon tetrachloride ($SiCl_4$) or germanium tetrachloride ($GeCl_4$) are injected with oxygen in the end of the tube. The gases are then heated by means of an external hydrogen burner, bringing the tem-perature of the gas up to 1,900 K (1,600 °C, 3,000 °F), where the tetrachlorides react with oxygen to produce silica or germania (germanium dioxide) particles. When the reaction conditions are chosen to allow this reaction to occur in the gas phase throughout the tube volume, in contrast to earlier techniques where the reaction occurred only on the glass surface, this technique is called *modified chemical vapor deposition (MCVD)*.

The oxide particles then agglomerate to form large particle chains, which subsequently deposit on the walls of the tube as soot. The deposition is due to the large difference in temperature between the gas core and the wall causing the gas to push the particles outwards (this is known as thermo-phoresis). The torch is then traversed up and down the length of the tube to deposit the material evenly. After the torch has reached the end of the tube, it is then brought back to the beginning of the tube and the deposited particles are then melted to form a solid layer. This process is repeated until a sufficient amount of material has been deposited. For each layer the composition can be modified by varying the gas composition, resulting in precise control of the finished fiber's optical properties.

In outside vapor deposition or vapor axial deposition, the glass is formed by *flame hydrolysis*, a reaction in which silicon tetrachloride and germanium tetrachloride are oxidized by reaction with water (H_2O) in an oxyhydrogen flame. In outside vapor deposition the glass is deposited onto a solid rod, which is removed before further processing. In vapor axial deposition, a short *seed rod* is used, and a porous preform, whose length is not limited by the size of the source rod, is built up on its end. The porous preform is consolidated into a transparent, solid preform by heating to about 1,800 K (1,500 °C, 2,800 °F).

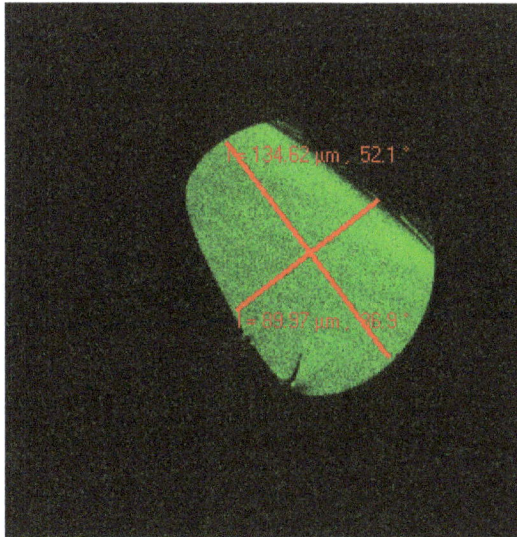

Cross-section of a fiber drawn from a D-shaped preform

Typical communications fiber uses a circular preform. For some applications such as double-clad fibers another form is preferred. In fiber lasers based on double-clad fiber, an asymmetric shape improves the filling factor for laser pumping.

Because of the surface tension, the shape is smoothed during the drawing process, and the shape of the resulting fiber does not reproduce the sharp edges of the preform. Nevertheless, careful polishing of the preform is important, since any defects of the preform surface affect the optical and mechanical properties of the resulting fiber. In particular, the preform for the test-fiber shown in the figure was not polished well, and cracks are seen with the confocal optical microscope.

Drawing

The preform, however constructed, is placed in a device known as a drawing tower, where the preform tip is heated and the optical fiber is pulled out as a string. By measuring the resultant fiber width, the tension on the fiber can be controlled to maintain the fiber thickness.

Coatings

The light is guided down the core of the fiber by an optical cladding with a lower refractive index that traps light in the core through total internal reflection.

The cladding is coated by a buffer that protects it from moisture and physical damage. The buffer coating is what gets stripped off the fiber for termination or splicing. These coatings are UV-cured urethane acrylate composite or polyimide materials applied to the outside of the fiber during the drawing process. The coatings protect the very delicate strands of glass fiber—about the size of a human hair—and allow it to survive the rigors of manufacturing, proof testing, cabling and installation.

Today's glass optical fiber draw processes employ a dual-layer coating approach. An inner primary coating is designed to act as a shock absorber to minimize attenuation caused by microbending. An outer secondary coating protects the primary coating against mechanical damage and acts as a barrier to lateral forces. Sometimes a metallic armor layer is added to provide extra protection.

These fiber optic coating layers are applied during the fiber draw, at speeds approaching 100 kilometers per hour (60 mph). Fiber optic coatings are applied using one of two methods: *wet-on-dry* and *wet-on-wet*. In wet-on-dry, the fiber passes through a primary coating application, which is then UV cured—then through the secondary coating application, which is subsequently cured. In wet-on-wet, the fiber passes through both the primary and secondary coating applications, then goes to UV curing.

Fiber optic coatings are applied in concentric layers to prevent damage to the fiber during the drawing application and to maximize fiber strength and microbend resistance. Unevenly coated fiber will experience non-uniform forces when the coating expands or contracts, and is susceptible to greater signal attenuation. Under proper drawing and coating processes, the coatings are concentric around the fiber, continuous over the length of the application and have constant thickness.

Fiber optic coatings protect the glass fibers from scratches that could lead to strength degradation. The combination of moisture and scratches accelerates the aging and deterioration of fiber strength. When fiber is subjected to low stresses over a long period, fiber fatigue can occur. Over time or in extreme conditions, these factors combine to cause microscopic flaws in the glass fiber to propagate, which can ultimately result in fiber failure.

Three key characteristics of fiber optic waveguides can be affected by environmental conditions: strength, attenuation and resistance to losses caused by microbending. External fiber optic coatings protect glass optical fiber from environmental conditions that can affect the fiber's performance and long-term durability. On the inside, coatings ensure the reliability of the signal being carried and help minimize attenuation due to microbending.

Practical Issues

Cable Construction

In practical fibers, the cladding is usually coated with a tough resin *buffer* layer, which may be further surrounded by a *jacket* layer, usually glass. These layers add strength to the fiber but do not contribute to its optical wave guide properties. Rigid fiber assemblies sometimes put light-absorbing ("dark") glass between the fibers, to prevent light that leaks out of one fiber from entering another. This reduces cross-talk between the fibers, or reduces flare in fiber bundle imaging applications.

An optical fiber cable

Modern cables come in a wide variety of sheathings and armor, designed for applications such as direct burial in trenches, high voltage isolation, dual use as power lines, installation in conduit, lashing to aerial telephone poles, submarine installation, and insertion in paved streets. The cost of small fiber-count pole-mounted cables has greatly decreased due to the high demand for fiber to the home (FTTH) installations in Japan and South Korea.

Fiber cable can be very flexible, but traditional fiber's loss increases greatly if the fiber is bent with a radius smaller than around 30 mm. This creates a problem when the cable is bent around corners or wound around a spool, making FTTX installations more complicated. "Bendable fibers", targeted towards easier installation in home environments, have been standardized as ITU-T G.657. This type of fiber can be bent with a radius as low as 7.5 mm without adverse impact. Even more bendable fibers have been developed. Bendable fiber may also be resistant to fiber hacking, in which the signal in a fiber is surreptitiously monitored by bending the fiber and detecting the leakage.

Another important feature of cable is cable's ability to withstand horizontally applied force. It is technically called max tensile strength defining how much force can be applied to the cable during the installation period.

Some fiber optic cable versions are reinforced with aramid yarns or glass yarns as intermediary strength member. In commercial terms, usage of the glass yarns are more cost effective while no loss in mechanical durability of the cable. Glass yarns also protect the cable core against rodents and termites.

Termination and Splicing

Optical fibers are connected to terminal equipment by optical fiber connectors. These connectors are usually of a standard type such as *FC, SC, ST, LC, MTRJ,* or *SMA*. Optical fibers may be connected to each other by connectors or by *splicing*, that is, joining two fibers together to form a continuous optical waveguide. The generally accepted splicing method is arc fusion splicing, which melts the fiber ends together with an electric arc. For quicker fastening jobs, a "mechanical splice" is used.

ST connectors on multi-mode fiber.

Fusion splicing is done with a specialized instrument. The fiber ends are first stripped of their protective polymer coating (as well as the more sturdy outer jacket, if present). The ends are *cleaved* (cut) with a precision cleaver to make them perpendicular, and are placed into special holders in the fusion splicer. The splice is usually inspected via a magnified viewing screen to check the cleaves before and after the splice. The splicer uses small motors to align the end faces together, and emits a small spark between electrodes at the gap to burn off dust and moisture. Then the splicer generates a larger spark that raises the temperature above the melting point of the glass, fusing the ends together permanently. The location and energy of the spark is carefully controlled so that the molten core and cladding do not mix, and this minimizes optical loss. A splice loss estimate is measured by the splicer, by directing light through the cladding on one side and measuring the light leaking from the cladding on the other side. A splice loss under 0.1 dB is typical. The complexity of this process makes fiber splicing much more difficult than splicing copper wire.

Mechanical fiber splices are designed to be quicker and easier to install, but there is still the need for stripping, careful cleaning and precision cleaving. The fiber ends are aligned and held together by a precision-made sleeve, often using a clear index-matching gel that enhances the transmission of light across the joint. Such joints typically have higher optical loss and are less robust than fusion splices, especially if the gel is used. All splicing techniques involve installing an enclosure that protects the splice.

Fibers are terminated in connectors that hold the fiber end precisely and securely. A fiber-optic connector is basically a rigid cylindrical barrel surrounded by a sleeve that holds the barrel in its mating socket. The mating mechanism can be *push and click*, *turn and latch* (*bayonet mount*), or *screw-in* (*threaded*). The barrel is typically free to move within the sleeve, and may have a key that prevents the barrel and fiber from rotating as the connectors are mated.

A typical connector is installed by preparing the fiber end and inserting it into the rear of the connector body. Quick-set adhesive is usually used to hold the fiber securely, and a strain relief is secured to the rear. Once the adhesive sets, the fiber's end is polished to a mirror finish. Various polish profiles are used, depending on the type of fiber and the application. For single-mode fiber, fiber ends are typically polished with a slight curvature that makes the mated connectors touch only at their cores. This is called a *physical contact* (PC) polish. The curved surface may be polished at an angle, to make an *angled physical contact (APC)* connection. Such connections have higher loss than PC connections, but greatly reduced back reflection, because light that reflects from the angled surface leaks out of the fiber core. The resulting signal strength loss is called *gap loss*. APC fiber ends have low back reflection even when disconnected.

In the 1990s, terminating fiber optic cables was labor-intensive. The number of parts per connector, polishing of the fibers, and the need to oven-bake the epoxy in each connector made terminating fiber optic cables difficult. Today, many connectors types are on the market that offer easier, less labor-intensive ways of terminating cables. Some of the most popular connectors are pre-polished at the factory, and include a gel inside the connector. Those two steps help save money on labor, especially on large projects. A cleave is made at a required length, to get as close to the polished piece already inside the connector. The gel surrounds the point where the two pieces meet inside the connector for very little light loss

Free-space Coupling

It is often necessary to align an optical fiber with another optical fiber, or with an optoelectronic device such as a light-emitting diode, a laser diode, or a modulator. This can involve either carefully aligning the fiber and placing it in contact with the device, or can use a lens to allow coupling over an air gap. In some cases the end of the fiber is polished into a curved form that makes it act as a lens. Some companies can even shape the fiber into lenses by cutting them with lasers.

In a laboratory environment, a bare fiber end is coupled using a fiber launch system, which uses a microscope objective lens to focus the light down to a fine point. A precision translation stage (micro-positioning table) is used to move the lens, fiber, or device to allow the coupling efficiency to be optimized. Fibers with a connector on the end make this process much simpler: the connector is simply plugged into a pre-aligned fiberoptic collimator, which contains a lens that is either accurately positioned with respect to the fiber, or is adjustable. To achieve the best injection efficiency into single-mode fiber, the direction, position, size and divergence of the beam must all be optimized. With good beams, 70 to 90% coupling efficiency can be achieved.

With properly polished single-mode fibers, the emitted beam has an almost perfect Gaussian shape—even in the far field—if a good lens is used. The lens needs to be large enough to support the full numerical aperture of the fiber, and must not introduce aberrations in the beam. Aspheric lenses are typically used.

Fiber Fuse

At high optical intensities, above 2 megawatts per square centimeter, when a fiber is subjected to a shock or is otherwise suddenly damaged, a *fiber fuse* can occur. The reflection from the damage vaporizes the fiber immediately before the break, and this new defect remains reflective so that the damage propagates back toward the transmitter at 1–3 meters per second (4–11 km/h, 2–8 mph). The open fiber control system, which ensures laser eye safety in the event of a broken fiber, can also effectively halt propagation of the fiber fuse. In situations, such as undersea cables, where high power levels might be used without the need for open fiber control, a "fiber fuse" protection device at the transmitter can break the circuit to keep damage to a minimum.

Chromatic Dispersion

The refractive index of fibers varies slightly with the frequency of light, and light sources are not perfectly monochromatic. Modulation of the light source to transmit a signal also slightly widens the frequency band of the transmitted light. This has the effect that, over long distances and at high modulation speeds, the different frequencies of light can take different times to arrive at the receiver, ultimately making the signal impossible to discern, and requiring extra repeaters. This problem can be overcome in a number of ways, including the use of a relatively short length of fiber that has the opposite refractive index gradient.

Spectrometer

In physics, a spectrometer is an apparatus to measure a spectrum. Generally, a spectrum is a graph that shows intensity as a function of wavelength, of frequency, of energy, of momentum, or of mass.

Optical Spectrometer

Optical spectrometers (often simply called "spectrometers"), in particular, show the intensity of light as a function of wavelength or of frequency. The deflection is produced either by refraction in a prism or by diffraction in a diffraction grating.

Mass Spectrometer

A mass spectrometer is an analytical instrument that is used to identify the amount and type of chemicals present in a sample by measuring the mass-to-charge ratio and abundance of gas-phase ions.

Time-of-flight Spectrometer

The energy spectrum of particles of known mass can also be measured by determining the time of flight between two detectors (and hence, the velocity) in a time-of-flight spectrometer. Alternatively, if the velocity is known, masses can be determined in a time-of-flight mass spectrometer.

Magnetic Spectrometer

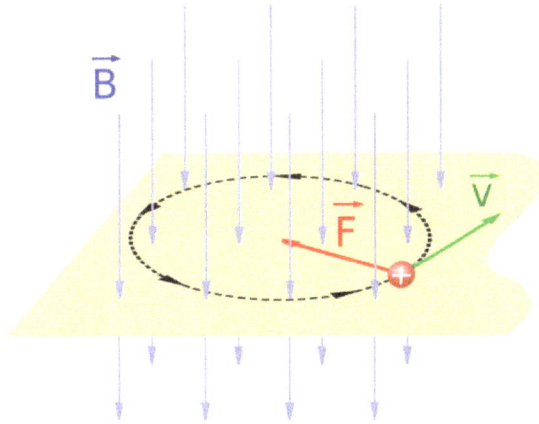

A positive charged particle moving in a circle under the influence of the Lorentz force F

When a fast charged particle (charge q, mass m) enters a constant magnetic field B at right angles, it is deflected into a circular path of radius r, due to the Lorentz force. The momentum p of the particle is then given by

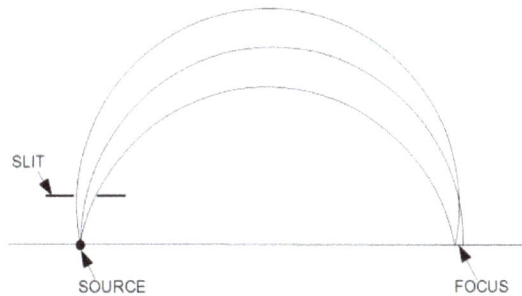

Focus of a magnetic semicircular spectrometer

where m and v are mass and velocity of the particle. The focussing principle of the oldest and simplest magnetic spectrometer, the semicircular spectrometer, invented by J. K. Danisz, is shown on the left. A constant magnetic field is perpendicular to the page. Charged particles of momentum p that pass the slit are deflected into circular paths of radius $r = p/qB$. It turns out that they all hit the horizontal line at nearly the same place, the focus; here a particle counter should be placed. Varying B, this makes possible to measure the energy spectrum of alpha particles in an alpha particle spectrometer, of beta particles in a beta particle spectrometer, of particles (e.g., fast ions) in a particle spectrometer, or to measure the relative content of the various masses in a mass spectrometer.

Since Danysz' time, many types of magnetic spectrometers more complicated than the semicircular type have been devised.

Resolution

Generally, the resolution of an instrument tells us how well two close-lying energies (or wavelengths, or frequencies, or masses) can be resolved. Generally, for an instrument with mechanical slits, higher resolution will mean lower intensity.

Light-emitting diode

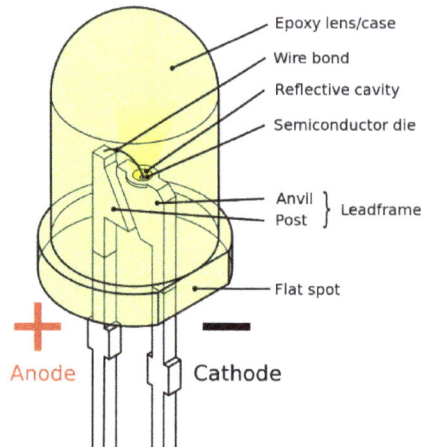

- Epoxy lens/case
- Wire bond
- Reflective cavity
- Semiconductor die
- Anvil } Leadframe
- Post
- Flat spot

Anode   Cathode

Parts of an LED. Although unlabeled, the flat bottom surfaces of the anvil and post embedded inside the epoxy act as anchors, to prevent the conductors from being forcefully pulled out via mechanical strain or vibration.

A bulb-shaped modern retrofit LED lamp with aluminium heat sink, a light diffusing dome and E27 screw base, using a built-in power supply working on mains voltage

A light-emitting diode (LED) is a two-lead semiconductor light source. It is a p–n junction diode, which emits light when activated. When a suitable voltage is applied to the leads, electrons are able to recombine with electron holes within the device, releasing energy in the form of photons. This effect is called electroluminescence, and the color of the light (corresponding to the energy of the photon) is determined by the energy band gap of the semiconductor.

An LED is often small in area (less than 1 mm²) and integrated optical components may be used to shape its radiation pattern.

Appearing as practical electronic components in 1962, the earliest LEDs emitted low-intensity infrared light. Infrared LEDs are still frequently used as transmitting elements in remote-control circuits, such as those in remote controls for a wide variety of consumer electronics. The first visible-light LEDs were also of low intensity, and limited to red. Modern LEDs are available across the visible, ultraviolet, and infrared wavelengths, with very high brightness.

Early LEDs were often used as indicator lamps for electronic devices, replacing small incandescent bulbs. They were soon packaged into numeric readouts in the form of seven-segment displays, and were commonly seen in digital clocks.

Recent developments in LEDs permit them to be used in environmental and task lighting. LEDs have many advantages over incandescent light sources including lower energy consumption, longer lifetime, improved physical robustness, smaller size, and faster switching. Light-emitting diodes are now used in applications as diverse as aviation lighting, automotive headlamps, advertising, general lighting, traffic signals, camera flashes and lighted wallpaper. As of 2016, LEDs powerful enough for room lighting remain somewhat more expensive, and require more precise current and heat management, than compact fluorescent lamp sources of comparable output. They are, however, significantly more energy efficient and, arguably, have less environmental concerns linked to their disposal. The governments of some countries are promoting the domestic use of LED-based lighting, and in some cases providing LED-based lighting solutions to the public at subsidized rates.

LEDs have allowed new displays and sensors to be developed, while their high switching rates are also used in advanced communications technology.

History

Discoveries and Early Devices

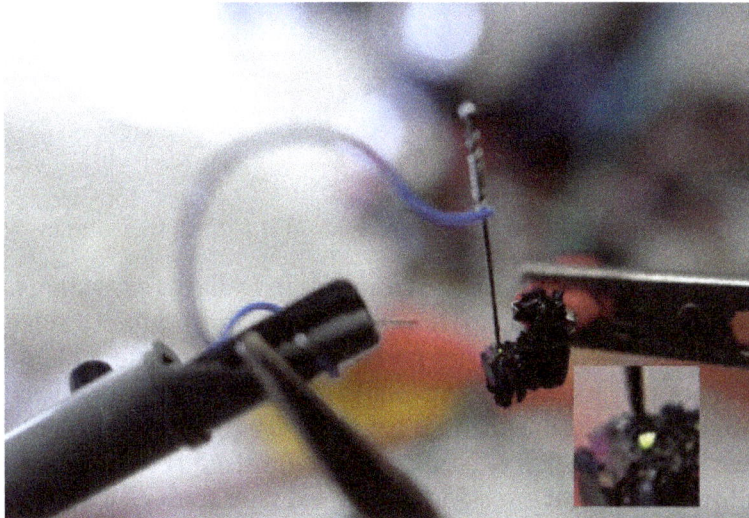

Green electroluminescence from a point contact on a crystal of SiC recreates Round's original experiment from 1907.

Electroluminescence as a phenomenon was discovered in 1907 by the British experimenter H. J. Round of Marconi Labs, using a crystal of silicon carbide and a cat's-whisker detector. Soviet inventor Oleg Losev reported creation of the first LED in 1927. His research was distributed in Soviet, German and British scientific journals, but no practical use was made of the discovery for several decades. Kurt Lehovec, Carl Accardo and Edward Jamgochian, explained these first light-emitting diodes in 1951 using an apparatus employing SiC crystals with a current source of battery or pulse generator and with a comparison to a variant, pure, crystal in 1953.

Rubin Braunstein of the Radio Corporation of America reported on infrared emission from gallium arsenide (GaAs) and other semiconductor alloys in 1955. Braunstein observed infrared emission generated by simple diode structures using gallium antimonide (GaSb), GaAs, indium phosphide (InP), and silicon-germanium (SiGe) alloys at room temperature and at 77 Kelvin.

In 1957, Braunstein further demonstrated that the rudimentary devices could be used for non-radio communication across a short distance. As noted by Kroemer Braunstein".. had set up a simple optical communications link: Music emerging from a record player was used via suitable electronics to modulate the forward current of a GaAs diode. The emitted light was detected by a PbS diode some distance away. This signal was fed into an audio amplifier, and played back by a loudspeaker. Intercepting the beam stopped the music. We had a great deal of fun playing with this setup." This setup presaged the use of LEDs for optical communication applications.

A Texas Instruments SNX-100 GaAs LED contained in a TO-18 transistor metal case.

In September 1961, while working at Texas Instruments in Dallas, Texas, James R. Biard and Gary Pittman discovered near-infrared (900 nm) light emission from a tunnel diode they had constructed on a GaAs substrate. By October 1961, they had demonstrated efficient light emission and signal coupling between a GaAs p-n junction light emitter and an electrically-isolated semiconductor photodetector. On August 8, 1962, Biard and Pittman filed a patent titled "Semiconductor Radiant Diode" based on their findings, which described a zinc diffused p–n junction LED with a spaced cathode contact to allow for efficient emission of infrared light under forward bias. After establishing the priority of their work based on engineering notebooks predating submissions from G.E. Labs, RCA Research Labs, IBM Research Labs, Bell Labs, and Lincoln Lab at MIT, the U.S. patent office issued the two inventors the patent for the GaAs infrared (IR) light-emitting diode (U.S. Patent US3293513), the first practical LED. Immediately after filing the patent, Texas Instruments (TI) began a project to manufacture infrared diodes. In October 1962, TI announced the first LED commercial product (the SNX-100), which employed a pure GaAs crystal to emit a 890 nm light output. In October 1963, TI announced the first commercial hemispherical LED, the SNX-110.

The first visible-spectrum (red) LED was developed in 1962 by Nick Holonyak, Jr., while working at General Electric. Holonyak first reported his LED in the journal *Applied Physics Letters* on the December 1, 1962. M. George Craford, a former graduate student of Holonyak, invented the first yellow LED and improved the brightness of red and red-orange LEDs by a factor of ten in 1972.

In 1976, T. P. Pearsall created the first high-brightness, high-efficiency LEDs for optical fiber tele-communications by inventing new semiconductor materials specifically adapted to optical fiber transmission wavelengths.

Initial Commercial Development

The first commercial LEDs were commonly used as replacements for incandescent and neon indicator lamps, and in seven-segment displays, first in expensive equipment such as laboratory and electronics test equipment, then later in such appliances as TVs, radios, telephones, calculators, as well as watches. Until 1968, visible and infrared LEDs were extremely costly, in the order of US$200 per unit, and so had little practical use. The Monsanto Company was the first organization to mass-produce visible LEDs, using gallium arsenide phosphide (GaAsP) in 1968 to produce red LEDs suitable for indicators. Hewlett Packard (HP) introduced LEDs in 1968, initially using GaAsP supplied by Monsanto. These red LEDs were bright enough only for use as indicators, as the light output was not enough to illuminate an area. Readouts in calculators were so small that plastic lenses were built over each digit to make them legible. Later, other colors became widely available and appeared in appliances and equipment. In the 1970s commercially successful LED devices at less than five cents each were produced by Fairchild Optoelectronics. These devices employed compound semiconductor chips fabricated with the planar process invented by Dr. Jean Hoerni at Fairchild Semiconductor. The combination of planar processing for chip fabrication and innovative packaging methods enabled the team at Fairchild led by optoelectronics pioneer Thomas Brandt to achieve the needed cost reductions. These methods continue to be used by LED producers.

LED display of a TI-30 scientific calculator (ca. 1978), which uses plastic lenses to increase the visible digit size

Most LEDs were made in the very common 5 mm T1¾ and 3 mm T1 packages, but with rising power output, it has grown increasingly necessary to shed excess heat to maintain reliability, so more complex packages have been adapted for efficient heat dissipation. Packages for state-of-the-art high-power LEDs bear little resemblance to early LEDs.

Blue LED

Blue LEDs were first developed by Herbert Paul Maruska at RCA in 1972 using gallium nitride (GaN) on a sapphire substrate. SiC-types were first commercially sold in the United States by Cree in 1989. However, neither of these initial blue LEDs were very bright.

The first high-brightness blue LED was demonstrated by Shuji Nakamura of Nichia Corporation in 1994 and was based on InGaN. In parallel, Isamu Akasaki and Hiroshi Amano in Nagoya were working on developing the important GaN nucleation on sapphire substrates and the demonstration of p-type doping of GaN. Nakamura, Akasaki and Amano were awarded the 2014 Nobel prize in physics for their work. In 1995, Alberto Barbieri at the Cardiff University Laboratory (GB) investigated the efficiency and reliability of high-brightness LEDs and demonstrated a "transparent contact" LED using indium tin oxide (ITO) on (AlGaInP/GaAs).

In 2001 and 2002, processes for growing gallium nitride (GaN) LEDs on silicon were successfully demonstrated. In January 2012, Osram demonstrated high-power InGaN LEDs grown on silicon substrates commercially.

White LEDs and the Illumination breakthrough

The attainment of high efficiency in blue LEDs was quickly followed by the development of the first white LED. In this device a Y3Al5O12:Ce (known as "YAG") phosphor coating on the emitter absorbs some of the blue emission and produces yellow light through fluorescence. The combination of that yellow with remaining blue light appears white to the eye. However, using different phosphors (fluorescent materials) it also became possible to instead produce green and red light through fluorescence. The resulting mixture of red, green and blue is not only perceived by humans as white light but is superior for illumination in terms of color rendering, whereas one cannot appreciate the color of red or green objects illuminated only by the yellow (and remaining blue) wavelengths from the YAG phosphor.

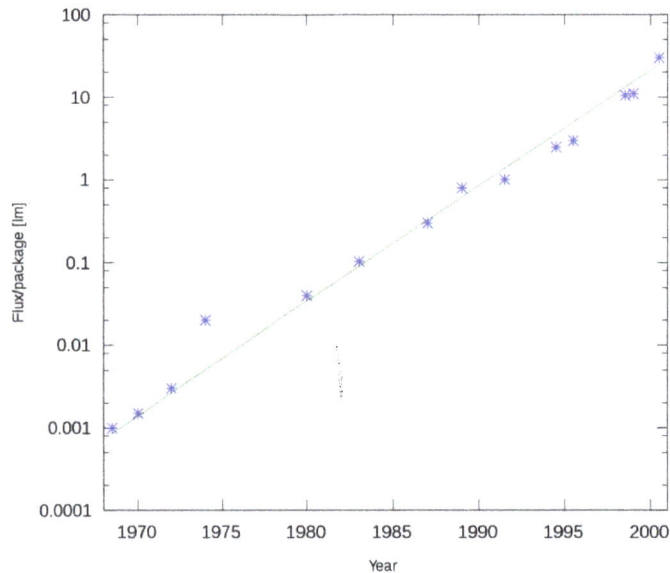

Illustration of Haitz's law, showing improvement in light output per LED over time, with a logarithmic scale on the vertical axis

The first white LEDs were expensive and inefficient. However, the light output of LEDs has increased exponentially, with a doubling occurring approximately every 36 months since the 1960s (similar to Moore's law). This trend is generally attributed to the parallel development of other semiconductor technologies and advances in optics and materials science, and has been called Haitz's law after Dr. Roland Haitz.

The light output and efficiency of blue and near-ultraviolet LEDs rose as the cost of reliable devices fell: this led to the use of (relatively) high-power white-light LEDs for the purpose of illumination which are replacing incandescent and fluorescent lighting.

White LEDs can now produce over 300 lumens per watt of electricity while lasting up to 100,000 hours. Compared to incandescent bulbs, this is not only a huge increase in electrical efficiency, but – over time – a similar or lower cost per bulb.

Working Principle

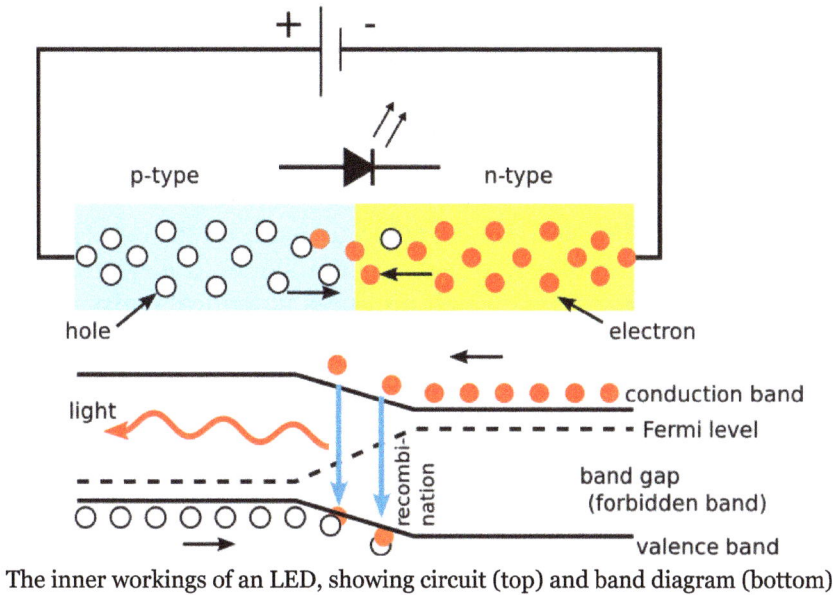

The inner workings of an LED, showing circuit (top) and band diagram (bottom)

A P-N junction can convert absorbed light energy into a proportional electric current. The same process is reversed here (i.e. the P-N junction emits light when electrical energy is applied to it). This phenomenon is generally called electroluminescence, which can be defined as the emission of light from a semi-conductor under the influence of an electric field. The charge carriers recombine in a forward-biased P-N junction as the electrons cross from the N-region and recombine with the holes existing in the P-region. Free electrons are in the conduction band of energy levels, while holes are in the valence energy band. Thus the energy level of the holes will be lesser than the energy levels of the electrons. Some portion of the energy must be dissipated in order to recombine the electrons and the holes. This energy is emitted in the form of heat and light.

The electrons dissipate energy in the form of heat for silicon and germanium diodes but in gallium arsenide phosphide (GaAsP) and gallium phosphide (GaP) semiconductors, the electrons dissipate energy by emitting photons. If the semiconductor is translucent, the junction becomes the source of light as it is emitted, thus becoming a light-emitting diode, but when the junction is reverse biased no light will be produced by the LED and, on the contrary, the device may also be damaged.

Technology

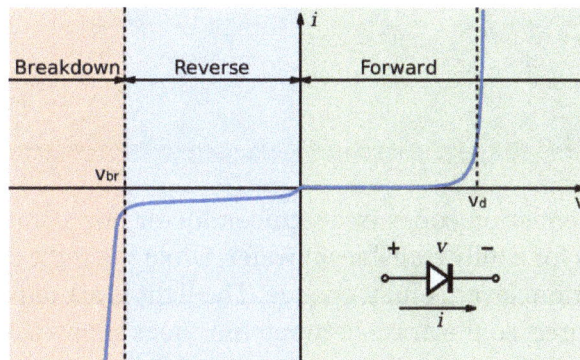

I-V diagram for a diode. An LED will begin to emit light when more than 2 or 3 volts is applied to it.

Physics

The LED consists of a chip of semiconducting material doped with impurities to create a *p-n junction*. As in other diodes, current flows easily from the p-side, or anode, to the n-side, or cathode, but not in the reverse direction. Charge-carriers—electrons and holes—flow into the junction from electrodes with different voltages. When an electron meets a hole, it falls into a lower energy level and releases energy in the form of a photon.

The wavelength of the light emitted, and thus its color, depends on the band gap energy of the materials forming the *p-n junction*. In silicon or germanium diodes, the electrons and holes usually recombine by a *non-radiative transition*, which produces no optical emission, because these are indirect band gap materials. The materials used for the LED have a direct band gap with energies corresponding to near-infrared, visible, or near-ultraviolet light.

LED development began with infrared and red devices made with gallium arsenide. Advances in materials science have enabled making devices with ever-shorter wavelengths, emitting light in a variety of colors.

LEDs are usually built on an n-type substrate, with an electrode attached to the p-type layer deposited on its surface. P-type substrates, while less common, occur as well. Many commercial LEDs, especially GaN/InGaN, also use sapphire substrate.

Most materials used for LED production have very high refractive indices. This means that much of the light will be reflected back into the material at the material/air surface interface. Thus, light extraction in LEDs is an important aspect of LED production, subject to much research and development.

Refractive Index

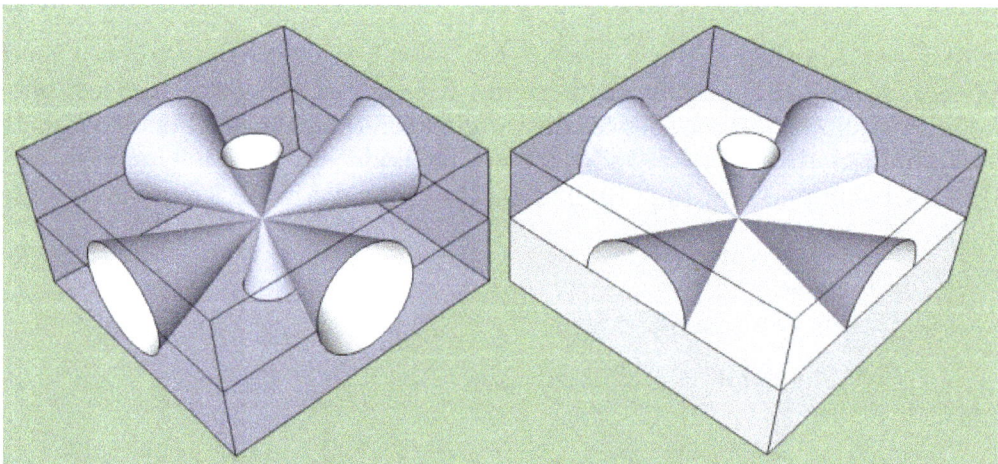

Idealized example of light emission cones in a semiconductor, for a single point-source emission zone. The left illustration is for a fully translucent wafer, while the right illustration shows the half-cones formed when the bottom layer is fully opaque. The light is actually emitted equally in all directions from the point-source, so the areas between the cones shows the large amount of trapped light energy that is wasted as heat.

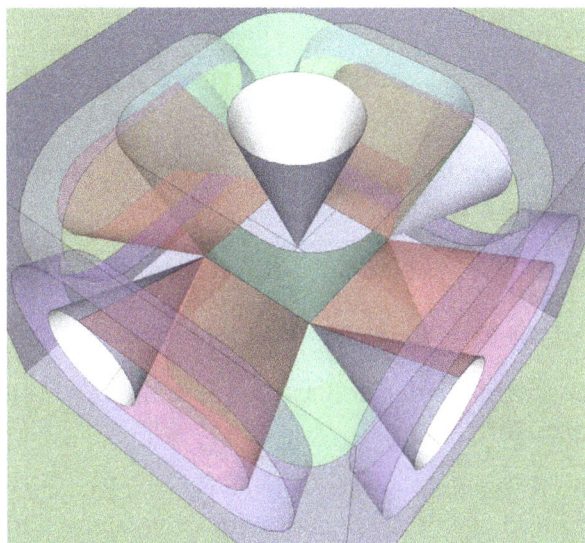

The light emission cones of a real LED wafer are far more complex than a single point-source light emission. The light emission zone is typically a two-dimensional plane between the wafers. Every atom across this plane has an individual set of emission cones. Drawing the billions of overlapping cones is impossible, so this is a simplified diagram showing the extents of all the emission cones combined. The larger side cones are clipped to show the interior features and reduce image complexity; they would extend to the opposite edges of the two-dimensional emission plane.

Bare uncoated semiconductors such as silicon exhibit a very high refractive index relative to open air, which prevents passage of photons arriving at sharp angles relative to the air-contacting surface of the semiconductor due to total internal reflection. This property affects both the light-emission efficiency of LEDs as well as the light-absorption efficiency of photovoltaic cells. The refractive index of silicon is 3.96 (at 590 nm), while air is 1.0002926.

In general, a flat-surface uncoated LED semiconductor chip will emit light only perpendicular to the semiconductor's surface, and a few degrees to the side, in a cone shape referred to as the *light cone, cone of light,* or the *escape cone.* The maximum angle of incidence is referred to as the critical angle. When this angle is exceeded, photons no longer escape the semiconductor but are instead reflected internally inside the semiconductor crystal as if it were a mirror.

Internal reflections can escape through other crystalline faces, if the incidence angle is low enough and the crystal is sufficiently transparent to not re-absorb the photon emission. But for a simple square LED with 90-degree angled surfaces on all sides, the faces all act as equal angle mirrors. In this case most of the light can not escape and is lost as waste heat in the crystal.

A convoluted chip surface with angled facets similar to a jewel or fresnel lens can increase light output by allowing light to be emitted perpendicular to the chip surface while far to the sides of the photon emission point.

The ideal shape of a semiconductor with maximum light output would be a microsphere with the photon emission occurring at the exact center, with electrodes penetrating to the center to contact at the emission point. All light rays emanating from the center would be perpendicular to the entire surface of the sphere, resulting in no internal reflections. A hemispherical semiconductor would also work, with the flat back-surface serving as a mirror to back-scattered photons.

Transition Coatings

After the doping of the wafer, it is cut apart into individual dies. Each die is commonly called a chip.

Many LED semiconductor chips are encapsulated or potted in clear or colored molded plastic shells. The plastic shell has three purposes:

1. Mounting the semiconductor chip in devices is easier to accomplish.

2. The tiny fragile electrical wiring is physically supported and protected from damage.

3. The plastic acts as a refractive intermediary between the relatively high-index semiconductor and low-index open air.

The third feature helps to boost the light emission from the semiconductor by acting as a diffusing lens, allowing light to be emitted at a much higher angle of incidence from the light cone than the bare chip is able to emit alone.

Efficiency and Operational Parameters

Typical indicator LEDs are designed to operate with no more than 30–60 milliwatts (mW) of electrical power. Around 1999, Philips Lumileds introduced power LEDs capable of continuous use at one watt. These LEDs used much larger semiconductor die sizes to handle the large power inputs. Also, the semiconductor dies were mounted onto metal slugs to allow for heat removal from the LED die.

One of the key advantages of LED-based lighting sources is high luminous efficacy. White LEDs quickly matched and overtook the efficacy of standard incandescent lighting systems. In 2002, Lumileds made five-watt LEDs available with luminous efficacy of 18–22 lumens per watt (lm/W). For comparison, a conventional incandescent light bulb of 60–100 watts emits around 15 lm/W, and standard fluorescent lights emit up to 100 lm/W.

As of 2012, Philips had achieved the following efficacies for each color. The efficiency values show the physics – light power out per electrical power in. The lumen-per-watt efficacy value includes characteristics of the human eye, and is derived using the luminosity function.

	Color	Wavelength range (nm)	Typical efficiency coefficient	Typical efficacy (lm/W)
	Red	$620 < \lambda < 645$	0.39	72
	Red-orange	$610 < \lambda < 620$	0.29	98
	Green	$520 < \lambda < 550$	0.15	93
	Cyan	$490 < \lambda < 520$	0.26	75
	Blue	$460 < \lambda < 490$	0.35	37

In September 2003, a new type of blue LED was demonstrated by Cree that consumes 24 mW at 20 milliamperes (mA). This produced a commercially packaged white light giving 65 lm/W at 20 mA, becoming the brightest white LED commercially available at the time, and more than four times as efficient as standard incandescents. In 2006, they demonstrated a prototype with a record white LED luminous efficacy of 131 lm/W at 20 mA. Nichia Corporation has developed a white LED with luminous efficacy of 150 lm/W at a forward current of 20 mA. Cree's XLamp XM-L LEDs, commercially available in 2011, produce 100 lm/W at their full power of 10 W, and up to 160 lm/W at around 2 W input power. In 2012, Cree announced a white LED giving 254 lm/W, and 303 lm/W in March 2014. Practical general lighting needs high-power LEDs, of one watt or more. Typical operating currents for such devices begin at 350 mA.

These efficiencies are for the light-emitting diode only, held at low temperature in a lab. Since LEDs installed in real fixtures operate at higher temperature and with driver losses, real-world efficiencies are much lower. United States Department of Energy (DOE) testing of commercial LED lamps designed to replace incandescent lamps or CFLs showed that average efficacy was still about 46 lm/W in 2009 (tested performance ranged from 17 lm/W to 79 lm/W).

Efficiency Droop

Efficiency droop is the decrease in luminous efficiency of LEDs as the electric current increases above tens of milliamperes.

This effect was initially theorized to be related to elevated temperatures. Scientists proved the opposite to be true that, although the life of an LED would be shortened, the efficiency droop is less severe at elevated temperatures. The mechanism causing efficiency droop was identified in 2007 as Auger recombination, which was taken with mixed reaction. In 2013, a study confirmed Auger recombination as the cause of efficiency droop.

In addition to being less efficient, operating LEDs at higher electric currents creates higher heat levels which compromise the lifetime of the LED. Because of this increased heating at higher currents, high-brightness LEDs have an industry standard of operating at only 350 mA, which is a compromise between light output, efficiency, and longevity.

Possible Solutions

Instead of increasing current levels, luminance is usually increased by combining multiple LEDs in one bulb. Solving the problem of efficiency droop would mean that household LED light bulbs would need fewer LEDs, which would significantly reduce costs.

Researchers at the U.S. Naval Research Laboratory have found a way to lessen the efficiency droop. They found that the droop arises from non-radiative Auger recombination of the injected carriers. They created quantum wells with a soft confinement potential to lessen the non-radiative Auger processes.

Researchers at Taiwan National Central University and Epistar Corp are developing a way to lessen the efficiency droop by using ceramic aluminium nitride (AlN) substrates, which are more thermally conductive than the commercially used sapphire. The higher thermal conductivity reduces self-heating effects.

Lifetime and Failure

Solid-state devices such as LEDs are subject to very limited wear and tear if operated at low currents and at low temperatures. Typical lifetimes quoted are 25,000 to 100,000 hours, but heat and current settings can extend or shorten this time significantly.

The most common symptom of LED (and diode laser) failure is the gradual lowering of light output and loss of efficiency. Sudden failures, although rare, can also occur. Early red LEDs were notable for their short service life. With the development of high-power LEDs the devices are subjected to higher junction temperatures and higher current densities than traditional devices. This causes stress on the material and may cause early light-output degradation. To quantitatively classify useful lifetime in a standardized manner it has been suggested to use L70 or L50, which are the runtimes (typically given in thousands of hours) at which a given LED reaches 70% and 50% of initial light output, respectively.

LED performance is temperature dependent. Most manufacturers' published ratings of LEDs are for an operating temperature of 25 °C (77 °F). LEDs used outdoors, such as traffic signals or in-pavement signal lights, and that are used in climates where the temperature within the light fixture gets very high, could result in low signal intensities or even failure.

Since LED efficacy is inversely proportional to operating temperature, LED technology is well suited for supermarket freezer lighting. Because LEDs produce less waste heat than incandescent lamps, their use in freezers can save on refrigeration costs as well. However, they may be more susceptible to frost and snow buildup than incandescent lamps, so some LED lighting systems have been designed with an added heating circuit. Additionally, research has developed heat sink technologies that will transfer heat produced within the junction to appropriate areas of the light fixture.

Colors and Materials

Conventional LEDs are made from a variety of inorganic semiconductor materials. The following table shows the available colors with wavelength range, voltage drop and material:

	Color	Wavelength [nm]	Voltage drop [ΔV]	Semiconductor material
	Infrared	$\lambda > 760$	$\Delta V < 1.63$	Gallium arsenide (GaAs) Aluminium gallium arsenide (AlGaAs)
	Red	$610 < \lambda < 760$	$1.63 < \Delta V < 2.03$	Aluminium gallium arsenide (AlGaAs) Gallium arsenide phosphide (GaAsP) Aluminium gallium indium phosphide (AlGaInP) Gallium(III) phosphide (GaP)
	Orange	$590 < \lambda < 610$	$2.03 < \Delta V < 2.10$	Gallium arsenide phosphide (GaAsP) Aluminium gallium indium phosphide (AlGaInP) Gallium(III) phosphide (GaP)
	Yellow	$570 < \lambda < 590$	$2.10 < \Delta V < 2.18$	Gallium arsenide phosphide (GaAsP) Aluminium gallium indium phosphide (AlGaInP) Gallium(III) phosphide (GaP)
	Green	$500 < \lambda < 570$	$1.9 < \Delta V < 4.0$	Traditional green: Gallium(III) phosphide (GaP) Aluminium gallium indium phosphide (AlGaInP) Aluminium gallium phosphide (AlGaP) Pure green: Indium gallium nitride (InGaN) / Gallium(III) nitride (GaN)
	Blue	$450 < \lambda < 500$	$2.48 < \Delta V < 3.7$	Zinc selenide (ZnSe) Indium gallium nitride (InGaN) Silicon carbide (SiC) as substrate Silicon (Si) as substrate—under development

Violet	$400 < \lambda < 450$	$2.76 < \Delta V < 4.0$	Indium gallium nitride (InGaN)
Purple	Multiple types	$2.48 < \Delta V < 3.7$	Dual blue/red LEDs, blue with red phosphor, or white with purple plastic
Ultraviolet	$\lambda < 400$	$3 < \Delta V < 4.1$	Indium gallium nitride (InGaN) (385-400 nm) Diamond (235 nm) Boron nitride (215 nm) Aluminium nitride (AlN) (210 nm) Aluminium gallium nitride (AlGaN) Aluminium gallium indium nitride (AlGaInN)—down to 210 nm
Pink	Multiple types	$\Delta V \sim 3.3$	Blue with one or two phosphor layers, yellow with red, orange or pink phosphor added afterwards, white with pink plastic, or white phosphors with pink pigment or dye over top.
White	Broad spectrum	$2.8 < \Delta V < 4.2$	Cool / Pure White: Blue/UV diode with yellow phosphor Warm White: Blue diode with orange phosphor

Blue and Ultraviolet

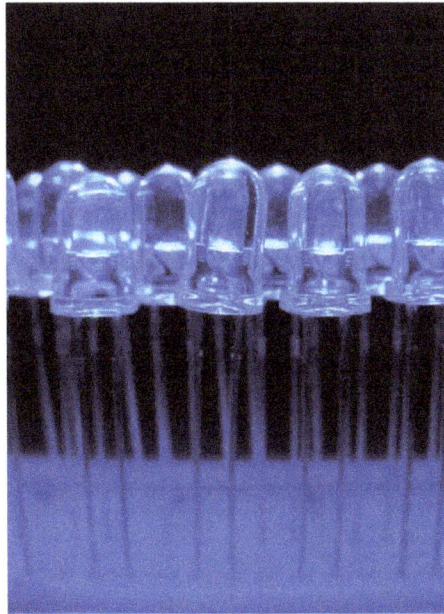

Blue LEDs

The first blue-violet LED using magnesium-doped gallium nitride was made at Stanford University in 1972 by Herb Maruska and Wally Rhines, doctoral students in materials science and engineering. At the time Maruska was on leave from RCA Laboratories, where he collaborated with Jacques Pankove on related work. In 1971, the year after Maruska left for Stanford, his RCA colleagues Pankove and Ed Miller demonstrated the first blue electroluminescence from zinc-doped gallium nitride, though the subsequent device Pankove and Miller built, the first actual gallium nitride light-emitting diode, emitted green light. In 1974 the U.S. Patent Office awarded Maruska, Rhines and Stanford professor David Stevenson a patent for their work in 1972 (U.S. Patent US3819974 A) and today magnesium-doping of gallium nitride continues to be the basis for all commercial blue LEDs and laser diodes. These devices built in the early 1970s had too little light output to be of practical use and research into gallium nitride devices slowed. In August 1989, Cree introduced the first commercially available blue LED based on the indirect bandgap semiconductor, silicon carbide (SiC). SiC LEDs had very low efficiency, no more than about 0.03%, but did emit in the blue portion of the visible light spectrum

In the late 1980s, key breakthroughs in GaN epitaxial growth and p-type doping ushered in the modern era of GaN-based optoelectronic devices. Building upon this foundation, Theodore Moustakas at Boston University patented a method for producing high-brightness blue LEDs using a new two-step process. Two years later, in 1993, high-brightness blue LEDs were demonstrated again by Shuji Nakamura of Nichia Corporation using a gallium nitride growth process similar to Moustakas's. Both Moustakas and Nakamura were issued separate patents, which confused the issue of who was the original inventor (partly because although Moustakas invented his first, Nakamura filed first) This new development revolutionized LED lighting, making high-power blue light sources practical, leading to the development of technologies like Blu-ray, as well as allowing the bright high resolution screens of modern tablets and phones

Nakamura was awarded the 2006 Millennium Technology Prize for his invention. Nakamura, Hiroshi Amano and Isamu Akasaki were awarded the Nobel Prize in Physics in 2014 for the invention of the blue LED. In 2015, a US court ruled that three companies (i.e. the litigants who had not previously settled out of court) that had licensed Nakamura's patents for production in the United States had infringed Moustakas's prior patent, and ordered them to pay licensing fees of not less than 13 million USD.

By the late 1990s, blue LEDs became widely available. They have an active region consisting of one or more InGaN quantum wells sandwiched between thicker layers of GaN, called cladding layers. By varying the relative In/Ga fraction in the InGaN quantum wells, the light emission can in theory be varied from violet to amber. Aluminium gallium nitride (AlGaN) of varying Al/Ga fraction can be used to manufacture the cladding and quantum well layers for ultraviolet LEDs, but these devices have not yet reached the level of efficiency and technological maturity of InGaN/GaN blue/green devices. If un-alloyed GaN is used in this case to form the active quantum well layers, the device will emit near-ultraviolet light with a peak wavelength centred around 365 nm. Green LEDs manufactured from the InGaN/GaN system are far more efficient and brighter than green LEDs produced with non-nitride material systems, but practical devices still exhibit efficiency too low for high-brightness applications

With nitrides containing aluminium, most often AlGaN and AlGaInN, even shorter wavelengths are achievable. Ultraviolet LEDs in a range of wavelengths are becoming available on the market. Near-UV emitters at wavelengths around 375–395 nm are already cheap and often encountered, for example, as black light lamp replacements for inspection of anti-counterfeiting UV watermarks in some documents and paper currencies. Shorter-wavelength diodes, while substantially more expensive, are commercially available for wavelengths down to 240 nm. As the photosensitivity of microorganisms approximately matches the absorption spectrum of DNA, with a peak at about 260 nm, UV LED emitting at 250–270 nm are to be expected in prospective disinfection and sterilization devices. Recent research has shown that commercially available UVA LEDs (365 nm) are already effective disinfection and sterilization devices. UV-C wavelengths were obtained in laboratories using aluminium nitride (210 nm), boron nitride (215 nm) and diamond (235 nm).

RGB

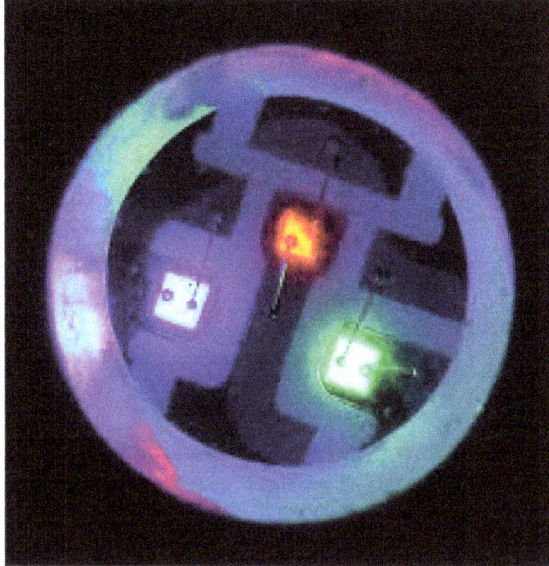

RGB LED

RGB LEDs consist of one red, one green, and one blue LED. By independently adjusting each of the three, RGB LEDs are capable of producing a wide color gamut. Unlike dedicated-color LEDs, however, these obviously do not produce pure wavelengths. Moreover, such modules as commercially available are often not optimized for smooth color mixing.

White

There are two primary ways of producing white light-emitting diodes (WLEDs), LEDs that generate high-intensity white light. One is to use individual LEDs that emit three primary colors—red, green, and blue—and then mix all the colors to form white light. The other is to use a phosphor material to convert monochromatic light from a blue or UV LED to broad-spectrum white light, much in the same way a fluorescent light bulb works. It is important to note that the 'whiteness' of the light produced is essentially engineered to suit the human eye, and depending on the situation it may not always be appropriate to think of it as white light.

There are three main methods of mixing colors to produce white light from an LED:

- blue LED + green LED + red LED (color mixing; can be used as backlighting for displays)

- near-UV or UV LED + RGB phosphor (an LED producing light with a wavelength shorter than blue's is used to excite an RGB phosphor)

- blue LED + yellow phosphor (two complementary colors combine to form white light; more efficient than first two methods and more commonly used)

Because of metamerism, it is possible to have quite different spectra that appear white. However, the appearance of objects illuminated by that light may vary as the spectrum varies.

RGB Systems

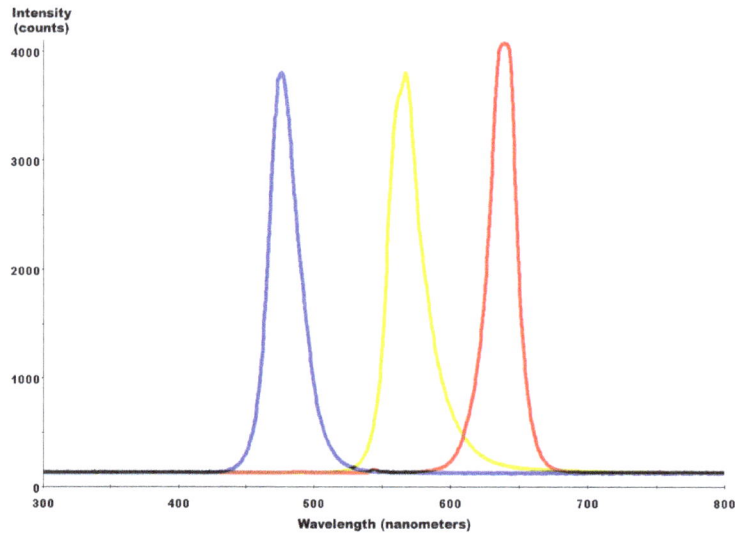

Combined spectral curves for blue, yellow-green, and high-brightness red solid-state semiconductor LEDs. FWHM spectral bandwidth is approximately 24–27 nm for all three colors.

RGB LED

White light can be formed by mixing differently colored lights; the most common method is to use red, green, and blue (RGB). Hence the method is called multi-color white LEDs (sometimes referred to as RGB LEDs). Because these need electronic circuits to control the blending and diffusion of different colors, and because the individual color LEDs typically have slightly different emission patterns (leading to variation of the color depending on direction) even if they are made as a single unit, these are seldom used to produce white lighting. Nonetheless, this method has many applications because of the flexibility of mixing different colors, and in principle, this mechanism also has higher quantum efficiency in producing white light

There are several types of multi-color white LEDs: di-, tri-, and tetrachromatic white LEDs. Several key factors that play among these different methods, include color stability, color rendering capability, and luminous efficacy. Often, higher efficiency will mean lower color rendering, presenting a trade-off between the luminous efficacy and color rendering. For example, the dichromatic white LEDs have the best luminous efficacy (120 lm/W), but the lowest color rendering capability. However, although tetrachromatic white LEDs have excellent color rendering capability, they often have poor luminous efficacy. Trichromatic white LEDs are in between, having both good luminous efficacy (>70 lm/W) and fair color rendering capability.

One of the challenges is the development of more efficient green LEDs. The theoretical maximum for green LEDs is 683 lumens per watt but as of 2010 few green LEDs exceed even 100 lumens per watt. The blue and red LEDs get closer to their theoretical limits.

Multi-color LEDs offer not merely another means to form white light but a new means to form light of different colors. Most perceivable colors can be formed by mixing different amounts of three primary colors. This allows precise dynamic color control. As more effort is devoted to investigating this method, multi-color LEDs should have profound influence on the fundamental method that we use to produce and control light color. However, before this type of LED can play a role on the market, several technical problems must be solved. These include that this type of LED's emission power decays exponentially with rising temperature, resulting in a substantial change in color stability. Such problems inhibit and may preclude industrial use. Thus, many new package designs aimed at solving this problem have been proposed and their results are now being reproduced by researchers and scientists.

Correlated color temperature (CCT) dimming for LED technology is regarded as a difficult task, since binning, age and temperature drift effects of LEDs change the actual color value output. Feedback loop systems are used for example with color sensors, to actively monitor and control the color output of multiple color mixing LEDs.

Phosphor-based LEDs

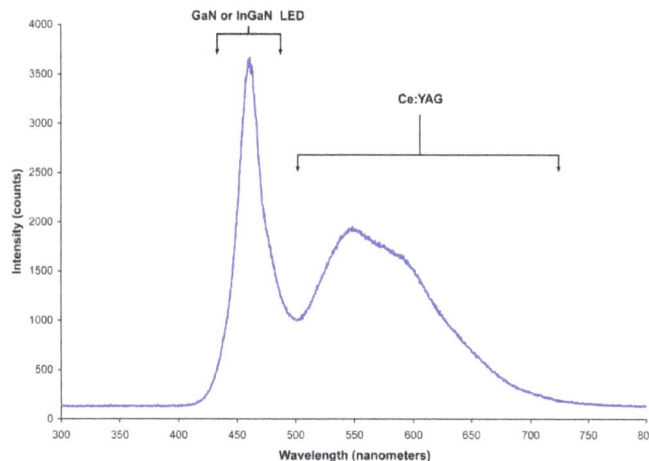

Spectrum of a white LED showing blue light directly emitted by the GaN-based LED (peak at about 465 nm) and the more broadband Stokes-shifted light emitted by the Ce^{3+}:YAG phosphor, which emits at roughly 500–700 nm

This method involves coating LEDs of one color (mostly blue LEDs made of InGaN) with phosphors of different colors to form white light; the resultant LEDs are called phosphor-based or phosphor-converted white LEDs (pcLEDs). A fraction of the blue light undergoes the Stokes shift being transformed from shorter wavelengths to longer. Depending on the color of the original LED, phosphors of different colors can be employed. If several phosphor layers of distinct colors are applied, the emitted spectrum is broadened, effectively raising the color rendering index (CRI) value of a given LED.

Phosphor-based LED efficiency losses are due to the heat loss from the Stokes shift and also other phosphor-related degradation issues. Their luminous efficacies compared to normal LEDs depend on the spectral distribution of the resultant light output and the original wavelength of the LED itself. For example, the luminous efficacy of a typical YAG yellow phosphor based white LED ranges from 3 to 5 times the luminous efficacy of the original blue LED because of the human eye's greater sensitivity to yellow than to blue (as modeled in the luminosity function). Due to the simplicity of manufacturing the phosphor method is still the most popular method for making high-intensity white LEDs. The design and production of a light source or light fixture using a monochrome emitter with phosphor conversion is simpler and cheaper than a complex RGB system, and the majority of high-intensity white LEDs presently on the market are manufactured using phosphor light conversion.

Among the challenges being faced to improve the efficiency of LED-based white light sources is the development of more efficient phosphors. As of 2010, the most efficient yellow phosphor is still the YAG phosphor, with less than 10% Stoke shift loss. Losses attributable to internal optical losses due to re-absorption in the LED chip and in the LED packaging itself account typically for another 10% to 30% of efficiency loss. Currently, in the area of phosphor LED development, much effort is being spent on optimizing these devices to higher light output and higher operation temperatures. For instance, the efficiency can be raised by adapting better package design or by using a more suitable type of phosphor. Conformal coating process is frequently used to address the issue of varying phosphor thickness.

Some phosphor-based white LEDs encapsulate InGaN blue LEDs inside phosphor-coated epoxy. Alternatively, the LED might be paired with a remote phosphor, a preformed polycarbonate piece coated with the phosphor material. Remote phosphors provide more diffuse light, which is desirable for many applications. Remote phosphor designs are also more tolerant of variations in the LED emissions spectrum. A common yellow phosphor material is cerium-doped yttrium aluminium garnet (Ce^{3+}:YAG).

White LEDs can also be made by coating near-ultraviolet (NUV) LEDs with a mixture of high-efficiency europium-based phosphors that emit red and blue, plus copper and aluminium-doped zinc sulfide (ZnS:Cu, Al) that emits green. This is a method analogous to the way fluorescent lamps work. This method is less efficient than blue LEDs with YAG:Ce phosphor, as the Stokes shift is larger, so more energy is converted to heat, but yields light with better spectral characteristics, which render color better. Due to the higher radiative output of the ultraviolet LEDs than of the blue ones, both methods offer comparable brightness. A concern is that UV light may leak from a malfunctioning light source and cause harm to human eyes or skin.

Other White LEDs

Another method used to produce experimental white light LEDs used no phosphors at all and was based on homoepitaxially grown zinc selenide (ZnSe) on a ZnSe substrate that simultaneously emitted blue light from its active region and yellow light from the substrate.

A new style of wafers composed of gallium-nitride-on-silicon (GaN-on-Si) is being used to produce white LEDs using 200-mm silicon wafers. This avoids the typical costly sapphire substrate in relatively small 100- or 150-mm wafer sizes. The sapphire apparatus must be coupled with a mirror-like collector to reflect light that would otherwise be wasted. It is predicted that by 2020, 40% of all GaN LEDs will be made with GaN-on-Si. Manufacturing large sapphire material is difficult, while large silicon material is cheaper and more abundant. LED companies shifting from using sapphire to silicon should be a minimal investment.

Organic Light-emitting Diodes (OLEDs)

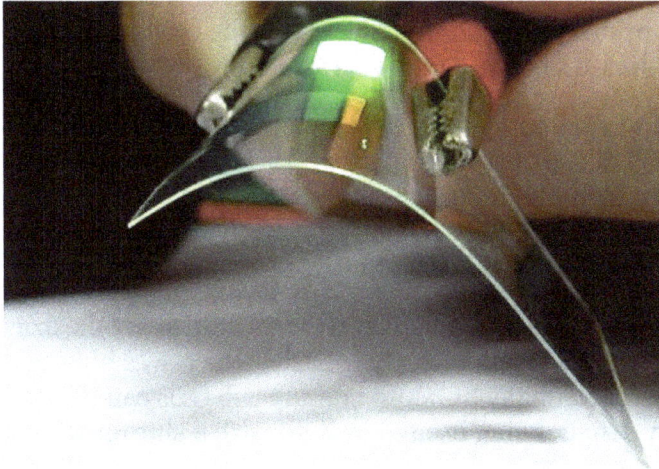

Demonstration of a flexible OLED device

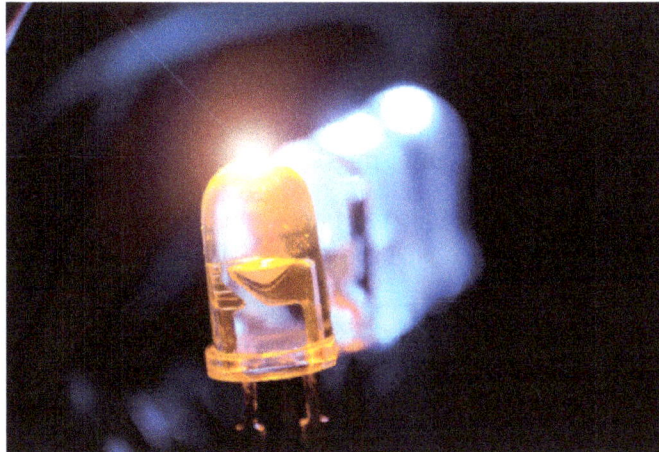

Orange light-emitting diode

In an organic light-emitting diode (OLED), the electroluminescent material comprising the emissive layer of the diode is an organic compound. The organic material is electrically conductive due to the delocalization of pi electrons caused by conjugation over all or part of the molecule, and the material therefore functions as an organic semiconductor. The organic materials can be small organic molecules in a crystalline phase, or polymers.

The potential advantages of OLEDs include thin, low-cost displays with a low driving voltage, wide viewing angle, and high contrast and color gamut. Polymer LEDs have the added benefit of printable and flexible displays. OLEDs have been used to make visual displays for portable electronic devices such as cellphones, digital cameras, and MP3 players while possible future uses include lighting and televisions.

Quantum Dot LEDs

Quantum dots (QD) are semiconductor nanocrystals that possess unique optical properties. Their emission color can be tuned from the visible throughout the infrared spectrum. This allows quantum dot LEDs to create almost any color on the CIE diagram. This provides more color options and better color rendering than white LEDs since the emission spectrum is much narrower, characteristic of quantum confined states. There are two types of schemes for QD excitation. One uses photo excitation with a primary light source LED (typically blue or UV LEDs are used). The other is direct electrical excitation first demonstrated by Alivisatos et al.

One example of the photo-excitation scheme is a method developed by Michael Bowers, at Vanderbilt University in Nashville, involving coating a blue LED with quantum dots that glow white in response to the blue light from the LED. This method emits a warm, yellowish-white light similar to that made by incandescent light bulbs. Quantum dots are also being considered for use in white light-emitting diodes in liquid crystal display (LCD) televisions.

In February 2011 scientists at PlasmaChem GmbH were able to synthesize quantum dots for LED applications and build a light converter on their basis, which was able to efficiently convert light from blue to any other color for many hundred hours. Such QDs can be used to emit visible or near infrared light of any wavelength being excited by light with a shorter wavelength.

The structure of QD-LEDs used for the electrical-excitation scheme is similar to basic design of OLEDs. A layer of quantum dots is sandwiched between layers of electron-transporting and hole-transporting materials. An applied electric field causes electrons and holes to move into the quantum dot layer and recombine forming an exciton that excites a QD. This scheme is commonly studied for quantum dot display. The tunability of emission wavelengths and narrow bandwidth is also beneficial as excitation sources for fluorescence imaging. Fluorescence near-field scanning optical microscopy (NSOM) utilizing an integrated QD-LED has been demonstrated.

In February 2008, a luminous efficacy of 300 lumens of visible light per watt of radiation (not per electrical watt) and warm-light emission was achieved by using nanocrystals.

Types

LEDs are produced in a variety of shapes and sizes. The color of the plastic lens is often the same as the actual color of light emitted, but not always. For instance, purple plastic is often used for infrared LEDs, and most blue devices have colorless housings. Modern high-power LEDs such as those used for lighting and backlighting are generally found in surface-mount technology (SMT) packages (not shown).

The main types of LEDs are miniature, high-power devices and custom designs such as alphanumeric or multi-color.

Miniature

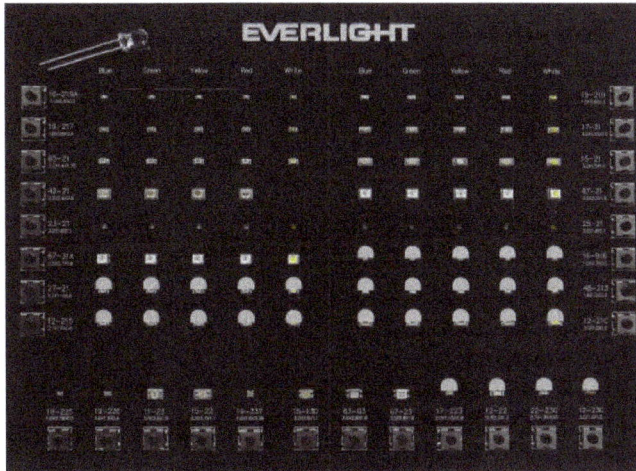

Photo of miniature surface mount LEDs in most common sizes. They can be much smaller than a traditional 5 mm lamp type LED which is shown on the upper left corner.

These are mostly single-die LEDs used as indicators, and they come in various sizes from 2 mm to 8 mm, through-hole and surface mount packages. They usually do not use a separate heat sink. Typical current ratings ranges from around 1 mA to above 20 mA. The small size sets a natural upper boundary on power consumption due to heat caused by the high current density and need for a heat sink. Often daisy chained as used in LED tapes.

Very small (1.6x1.6x0.35 mm) red, green, and blue surface mount miniature LED package with gold wire bonding details.

Common package shapes include round, with a domed or flat top, rectangular with a flat top (as used in bar-graph displays), and triangular or square with a flat top. The encapsulation may also be clear or tinted to improve contrast and viewing angle.

Researchers at the University of Washington have invented the thinnest LED. It is made of two-dimensional (2-D) flexible materials. It is three atoms thick, which is 10 to 20 times thinner than three-dimensional (3-D) LEDs and is also 10,000 times smaller than the thickness of a human hair. These 2-D LEDs are going to make it possible to create smaller, more energy-efficient lighting, optical communication and nano lasers.

There are three main categories of miniature single die LEDs:

Low-current

Typically rated for 2 mA at around 2 V (approximately 4 mW consumption)

Standard

20 mA LEDs (ranging from approximately 40 mW to 90 mW) at around:

- 1.9 to 2.1 V for red, orange, yellow, and traditional green

- 3.0 to 3.4 V for pure green and blue

- 2.9 to 4.2 V for violet, pink, purple and white

Ultra-high-output

20 mA at approximately 2 or 4–5 V, designed for viewing in direct sunlight

5 V and 12 V LEDs are ordinary miniature LEDs that incorporate a suitable series resistor for direct connection to a 5 V or 12 V supply.

High-power

High-power light-emitting diodes attached to an LED star base (Luxeon, Lumileds)

High-power LEDs (HP-LEDs) or high-output LEDs (HO-LEDs) can be driven at currents from hundreds of mA to more than an ampere, compared with the tens of mA for other LEDs. Some can emit over a thousand lumens. LED power densities up to 300 W/cm² have been achieved. Since overheating is destructive, the HP-LEDs must be mounted on a heat sink to allow for heat dissipation. If the heat from a HP-LED is not removed, the device will fail in seconds. One HP-LED can often replace an incandescent bulb in a flashlight, or be set in an array to form a powerful LED lamp.

Some well-known HP-LEDs in this category are the Nichia 19 series, Lumileds Rebel Led, Osram Opto Semiconductors Golden Dragon, and Cree X-lamp. As of September 2009, some HP-LEDs manufactured by Cree now exceed 105 lm/W.

Examples for Haitz's law, which predicts an exponential rise in light output and efficacy of LEDs over time, are the CREE XP-G series LED which achieved 105 lm/W in 2009 and the Nichia 19 series with a typical efficacy of 140 lm/W, released in 2010.

AC Driven

LEDs have been developed by Seoul Semiconductor that can operate on AC power without the need for a DC converter. For each half-cycle, part of the LED emits light and part is dark, and this is reversed during the next half-cycle. The efficacy of this type of HP-LED is typically 40 lm/W. A large number of LED elements in series may be able to operate directly from line voltage. In 2009, Seoul Semiconductor released a high DC voltage LED, named as 'Acrich MJT', capable of being driven from AC power with a simple controlling circuit. The low-power dissipation of these LEDs affords them more flexibility than the original AC LED design.

Application-specific Variations

Flashing

Flashing LEDs are used as attention seeking indicators without requiring external electronics. Flashing LEDs resemble standard LEDs but they contain an integrated multivibrator circuit that causes the LED to flash with a typical period of one second. In diffused lens LEDs, this circuit is visible as a small black dot. Most flashing LEDs emit light of one color, but more sophisticated devices can flash between multiple colors and even fade through a color sequence using RGB color mixing.

Bi-color

Bi-color LEDs contain two different LED emitters in one case. There are two types of these. One type consists of two dies connected to the same two leads antiparallel to each other. Current flow in one direction emits one color, and current in the opposite direction emits the other color. The other type consists of two dies with separate leads for both dies and another lead for common anode or cathode, so that they can be controlled independently.

Tri-color

Tri-color LEDs contain three different LED emitters in one case. Each emitter is connected to a separate lead so they can be controlled independently. A four-lead arrangement is typical with one common lead (anode or cathode) and an additional lead for each color.

RGB

RGB LEDs are tri-color LEDs with red, green, and blue emitters, in general using a four-wire connection with one common lead (anode or cathode). These LEDs can have either common positive or common negative leads. Others however, have only two leads (positive and negative) and have a built-in tiny electronic control unit.

Decorative-multicolor

Decorative-multicolor LEDs incorporate several emitters of different colors supplied by only two lead-out wires. Colors are switched internally by varying the supply voltage.

Alphanumeric

Alphanumeric LEDs are available in seven-segment, starburst and dot-matrix format. Seven-segment displays handle all numbers and a limited set of letters. Starburst displays can display all letters. Dot-matrix displays typically use 5x7 pixels per character. Seven-segment LED displays were in widespread use in the 1970s and 1980s, but rising use of liquid crystal displays, with their lower power needs and greater display flexibility, has reduced the popularity of numeric and alphanumeric LED displays.

Digital-RGB

Digital-RGB LEDs are RGB LEDs that contain their own "smart" control electronics. In addition to power and ground, these provide connections for data-in, data-out, and sometimes a clock or strobe signal. These are connected in a daisy chain, with the data in of the first LED sourced by a microprocessor, which can control the brightness and color of each LED independently of the others. They are used where a combination of maximum control and minimum visible electronics are needed such as strings for Christmas and LED matrices. Some even have refresh rates in the kHz range, allowing for basic video applications.

Filament

An LED filament consists of multiple LED dice connected in series on a common longitudinal substrate that form a thin rod reminiscent of a traditional incandescent filament. These are being used as a low cost decorative alternative for traditional light bulbs that are being phased out in many countries. The filaments require a rather high voltage to light to nominal brightness, allowing them to work efficiently and simply with mains voltages. Often a simple rectifier and capacitive current limiting are employed to create a low-cost replacement for a traditional light bulb without the complexity of creating a low voltage, high current converter which is required by single die LEDs. Usually they are packaged in a sealed enclosure with a shape similar to lamps they were designed to replace (e.g. a bulb), and filled with inert nitrogen or carbon dioxide gas to remove heat efficiently.

Considerations for Use

Power Sources

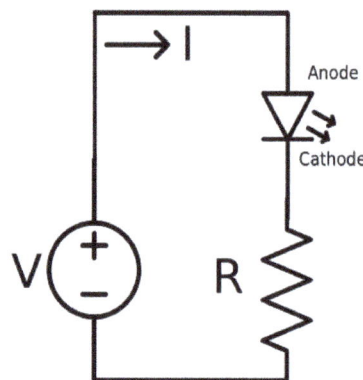

Simple LED circuit with resistor for current limiting

The current–voltage characteristic of an LED is similar to other diodes, in that the current is dependent exponentially on the voltage (see Shockley diode equation). This means that a small change in voltage can cause a large change in current. If the applied voltage exceeds the LED's forward voltage drop by a small amount, the current rating may be exceeded by a large amount, potentially damaging or destroying the LED. The typical solution is to use constant-current power supplies to keep the current below the LED's maximum current rating. Since most common power sources (batteries, mains) are constant-voltage sources, most LED fixtures must include a power converter, at least a current-limiting resistor. However, the high resistance of three-volt coin cells combined with the high differential resistance of nitride-based LEDs makes it possible to power such an LED from such a coin cell without an external resistor.

Electrical Polarity

As with all diodes, current flows easily from p-type to n-type material. However, no current flows and no light is emitted if a small voltage is applied in the reverse direction. If the reverse voltage grows large enough to exceed the breakdown voltage, a large current flows and the LED may be damaged. If the reverse current is sufficiently limited to avoid damage, the reverse-conducting LED is a useful noise diode.

Safety and Health

The vast majority of devices containing LEDs are "safe under all conditions of normal use", and so are classified as "Class 1 LED product"/"LED Klasse 1". At present, only a few LEDs—extremely bright LEDs that also have a tightly focused viewing angle of 8° or less—could, in theory, cause temporary blindness, and so are classified as "Class 2". The opinion of the French Agency for Food, Environmental and Occupational Health & Safety (ANSES) of 2010, on the health issues concerning LEDs, suggested banning public use of lamps which were in the moderate Risk Group 2, especially those with a high blue component in places frequented by children. In general, laser safety regulations—and the "Class 1", "Class 2", etc. system—also apply to LEDs.

While LEDs have the advantage over fluorescent lamps that they do not contain mercury, they may contain other hazardous metals such as lead and arsenic. Regarding the toxicity of LEDs when treated as waste, a study published in 2011 stated: "According to federal standards, LEDs are not hazardous except for low-intensity red LEDs, which leached Pb [lead] at levels exceeding regulatory limits (186 mg/L; regulatory limit: 5). However, according to California regulations, excessive levels of copper (up to 3892 mg/kg; limit: 2500), lead (up to 8103 mg/kg; limit: 1000), nickel (up to 4797 mg/kg; limit: 2000), or silver (up to 721 mg/kg; limit: 500) render all except low-intensity yellow LEDs hazardous."

Advantages

- Efficiency: LEDs emit more lumens per watt than incandescent light bulbs. The efficiency of LED lighting fixtures is not affected by shape and size unlike fluorescent light bulbs or tubes.

- Color: LEDs can emit light of an intended color without using any color filters as traditional lighting methods need. This is more efficient and can lower initial costs.

- Size: LEDs can be very small (smaller than 2 mm^2) and are easily attached to printed circuit boards.

- Warmup time: LEDs light up very quickly. A typical red indicator LED will achieve full brightness in under a microsecond. LEDs used in communications devices can have even faster response times.

- Cycling: LEDs are ideal for uses subject to frequent on-off cycling, unlike incandescent and fluorescent lamps that fail faster when cycled often, or high-intensity discharge lamps (HID lamps) that require a long time before restarting.

- Dimming: LEDs can very easily be dimmed either by pulse-width modulation or lowering the forward current. This pulse-width modulation is why LED lights, particularly headlights on cars, when viewed on camera or by some people, appear to be flashing or flickering. This is a type of stroboscopic effect.

- Cool light: In contrast to most light sources, LEDs radiate very little heat in the form of IR that can cause damage to sensitive objects or fabrics. Wasted energy is dispersed as heat through the base of the LED.

- Slow failure: LEDs mostly fail by dimming over time, rather than the abrupt failure of incandescent bulbs.

- Lifetime: LEDs can have a relatively long useful life. One report estimates 35,000 to 50,000 hours of useful life, though time to complete failure may be longer. Fluorescent tubes typically are rated at about 10,000 to 15,000 hours, depending partly on the conditions of use, and incandescent light bulbs at 1,000 to 2,000 hours. Several DOE demonstrations have shown that reduced maintenance costs from this extended lifetime, rather than energy savings, is the primary factor in determining the payback period for an LED product.

- Shock resistance: LEDs, being solid-state components, are difficult to damage with external shock, unlike fluorescent and incandescent bulbs, which are fragile.

- Focus: The solid package of the LED can be designed to focus its light. Incandescent and fluorescent sources often require an external reflector to collect light and direct it in a usable manner. For larger LED packages total internal reflection (TIR) lenses are often used to the same effect. However, when large quantities of light are needed many light sources are usually deployed, which are difficult to focus or collimate towards the same target.

Disadvantages

- Initial price: LEDs are currently slightly more expensive (price per lumen) on an initial capital cost basis, than other lighting technologies. As of March 2014, at least one manufacturer claims to have reached $1 per kilolumen. The additional expense partially stems from the relatively low lumen output and the drive circuitry and power supplies needed.

- Temperature dependence: LED performance largely depends on the ambient temperature of the operating environment – or thermal management properties. Overdriving an LED in high ambient temperatures may result in overheating the LED package, eventually leading to device failure. An adequate heat sink is needed to maintain long life. This is especially important in automotive, medical, and military uses where devices must operate over a wide range of temperatures, which require low failure rates. Toshiba has produced LEDs with an operating temperature range of −40 to 100 °C, which suits the LEDs for both indoor and outdoor use in applications such as lamps, ceiling lighting, street lights, and floodlights.

- Voltage sensitivity: LEDs must be supplied with a voltage above their threshold voltage and a current below their rating. Current and lifetime change greatly with a small change in applied voltage. They thus require a current-regulated supply (usually just a series resistor for indicator LEDs).

- Color rendition: Most cool-white LEDs have spectra that differ significantly from a black body radiator like the sun or an incandescent light. The spike at 460 nm and dip at 500 nm can cause the color of objects to be perceived differently under cool-white LED illumination than sunlight or incandescent sources, due to metamerism, red surfaces being rendered particularly poorly by typical phosphor-based cool-white LEDs.

- Area light source: Single LEDs do not approximate a point source of light giving a spherical light distribution, but rather a lambertian distribution. So LEDs are difficult to apply to uses needing a spherical light field; however, different fields of light can be manipulated by the application of different optics or "lenses". LEDs cannot provide divergence below a few degrees. In contrast, lasers can emit beams with divergences of 0.2 degrees or less.

- Electrical polarity: Unlike incandescent light bulbs, which illuminate regardless of the electrical polarity, LEDs will only light with correct electrical polarity. To automatically match source polarity to LED devices, rectifiers can be used.

- Blue hazard: There is a concern that blue LEDs and cool-white LEDs are now capable of exceeding safe limits of the so-called blue-light hazard as defined in eye safety specifications such as ANSI/IESNA RP-27.1–05: Recommended Practice for Photobiological Safety for Lamp and Lamp Systems.

- Light pollution: Because white LEDs, especially those with high color temperature, emit much more short wavelength light than conventional outdoor light sources such as high-pressure sodium vapor lamps, the increased blue and green sensitivity of scotopic vision means that white LEDs used in outdoor lighting cause substantially more sky glow. The American Medical Association warned on the use of high blue content white LEDs in street lighting, due to their higher impact on human health and environment, compared to low blue content light sources (e.g. High Pressure Sodium, PC amber LEDs, and low CCT LEDs).

- Efficiency droop: The efficiency of LEDs decreases as the electric current increases. Heating also increases with higher currents which compromises the lifetime of the LED. These effects put practical limits on the current through an LED in high power applications.

- Impact on insects: LEDs are much more attractive to insects than sodium-vapor lights, so much so that there has been speculative concern about the possibility of disruption to food webs.

- Use in winter conditions: Since they do not give off much heat in comparison to incandescent lights, LED lights used for traffic control can have snow obscuring them, leading to accidents.

Applications

LED uses fall into four major categories:

- Visual signals where light goes more or less directly from the source to the human eye, to convey a message or meaning

- Illumination where light is reflected from objects to give visual response of these objects

- Measuring and interacting with processes involving no human vision

- Narrow band light sensors where LEDs operate in a reverse-bias mode and respond to incident light, instead of emitting light

Indicators and Signs

The low energy consumption, low maintenance and small size of LEDs has led to uses as status indicators and displays on a variety of equipment and installations. Large-area LED displays are used as stadium displays and as dynamic decorative displays. Thin, lightweight message displays are used at airports and railway stations, and as destination displays for trains, buses, trams, and ferries.

Red and green LED traffic signals

One-color light is well suited for traffic lights and signals, exit signs, emergency vehicle lighting, ships' navigation lights or lanterns (chromacity and luminance standards being set under the Convention on the International Regulations for Preventing Collisions at Sea 1972, Annex I and the CIE) and LED-based Christmas lights. In cold climates, LED traffic lights may remain snow-covered. Red or yellow LEDs are used in indicator and alphanumeric displays in environments where night vision must be retained: aircraft cockpits, submarine and ship bridges, astronomy observatories, and in the field, e.g. night time animal watching and military field use.

Automotive applications for LEDs continue to grow.

Because of their long life, fast switching times, and their ability to be seen in broad daylight due to their high output and focus, LEDs have been used in brake lights for cars' high-mounted brake lights, trucks, and buses, and in turn signals for some time, but many vehicles now use LEDs for their rear light clusters. The use in brakes improves safety, due to a great reduction in the time needed to light fully, or faster rise time, up to 0.5 second faster than an incandescent bulb. This gives drivers behind more time to react. In a dual intensity circuit (rear markers and brakes) if the LEDs are not pulsed at a fast enough frequency, they can create a phantom array, where ghost images of the LED will appear if the eyes quickly scan across the array. White LED headlamps are starting to be used. Using LEDs has styling advantages because LEDs can form much thinner lights than incandescent lamps with parabolic reflectors.

Due to the relative cheapness of low output LEDs, they are also used in many temporary uses such as glowsticks, throwies, and the photonic textile Lumalive. Artists have also used LEDs for LED art.

Weather and all-hazards radio receivers with Specific Area Message Encoding (SAME) have three LEDs: red for warnings, orange for watches, and yellow for advisories and statements whenever issued.

Lighting

With the development of high-efficiency and high-power LEDs, it has become possible to use LEDs in lighting and illumination. To encourage the shift to LED lamps and other high-efficiency lighting, the US Department of Energy has created the L Prize competition. The Philips Lighting North America LED bulb won the first competition on August 3, 2011 after successfully completing 18 months of intensive field, lab, and product testing.

LEDs are used as street lights and in other architectural lighting. The mechanical robustness and long lifetime is used in automotive lighting on cars, motorcycles, and bicycle lights. LED light emission may be efficiently controlled by using nonimaging optics principles.

LED street lights are employed on poles and in parking garages. In 2007, the Italian village of Torraca was the first place to convert its entire illumination system to LEDs.

LEDs are used in aviation lighting. Airbus has used LED lighting in its Airbus A320 Enhanced since 2007, and Boeing uses LED lighting in the 787. LEDs are also being used now in airport and heliport lighting. LED airport fixtures currently include medium-intensity runway lights, runway centerline lights, taxiway centerline and edge lights, guidance signs, and obstruction lighting.

LEDs are also used as a light source for DLP projectors, and to backlight LCD televisions (referred to as LED TVs) and laptop displays. RGB LEDs raise the color gamut by as much as 45%. Screens for TV and computer displays can be made thinner using LEDs for backlighting.

The lack of IR or heat radiation makes LEDs ideal for stage lights using banks of RGB LEDs that can easily change color and decrease heating from traditional stage lighting, as well as medical lighting where IR-radiation can be harmful. In energy conservation, the lower heat output of LEDs also means air conditioning (cooling) systems have less heat in need of disposal.

LEDs are small, durable and need little power, so they are used in handheld devices such as flashlights. LED strobe lights or camera flashes operate at a safe, low voltage, instead of the 250+ volts commonly found in xenon flashlamp-based lighting. This is especially useful in cameras on mobile phones, where space is at a premium and bulky voltage-raising circuitry is undesirable.

LEDs are used for infrared illumination in night vision uses including security cameras. A ring of LEDs around a video camera, aimed forward into a retroreflective background, allows chroma keying in video productions.

LED to be used for miners, to increase visibility inside mines

LEDs are used in mining operations, as cap lamps to provide light for miners. Research has been done to improve LEDs for mining, to reduce glare and to increase illumination, reducing risk of injury to the miners.

LEDs are now used commonly in all market areas from commercial to home use: standard lighting, AV, stage, theatrical, architectural, and public installations, and wherever artificial light is used.

LEDs are increasingly finding uses in medical and educational applications, for example as mood enhancement, and new technologies such as AmBX, exploiting LED versatility. NASA has even sponsored research for the use of LEDs to promote health for astronauts.

Data Communication and Other Signalling

Light can be used to transmit data and analog signals. For example, lighting white LEDs can be used in systems assisting people to navigate in closed spaces while searching necessary rooms or objects.

Assistive listening devices in many theaters and similar spaces use arrays of infrared LEDs to send sound to listeners' receivers. Light-emitting diodes (as well as semiconductor lasers) are used to send data over many types of fiber optic cable, from digital audio over TOSLINK cables to the very high bandwidth fiber links that form the Internet backbone. For some time, computers were commonly equipped with IrDA interfaces, which allowed them to send and receive data to nearby machines via infrared.

Because LEDs can cycle on and off millions of times per second, very high data bandwidth can be achieved.

Sustainable Lighting

Efficient lighting is needed for sustainable architecture. In 2009, US Department of Energy testing results on LED lamps showed an average efficacy of 35 lm/W, below that of typical CFLs, and as low as 9 lm/W, worse than standard incandescent bulbs. A typical 13-watt LED lamp emitted 450 to 650 lumens, which is equivalent to a standard 40-watt incandescent bulb.

However, as of 2011, there are LED bulbs available as efficient as 150 lm/W and even inexpensive low-end models typically exceed 50 lm/W, so that a 6-watt LED could achieve the same results as a standard 40-watt incandescent bulb. The latter has an expected lifespan of 1,000 hours, whereas an LED can continue to operate with reduced efficiency for more than 50,000 hours.

See the chart below for a comparison of common light types:

	LED	CFL	Incandescent
Lightbulb Projected Lifespan	50,000 hours	10,000 hours	1,200 hours
Watts Per Bulb (equiv. 60 watts)	10	14	60
Cost Per Bulb	Approx. $19.00	$7.00	$1.25
KWh of Electricity Used Over 50,000 Hours	500	700	3000
Cost of Electricity (@ 0.10 per KWh)	$50	$70	$300
Bulbs Needed for 50,000 Hours of Use	1	5	42
Equivalent 50,000 Hours Bulb Expense	$19.00	$35.00	$52.50
TOTAL Cost for 50,000 Hours	$69.00	$105.00	$352.50

Energy Consumption

In the US, one kilowatt-hour (3.6 MJ) of electricity currently causes an average 1.34 pounds (610 g) of CO_2 emission. Assuming the average light bulb is on for 10 hours a day, a 40-watt bulb will cause 196 pounds (89 kg) of CO_2 emission per year. The 6-watt LED equivalent will only cause 30 pounds (14 kg) of CO_2 over the same time span. A building's carbon footprint from lighting can therefore be reduced by 85% by exchanging all incandescent bulbs for new LEDs if a building previously used only incandescent bulbs.

In practice, most buildings that use a lot of lighting use fluorescent lighting, which has 22% luminous efficiency compared with 5% for filaments, so changing to LED lighting would still give a 34% reduction in electrical power use and carbon emissions.

The reduction in carbon emissions depends on the source of electricity. Nuclear power in the United States produced 19.2% of electricity in 2011, so reducing electricity consumption in the U.S. reduces carbon emissions more than in France (75% nuclear electricity) or Norway (almost entirely hydroelectric).

Replacing lights that spend the most time lit results in the most savings, so LED lights in infrequently used locations bring a smaller return on investment.

Light Sources for Machine Vision Systems

Machine vision systems often require bright and homogeneous illumination, so features of interest are easier to process. LEDs are often used for this purpose, and this is likely to remain one of their major uses until the price drops low enough to make signaling and illumination uses more widespread. Barcode scanners are the most common example of machine vision, and many low cost products use red LEDs instead of lasers. Optical computer mice are an example of LEDs in machine vision, as it is used to provide an even light source on the surface for the miniature camera within the mouse. LEDs constitute a nearly ideal light source for machine vision systems for several reasons:

- The size of the illuminated field is usually comparatively small and machine vision systems are often quite expensive, so the cost of the light source is usually a minor concern. However, it might not be easy to replace a broken light source placed within complex machinery, and here the long service life of LEDs is a benefit.

- LED elements tend to be small and can be placed with high density over flat or even-shaped substrates (PCBs etc.) so that bright and homogeneous sources that direct light from tightly controlled directions on inspected parts can be designed. This can often be obtained with small, low-cost lenses and diffusers, helping to achieve high light densities with control over lighting levels and homogeneity. LED sources can be shaped in several configurations (spot lights for reflective illumination; ring lights for coaxial illumination; back lights for contour illumination; linear assemblies; flat, large format panels; dome sources for diffused, omnidirectional illumination).

- LEDs can be easily strobed (in the microsecond range and below) and synchronized with imaging. High-power LEDs are available allowing well-lit images even with very short light pulses. This is often used to obtain crisp and sharp "still" images of quickly moving parts.

- LEDs come in several different colors and wavelengths, allowing easy use of the best color for each need, where different color may provide better visibility of features of interest. Having a precisely known spectrum allows tightly matched filters to be used to separate informative bandwidth or to reduce disturbing effects of ambient light. LEDs usually operate at comparatively low working temperatures, simplifying heat management and dissipation. This allows using plastic lenses, filters, and diffusers. Waterproof units can also easily be designed, allowing use in harsh or wet environments (food, beverage, oil industries).

Other Applications

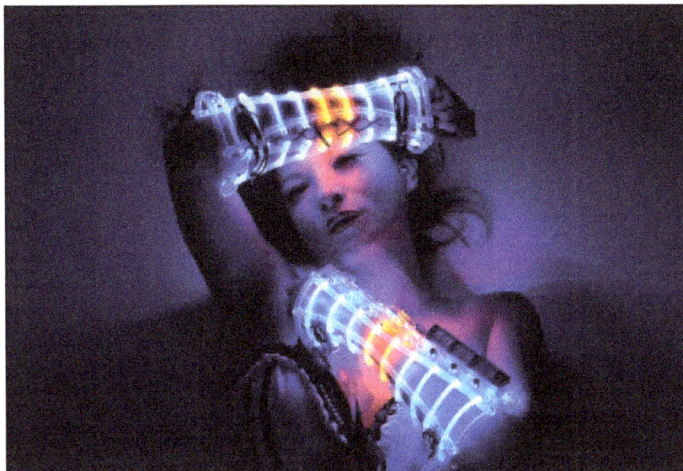

LED costume for stage performers

The light from LEDs can be modulated very quickly so they are used extensively in optical fiber and free space optics communications. This includes remote controls, such as for TVs, VCRs, and LED Computers, where infrared LEDs are often used. Opto-isolators use an LED combined with a photodiode or phototransistor to provide a signal path with electrical isolation between two circuits. This is especially useful in medical equipment where the signals from a low-voltage sensor circuit (usually battery-powered) in contact with a living organism must be electrically isolated from any possible electrical failure in a recording or monitoring device operating at potentially dangerous voltages. An optoisolator also allows information to be transferred between circuits not sharing a common ground potential.

Many sensor systems rely on light as the signal source. LEDs are often ideal as a light source due to the requirements of the sensors. LEDs are used as motion sensors, for example in optical computer mice. The Nintendo Wii's sensor bar uses infrared LEDs. Pulse oximeters use them for measuring oxygen saturation. Some flatbed scanners use arrays of RGB LEDs rather than the typical cold-cathode fluorescent lamp as the light source. Having independent control of three illuminated colors allows the scanner to calibrate itself for more accurate color balance, and there is no need for warm-up. Further, its sensors only need be monochromatic, since at any one time the page being scanned is only lit by one color of light. Since LEDs can also be used as photodiodes, they can be used for both photo emission and detection. This could be used, for example, in a touchscreen that registers reflected light from a finger or stylus. Many materials and biological systems are sensitive to, or dependent on, light. Grow lights use LEDs to increase photosynthesis in plants, and bacteria and viruses can be removed from water and other substances using UV LEDs for sterilization.

LEDs have also been used as a medium-quality voltage reference in electronic circuits. The forward voltage drop (e.g. about 1.7 V for a normal red LED) can be used instead of a Zener diode in low-voltage regulators. Red LEDs have the flattest I/V curve above the knee. Nitride-based LEDs have a fairly steep I/V curve and are useless for this purpose. Although LED forward voltage is far more current-dependent than a Zener diode, Zener diodes with breakdown voltages below 3 V are not widely available.

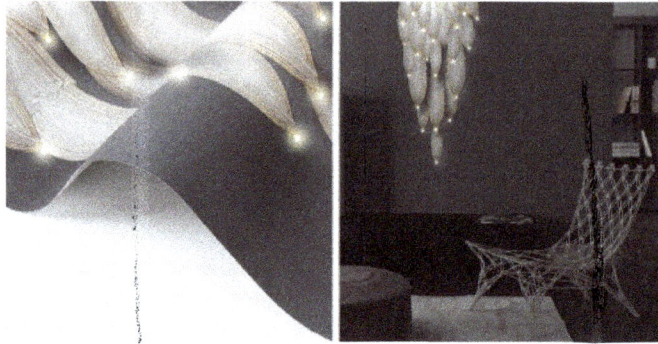

LED wallpaper by Meystyle

The progressive miniaturization of low-voltage lighting technology, such as LEDs and OLEDs, suitable to be incorporated into low-thickness materials has fostered in recent years the experimentation on combining light sources and wall covering surfaces to be applied onto interior walls. The new possibilities offered by these developments have prompted some designers and companies, such as Meystyle, Ingo Maurer, Lomox and Philips, to research and develop proprietary LED wallpaper technologies, some of which are currently available for commercial purchase. Other solutions mainly exist as prototypes or are in the process of being further refined.

Photonic Integrated Circuit

A photonic integrated circuit (PIC) or integrated optical circuit is a device that integrates multiple (at least two) photonic functions and as such is similar to an electronic integrated circuit. The major difference between the two is that a photonic integrated circuit provides functions for information signals imposed on optical wavelengths typically in the visible spectrum or near infrared 850 nm-1650 nm.

The most commercially utilized material platform for photonic integrated circuits is indium phosphide, which allows for the integration of various optically active and passive functions on the same chip. Initial examples of photonic integrated circuits were simple 2 section distributed Bragg reflector (DBR) lasers, consisting of two independently controlled device sections - a gain section and a DBR mirror section. Consequently, all modern monolithic tunable lasers, widely tunable lasers, externally modulated lasers and transmitters, integrated receivers, etc. are examples of photonic integrated circuits. Current state-of-the-art devices integrate hundreds of functions onto single chip. Pioneering work in this arena was performed at Bell Laboratories. Most notable academic centers of excellence of photonic integrated circuits in InP are the University of California at Santa Barbara, USA, and the Eindhoven University of Technology in the Netherlands.

A 2005 development showed that silicon can, even though it is an indirect bandgap material, still be used to generate laser light via the Raman nonlinearity. Such lasers are not electrically driven but optically driven and therefore still necessitate a further optical pump laser source.

Comparison to Electronic Integration

Unlike electronic integration where silicon is the dominant material, system photonic integrated circuits have been fabricated from a variety of material systems, including electro-optic crystals such as lithium niobate, silica on silicon, Silicon on insulator, various polymers and semiconductor materials which are used to make semiconductor lasers such as GaAs and InP. The different material systems are used because they each provide different advantages and limitations depending on the function to be integrated. For instance, silica (silicon dioxide) based PICs have very desirable properties for passive photonic circuits such as AWGs due to their comparatively low losses and low thermal sensitivity, GaAs or InP based PICs allow the direct integration of light sources and Silicon PICs enable co-integration of the photonics with transistor based electronics.

The fabrication techniques are similar to those used in electronic integrated circuits in which photolithography is used to pattern wafers for etching and material deposition. Unlike electronics where the primary device is the transistor, there is no single dominant device. The range of devices required on a chip includes low loss interconnect waveguides, power splitters, optical amplifiers, optical modulators, filters, lasers and detectors. These devices require a variety of different materials and fabrication techniques making it difficult to realize all of them on a single chip.

Newer techniques using resonant photonic interferometry is making way for UV LEDs to be used for optical computing requirements with much cheaper costs leading the way to PHz consumer electronics.

Examples of Photonic Integrated Circuits

The primary application for photonic integrated circuits is in the area of fiber-optic communication though applications in other fields such as biomedical and photonic computing are also possible.

The arrayed waveguide grating (AWG) which are commonly used as optical (de)multiplexers in wavelength division multiplexed (WDM) fiber-optic communication systems are an example of a photonic integrated circuit which has replaced previous multiplexing schemes which utilized multiple discrete filter elements. Since separating optical modes is a need for quantum computing, this technology may be helpful to miniaturize quantum computers.

Another example of a photonic integrated chip in wide use today in fiber-optic communication systems is the externally modulated laser (EML) which combines a distributed feed back laser diode with an electro-absorption modulator on a single InP based chip.

Current Status

Photonic integration is currently an active topic in U.S. Defense contracts. It is included by the Optical Internetworking Forum for inclusion in 100 gigahertz optical networking standards.

Optical Computing

Optical or photonic computing uses photons produced by lasers or diodes for computation. For decades, photons have promised to allow a higher bandwidth than the electrons used in conventional computers.

Most research projects focus on replacing current computer components with optical equivalents, resulting in an optical digital computer system processing binary data. This approach appears to offer the best short-term prospects for commercial optical computing, since optical components could be integrated into traditional computers to produce an optical-electronic hybrid. However, optoelectronic devices lose 30% of their energy converting electronic energy into photons and back; this conversion also slows the transmission of messages. All-optical computers eliminate the need for optical-electrical-optical (OEO) conversions.

Application-specific devices, such as *optical correlators*, have been designed to use the principles of optical computing. Such devices can be used, for example, to detect and track objects, and to classify serial time-domain optical data.

Optical Components for Binary Digital Computer

The fundamental building block of modern electronic computers is the transistor. To replace electronic components with optical ones, an equivalent optical transistor is required. This is achieved using materials with a non-linear refractive index. In particular, materials exist where the intensity of incoming light affects the intensity of the light transmitted through the material in a similar manner to the current response of a bipolar transistor. Such an 'optical transistor' can be used to create optical logic gates, which in turn are assembled into the higher level components of the computer's CPU. These will be non linear crystals used to manipulate light beams into controlling others.

Controversy

There are disagreements between researchers about the future capabilities of optical computers: will they be able to compete with semiconductor-based electronic computers on speed, power consumption, cost, and size. Critics note that real-world logic systems require "logic-level restoration, cascadability, fan-out and input–output isolation", all of which are currently provided by electronic transistors at low cost, low power, and high speed. For optical logic to be competitive beyond a few niche applications, major breakthroughs in non-linear optical device technology would be required, or perhaps a change in the nature of computing itself.

Misconceptions, Challenges and Prospects

A significant challenge to optical computing is that computation is a nonlinear process in which multiple signals must interact. Light, which is an electromagnetic wave, can only interact with another electromagnetic wave in the presence of electrons in a material, and the strength of this interaction is much weaker for electromagnetic waves, such as light, than for the electronic signals in a conventional computer. This may result in the processing elements for an optical computer requiring more power and larger dimensions than those for a conventional electronic computer using transistors

Photonic Logic

Photonic logic is the use of photons (light) in logic gates (NOT, AND, OR, NAND, NOR, XOR, XNOR). Switching is obtained using nonlinear optical effects when two or more signals are combined.

Resonators are especially useful in photonic logic, since they allow a build-up of energy from constructive interference, thus enhancing optical nonlinear effects.

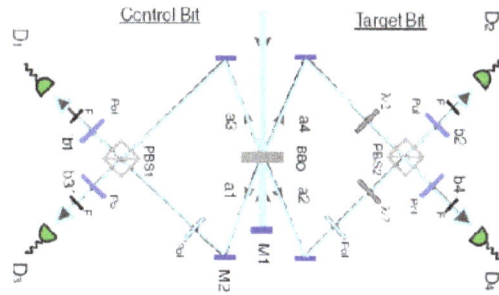

Realization of a Photonic Controlled-NOT Gate for use in Quantum Computing

Other approaches currently being investigated include photonic logic at a molecular level, using photoluminescent chemicals. In a recent demonstration, Witlicki et al. performed logical operations using molecules and SERS.

References

- *Mary Kay Carson (2007). Alexander Graham Bell: Giving Voice To The World*. Sterling Biographies. New York: Sterling Publishing. pp. 76–78. *ISBN 978-1-4027-3230-0.*

- *Novotny, Lukas,* (November 2007). *"Adapted from "The History of Near-field Optics""* (PDF). In Wolf, Emil. Progress in Optics. Progress In Optics series. 50. Amsterdam: Elsevier. pp. 142–150. *ISBN 978-0-444-53023-3.*

- *Nishizawa, Jun-ichi & Suto, Ken (2004). "Terahertz wave generation and light amplification using Raman effect". In Bhat, K. N. & DasGupta, Amitava. Physics of semiconductor devices. New Delhi, India: Narosa Publishing House. p. 27. ISBN 81-7319-567-6.*

- *Bănică, Florinel-Gabriel (2012). Chemical Sensors and Biosensors: Fundamentals and Applications. Chichester: John Wiley and Sons. Ch. 18–20. ISBN 978-0-470-71066-1.*

- *Williams, E. A. (ed.) (2007). National Association of Broadcasters Engineering Handbook (Tenth Edition). Taylor & Francis. pp. 1667–1685 (Chapter 6.10 – Fiber–Optic Transmission Systems, Jim Jachetta). ISBN 978-0-240-80751-5.*

- *Narra, Prathyusha; Zinger, D.S. (2004). "An effective LED dimming approach". Industry Applications Conference, 2004. 39th IAS Annual Meeting. Conference Record of the 2004 IEEE. 3: 1671–1676. doi:10.1109/ IAS.2004.1348695. ISBN 0-7803-8486-5.*

- *Feitelson, Dror G. (1988). "Chapter 3: Optical Image and Signal Processing". Optical Computing: A Survey for Computer Scientists. Cambridge, MA: MIT Press. ISBN 0-262-06112-0.*

- *"AMA Adopts Community Guidance to Reduce the Harmful Human and Environmental Effects of High Intensity Street Lighting".* www.ama-assn.org. Retrieved 2016-08-01.

- *"Philips Announces Partnership with Kvadrat Soft Cells to Bring Spaces Alive with luminous textile".* Philips. com. Philips. 2011. Retrieved 31 March 2016.

- *"The Nobel Prize in Physics 2014 Isamu Akasaki, Hiroshi Amano, Shuji Nakamura".* nobelprize.org. The Royal Swedish Academy of Sciences. Retrieved 17 March 2015.

- *"All in 1 LED Lighting Solutions Guide".* PhilipsLumileds.com. *Philips.* 2012-10-04. p. 15. Archived from *the original* (PDF) on March 14, 2013. Retrieved 2015-11-18.

- *Neidhardt, H.; Wilhelm, L.; Zagrebnov, V. A. (February 2015). "A New Model for Quantum Dot Light Emitting-Absorbing Bevices: Proofs and Supplements". Nanosystems: Physics, Chemistry, Mathematics. 6 (1): 6–45. doi:10.17586/2220-8054-2015-6-1-6-45. Retrieved 2015-10-31.*

Permissions

Index

www.ingramcontent.com/pod-product-compliance
Lightning Source LLC
Chambersburg PA
CBHW061313190326
41458CB00011B/3798